THE SWEET PEA MAN

The life and times of the
Victorian plant hybridist
Henry Eckford VMH
1823 – 1905

by

GRAHAM MARTIN

Scotland Street Press
EDINBURGH

2 5 7 3 1 2 9 7 5 3

First published in Great Britain in 2017 by
Scotland Street Press
Edinburgh
scotlandstreetpress@gmail.com

Cover Design by Bookmark Studio
ISBN: 978-1-91089518-4

Typeset in Scotland by Bookmark Studio
Printed in Poland

Contents

UPPLEMENT TO THE GARDENERS' MAGAZINE, DECEMBER
The
GARDENERS'
MAGAZINE.
TWOPENCE EVERY SATURDAY.
CALENDAR FOR
1900.
JANUARY.
FEBRUARY.
MARCH.
APRIL.
MAY.
JUNE.
JULY.
AUGUST.
SEPTEMBER.
OCTOBER.
NOVEMBER.
DECEMBER.

Chapter One

Monday the third of December, eighteen-hundred and thirty-eight. Edinburgh braced itself for winter. *The Edinburgh Evening Courant* was reporting storms off the coast, with the loss of many valuable lives, cargoes and ships. The month was to prove colder, too, than last year's, whose winter had been one of the coldest on record.[(1)] And in the winter before that one the city had experienced forty-nine days of snow.[(2)]

A snow-storm was already brewing as the passengers alighted the Defiance stage-coach at the east end of Princes Street, Edinburgh's beautifully set thoroughfare that runs east-west along the southern edge of the 'new town'. The journey began opposite The Crown hotel, where the Defiance was waiting. Here was the crowded travel hub of the city, which it still is. A few yards south from where the stage-coach stood now lies the main railway station, with the bus station just north across St. Andrew Square. Some buses also leave from Waterloo Place, the road that Princes Street runs into going east. In the 1830s, things were even closer at hand. The Crown, its proprietor Archibald Stewart and the Defiance were surrounded on all sides by coach-hirers, coach-builders, coach-makers, saddlers and coach-offices.[(3)]

Henry Eckford, fifteen years old, had boarded a cheaper outside seat on the Defiance for the two day journey to Inverness and horticultural apprenticeship at Beaufort Castle, the seat of Lord Lovat at Beauly. Safely in his pocket lay a silver sixpence, a good-luck token from his mother to ensure that he'd never be short of money. He was to keep it all his life, together with the memory of the blinding white morning sky, and of the extreme coldness of the journey. The party spent the first night in Aberdeen, arriving safely in Inverness around 7 o'clock the following evening. The journey loomed so large in his memory that in later life, where he'd written 'two days' in his *Curriculum Vitae* he pencilled 'three' over it. Surely, he thought of the impressionable event, it must have taken longer? By 1830 however, the Edinburgh – Aberdeen coach could cover the one hundred and thirty-four miles in, at a pinch, fourteen hours and twenty-two minutes, including the ferry crossing of the Firth of Forth at Queensferry. Even slowed by the snow, the shorter distance from Aberdeen to Inverness is unlikely to have taken twice as long as that.

~~~

Henry Eckford was born on Saturday 17th May, 1823, the seventh of the eight children of James and Isabel Eckford. In 1810, with Isabel possibly already pregnant the two had run off to nearby Edinburgh to marry. There was, however, a problem...

'Married irregularly at Edinburgh July 31st, 1810 James Eckford and Isabel Perie both of this parish they compeared before the Session September 9th 1810 paid a fine for their irregularity and the modr. [moderator] declared them married persons.' So runs the Cockpen Parish Register for 1810. An irregular marriage was one that was, or was considered to be, contrary to Church of Scotland rules. Perhaps the banns hadn't been read to the congregation on consecutive Sundays. Pregnancy is an unlikely cause. If it had occurred it was extremely recent, was probably unknown and could be hidden if it was. Although it was customary to marry in one's parish, it was common enough here for couples to go to Edinburgh to be married. For one thing it was cheaper. It was usual to seek permission to marry from one's parents or employers, permission that James and Isabel may not have had. James was thirty and Isabel sixteen, just, which gives us a clue; perhaps James had a past that he hadn't quite managed to keep covered? A few years before, one James Eckford had been fathering illegitimate children down in the borders, at Traquair Mill. Perhaps too, the gentle loosening of the church's grip on the nation's affairs saw the church zealous where its writ remained.

The couple's first child, Elizabeth, was born the following spring, at Shiels in Cockpen. They had a further three boys before moving to Stenhouse in Liberton, (not to be confused with the larger Edinburgh suburb of Stenhouse) in 1818. Here they had another four children, Henry being the third. Their last child, Isabella, was born in January 1826. If all the children survived, Henry as one of the youngest would have had a lot of siblings to make a fuss over him. Unfortunately, his father couldn't have done. At some time between April 1825 and January 1826, James Eckford died. Isabel now had eight children to bring up alone, though almost
~~~

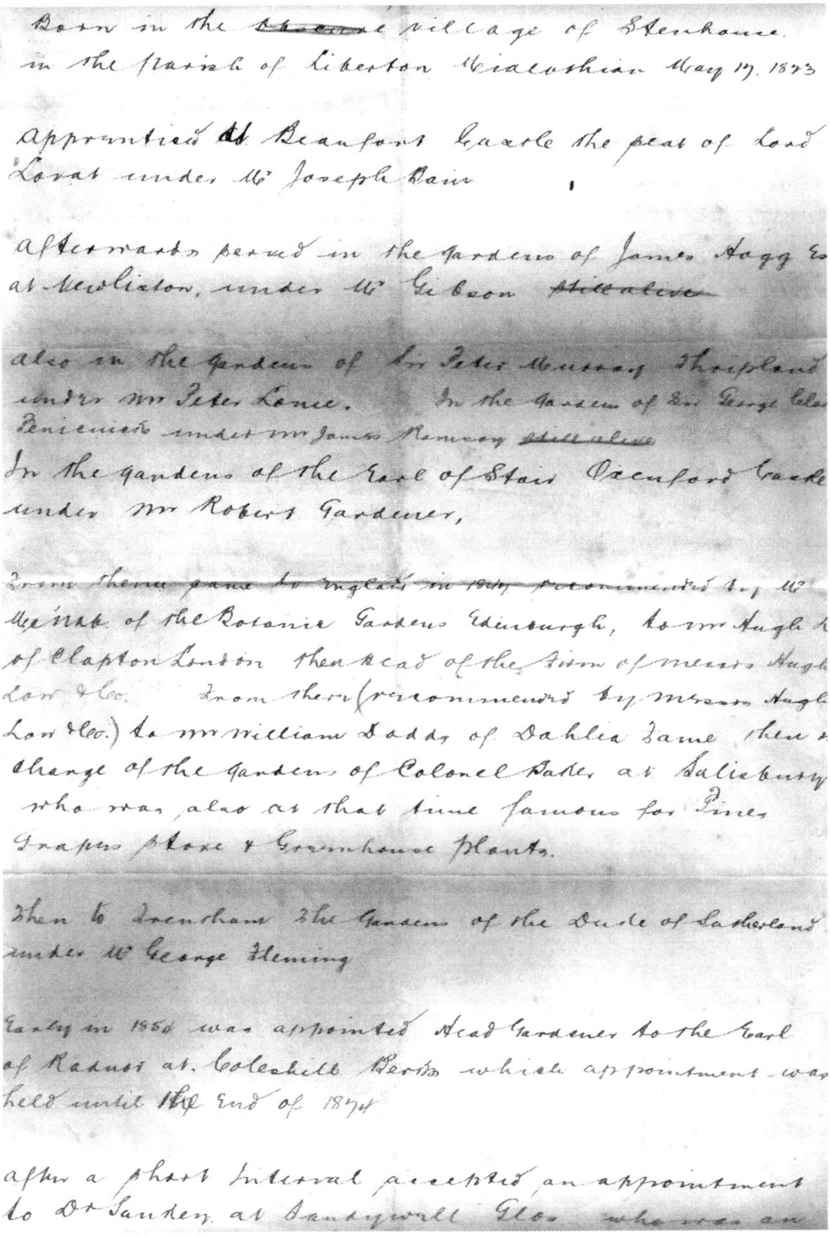
Born in the [struck through] village of Stenhouse
in the Parish of Liberton Midlothian May 17, 1823

Apprenticed at Beaufort Castle the seat of Lord
Lovat under Mr Joseph Bain

Afterwards served in the Gardens of James Hogg Es
at Newliston, under Mr Gibson ~~still alive~~

also in the Gardens of Sir Peter Murray Threipland
under Mr Peter Lonie. In the Garden of Sir George Cla
Penicuick under Mr James Ramsay ~~still alive~~
In the Gardens of the Earl of Stair Oxenford Castle
under Mr Robert Gardener,

~~From thence came to England in 1847 recommended by Mr~~
McNab of the Botanic Gardens Edinburgh, to Mr Hugh L
of Clapton London then head of the firm of Messrs Hugh
Low & Co. From there (recommended by Messrs Hugh
Low & Co.) to Mr William Dodds of Dahlia fame then i
charge of the Gardens of Colonel Baker at Salisbury
who was also at that time famous for Pines
Grapes Stove & Greenhouse Plants.

Then to Trentham The Gardens of the Duke of Sutherland
under Mr George Fleming

Early in 1854 was appointed Head Gardener to the Earl
of Radnor at Coleshill Berks which appointment was
held until the End of 1874

after a short Interval accepted an appointment
to Dr Sankey at Sandywell Glos who was an

Excerpt from Henry Eckford's CV

certainly with the help of the extended family. Henry must have exhibited an early fascination with flowers. Still in his cradle, his grandfather told his parents "if you make that boy anything but a gardener you'll spoil him!"[(4)]

Henry's father had been a farmer. Henry called his father a farmer on one occasion and a butcher (or flesher as they were more commonly known) on another and he was probably both. He may also have been a publican, though perhaps not all three at the same time! Farming could be a very profitable business. The agricultural advances of neighbouring East Lothian were now noted in mainland Europe. The historian T.C. Smout speaks of the 'enormous pride and self-confidence of the Lothian farming class' at this time.[(5)] The thirty-odd farmers in Liberton were considered by the local minister to be 'a highly respectable class of men – men of great skill and capital'.[(6)] If James had had to work in the deep Gilmerton mines instead, as many in the parish did, he would have been joined in them by his family. Wives and children climbed ladders with back-deforming loads, work for which they appear to have received no direct financial reward. The men were on 10d. to 1/- a day. When the carters delivered the coal to Edinburgh it fetched '7/6d a ton, small coal 2/6d a ton'.[(7)] As a profession, carters had a reputation for thievery, but those that stole people's lives away lay higher up the social scale. Henry was nineteen before such terrible child labour was outlawed in 1842.

As it was, Henry had the good fortune to be born to a farmer and the family were unlikely to have been left penniless by James's demise. Money however had to be found. The 1841 census describes Isabel as a 'grocer' as does the Post Office directory for 1842-43. The 1851 census, a little more enlighteningly, expands this to a 'grocer in spirit dealer'. Still living with Isabel in 1841 was Henry's sister Isabella, now fifteen, and his twenty-three year old brother Peter, well on his way to becoming a master-tailor. In a separate household within the house were Henry's uncle and aunt, John and Elizabeth Pirrie (spelling of names was quite haphazard) and a little boy of seven, James Kery. (Again, Kery could be Kerr. The witnesses at Henry's baptism on 13th June 1823 were William Kerr and Alexander Marr).

Henry's mother had probably started selling spirits as soon as her husband died. It could have been a pub she was running. There was one in Stenhouse called 'The Robins Nest'.[(8)] It lay on the north-west side of the staggered cross-roads there, a crossing formed when the present little Hyvot Loan road was a main thoroughfare, bisecting Elens Glen Road to run on north-westwards to Edinburgh. In 1861, Henry's brother Peter is listed as running a public house in Upper Stenhouse, having thrown over tailoring. So it's quite likely that Henry grew up in one. Statistically though, it's more likely to have been one of the smaller-scale whisky-shops, popularly known as 'dram-drinking shops'.

When Henry left for Inverness, the parish of Liberton had thirty-two of them. As well as taking away beer and spirits you could, if you had 'a penny to squander' buy spirits in single glasses, in very small quantities, over the counter. This was seen as 'a frightful source of intemperance among the lower orders'.[9] For the Reverend James Begg, the young parish minister, these thirty-two shops were 'just thirty too many, and the effect is as pernicious as possible. It is just so many persons scattered over the parish with their families and relations, whose living depends on the success with which they can prevail upon their neighbours to drink. One man is paid for teaching sobriety, but thirty-two have an interest in defeating his efforts, and human nature is on their side. At the same time' he concedes, well aware that someone somewhere would read what he'd written and tell everyone, 'some of these publicans are very respectable people…'.[10]

If you walk down to Liberton from Edinburgh today you'll notice a peculiar topography. The land rises up and down continually as though you were riding waves at sea. It's like that at Borthwick, too, where Henry's father came from; a common feature of this part of the country, and often commented on. Begg called it 'beautifully diversified with plains and rising grounds'.[11] A later writer saw the white Lothian farmhouses poetically as 'beacons in a slow-heaving sea'.[12] Travelling from Shrewsbury to Wem, the small town in Shropshire where Henry would end his days is another place you notice it. Shropshire was where another famous son of Liberton's Stenhouse ended up too, the architect John Simpson (1755-1815).[13] But we are getting ahead of ourselves. Beyond Liberton the wider area is bordered north-east and north above Edinburgh by the Firth of Forth. The northern tip of the Pentland Hills juts in, cliff-like, from the south-west, while away to the south the Moorfoot Hills rise away steeply from this gentle land.

Stenhouse is an attractive part of what is now otherwise nondescript outer - Edinburgh suburb. Its centre is the wooded hollow of Elen's Glen, where the Burdiehouse Burn flows north-eastwards to the sea. In Henry's time, Liberton was as noted for its gardens as much as its farms. The circumference of a local sycamore was measured at 19 feet (5.79m) which shows the depth and fertility of the soil. Stenhouse gardens sometimes raised the earliest strawberries for the Edinburgh market. In 1832, they were ripe on the 5th June. The large garden of Moredun House can hardly have been unknown to Henry. It was bigger than his village, and ran along its eastern side. 'Besides moveable glass frames', James Begg enthused in 1839, 'there is exposed to the light in the vineries, peach-houses and pine-pits [i.e. pineapple pits] .... upwards of 8223 square feet of glass. Hollies thrive admirably…'.[14] Hollies will thrive admirably almost anywhere but those expensive frames and glasshouses, unusual enough to be noted, are significant in view of Henry's later development.

We don't know where Henry went to school. In 1839 there were ten schools in the Liberton area, all within easy walking distance, all fee-paying (1/- to 1/6 a month)[15] with eminent schools in Edinburgh for those that could afford them. Of those in Liberton, one would have been a parochial school, as required by Acts of 1696 and 1803. Parochial schools were schools run by the parish and subsidised by the local landowners. The teacher was chosen by the Kirk Session of the local church to teach academic subjects and the Christian religion.[16] None of the schools in Liberton were as well-funded and attended as the parochial school, which Henry presumably attended. (In 1839, this was in Burdiehouse, though this was not Henry's; it had only commenced the year before, when Henry was fifteen). No mention is made of schools in the previous *Statistical Accounts* of Liberton in the 1790s.

We are speaking of a time, before universal education, when children were not required to attend school. Most poor children didn't. Although mixed education was the norm here in Scotland, the system favoured boys, especially at secondary level. 'The special characteristic of the Scottish parochial is this', ran an 1871 church report, 'that it is a *lower secondary* as well as a primary school, and thus connects [some of] the clever and aspiring boys of the humblest classes with the Universities and higher walks of life…'.[17] If Henry went to school in 1828, and if he attended a parochial school like the one described, then you could say he had a head start in life. Which is always an advantage!

Although the city of Edinburgh expanded to engulf Liberton in the nineteenth century, the 1852 Ordnance Survey map shows the city still two miles off from Stenhouse; it was a rural environment that the young Henry grew up in. Apart from short stays in London and Salisbury he stayed with the countryside, living in or around villages and small towns. In this way he went against the urbanising trend of the times. In the broad view of history such people can get overlooked, yet in their stories we learn that the great tides of change flow into numerous gulfs and inlets.

So, a rural environment, yes; but for diversion Edinburgh was within easy walking distance. Easier still, if you hitched a ride on the carters' wagons taking lime, and, until Henry was eight, coal to the city from the mine and quarry at Gilmerton. Nether Liberton, in the north of the parish, was large enough to hold a weekly market and once a year a 'Carters' Play'. Gilmerton had anciently had an annual May Day play, 'Robin Hood and Little John', unfortunately suppressed in 1600 for 'excessive lewdness'[18] the consequences of post-Reformation zealotry. Revived in an expurgated version, it had in Henry's time become a procession of decorated carts, much like a modern festival parade. Traces of the old days remained, however. James Begg again:

'The carters have friendly societies for the purpose of supporting each other in old age or during ill health, and with the view partly of securing a day's recreation, and partly of recruiting their numbers and funds, they have an annual procession. Every man decorates his cart-horses with flowers and ribbons, and a regular procession is made, accompanied by a band of music, through this and some of the neighbouring parishes. To crown all, there is an uncouth uproarious race with cart-horses on the public road, which draws forth a crowd of Edinburgh idlers, and all ends in a dinner, for which a fixed sum is paid. Much rioting and profligacy often take place in connection with these amusements, and the whole scene is melancholy. There are other societies in the parish which have also annual parades with a similar result'.[19]

They weren't to everyone's taste, but the Reverend Begg was, perhaps, being slightly disingenuous. Another observer in the 1830s wrote of this season, whitsuntide, as 'the only ancient religious festival that has become a popular one since the Reformation, through the addition of a modern circumstance. Clubs, or Friendly Societies, have *substituted for the old church ceremonies*, a strong motive to assemble in the early days of this week as their anniversary, and the time of the year being so delightful, this holiday has, in fact, become more than any other, what May-day was to the people.'[20] (My italics). By the by, in the 1851 census James Begg's name is spelt *Jeynes* Begg. The census-taker, chosen by the parochial school-master[21] for his competence and local knowledge, was unlikely to be thrown by the regions' dialects for such a common name. So we can see, surely, how James pronounced his name and from that perhaps, his demeanour.

James Begg went on to become a well-known figure in Scotland[22] Politically, he mixed a laudable reformist agenda (better city housing, reform of the rural bothy system, temperance, education, a wider franchise) with a religious one. Here, an intense anti-Catholicism and opposition to attempts to reunite the Presbyterian churches seem, to most twenty-first century Britons, as obscure arguments and hatreds from the mists of history. When Henry eventually moved to the south of England, even then, in the mid-nineteenth century, he would have noticed a change in the religious climate to match that in the weather.

Henry's father, as a farmer, would have practised a four to five year crop rotation of potatoes or turnips, wheat, barley, grass and oats. [(23)] Hard physical work required a lot of carbohydrate and the farm crops produced it. The main difference with England was the predominance of oatmeal (for porridge) and potatoes, compared to wheat (for bread). The diet of the people of the rural south-east was extremely good. They had a better and more varied diet than anywhere else in Scotland, if bland by today's tastes. The diet of someone of Henry's place, time and background is quite well documented. Breakfast would have been porridge, usually with milk, sometimes with beer. There was no sugar. Tea, if taken, was drunk here. Lunch was generally broth and potatoes, sometimes taken with bread. Alternatively, there might be fish, cheese or meat (beef or ham). The broth would have been vegetables from the garden, or barleymeal or peasemeal from the mill at Moredun, with pot barley and small quantities of meat-fat or meat-stock. Supper was either porridge or potatoes again, usually with milk. [(24)]

Henry may have had the odd ginger-beer. His mother, being a grocer as she called herself, may have stocked them, and if made by herself or a neighbour, sold at less than the 4d. a bottle charged in the proper posh shops. [(25)] Any fruit from the garden would have typically included strawberries, raspberries and gooseberries.[(26)] The volume of porridge eaten was enormous, a working man having some five times more on his plate than he sensibly would today. No wonder that breakfast 'was traditionally eaten standing', says T.C. Smout, 'on the grounds that an upright sack fills fullest'. [(27)] The average Scotsman was 1½" (3.8cm) taller than the average Englishman in the nineteenth century. Perhaps we can put Henry's eventual height of 6ft+ (1.8m+) down to porridge!

That closeness to Edinburgh raised the possibility of an apprenticeship for Henry at the Botanic Gardens there. There is no mention of it on Henry's *Curriculum Vitae*, but there's a reference to a 'Royal Park' in the papers of the late sweet pea expert and collector of Eckford memorabilia, Bernard Jones. Henry would probably have left school at the latest in 1837, which leaves us with a year before he arrived at Beauly. When Henry went to work in London in 1847, he had a recommendation from 'W. McNab'[(28)] with him. William McNab was the curator of the Edinburgh Botanic

Gardens in 1837. So the 'Royal Park' is possibly the Royal Botanic Gardens and not the Royal Park around Holyrood Palace. The epithet 'royal' seems to have followed the Botanic Gardens around its various sites from its origins in the grounds of Holyrood Palace in 1670. Incidentally, the Regius Keeper, McNab's boss at the gardens until 1819 was Daniel Rutherford,(29) uncle of the novelist Sir Walter Scott, a frequent visitor to Liberton...

Be this as it may, shortly before Henry's death in 1905 a garden magazine article stated that Henry had merely 'come under the notice' of McNab while working in the Lothians in the 1840s, McNab giving Henry 'letters of introduction' to take with him to London.(30) Henry would at least have visited the Gardens and was probably known to the staff there, one of whom, in 1838, was the future plant hunter Robert Fortune.(31) Determined on a successful career in horticulture, contact with the likes of Fortune could hardly be bettered. Robert Fortune helped to make us a tea-drinking nation by appropriating tea plants from China to grow in India and Ceylon. He also gave us our now popular garden plants like *Weigela*, Japanese anemone, winter jasmine and winter honeysuckle. Such wanderlust didn't rub off on Henry but the determination to succeed did.

At this time, the Botanic Gardens were on a 14.5 acre (5.9 ha) site, later expanded, at their present site in Inverleith. The construction of a palm house, for which the government had advanced over £1500 had just been completed, in 1834.(32) Go into the glasshouses here on a quiet day in January. If you're not fourteen imagine you are and in a new, pre-Victorian world. A world warm, clammy and still, so unlike Edinburgh. A world of enormous exotic plants.

In 1838, Edinburgh was a growing city of 140,000 people. As elsewhere, urbanisation had brought problems. Waves of infectious diseases had hit Edinburgh throughout recorded history and there was now a typhoid epidemic. It culminated this year with 2,244 cases being admitted into the Infirmary. More than half of them probably died. A cholera epidemic of just six years before saw 1,886 people infected, of whom 1,065 died; a survival rate of 44%.(33) In an epidemic, as in a war, their end is unknown. With children particularly at risk, Edinburgh was to be avoided if at all possible. Isabel Eckford, her life prescribed by early motherhood, was determined that her sons would succeed. Henry would one day be a head gardener, Peter a master tailor. And so it was that Henry came to be stamping his feet outside The Crown in Princes Street, on that bitterly cold Monday, the third of December, eighteen-hundred and thirty-eight.

References to Chapter One

(1) A.B. Thomson, *Mean winter temperatures in Edinburgh, 1764 – 1963*, (1971), The Meteorological Office, p.15. It was the sixth coldest winter between these dates.

(2) Robert C. Mossman, 'The meteorology of Edinburgh, part 2', *Transactions of the Royal Society of Edinburgh*, Vol. 39, (1900), p.80.

(3) *The Post Office Annual Directory and Calendar for Edinburgh and Leith* for 1838-39, (1838), p.221.

(4) 'Famous gardeners at home', *Garden Life*, (26-9-1903), pp. 463-465.

(5) T.C. Smout, *A century of the Scottish people 1830-1950*, 1986, (1997), p.310.

(6) Rev. James Begg, Parish of Liberton, *The new statistical account of Scotland*, Vol. I, Edinburgh, (1845), p.12. (Written in 1839).

(7) Rev. C. Ferenbach, *Annals of Liberton*, (1975), p.38.

(8) Ibid, p.39.

(9) Rev. James Millar, Parish of Perth, *The new statistical account of Scotland*, Vol. X, Perth, (1845), p.139. (Written in 3-1837).

(10) J. Begg, op.cit., p.27.

(11) Ibid, p.1.

(12) George Blake, in J. Dixon-Scott, *Scottish counties*, (1950), p.xi.

(13) James Grant, *Old and new Edinburgh*, Vol. 3, (1882), pp.339-340.

(14) J. Begg, op. cit., p.18.

(15) Ibid, p.25.

(16) R.D. Anderson, *Education and the Scottish people, 1750-1918*, (1995), passim.

(17) 'Parochial schools (Scotland) bill', Special report, Education committee of the Church of Scotland, (1871), pp.13-14.

(18) C.Ferenbach, op.cit., p.37.

(19) J. Begg, op.cit., p.12.

(20) William Howitt, *Rural life of England*, (1845), p.444.

(21) Parliamentary papers, Vol. 85, (1852-53), p.11.

(22) John & Julia Keay, eds., 'Begg, James', *Collins Encyclopaedia of Scotland*, 1994, (2000), p.73; T.C. Smout, op. cit., pp.166, 181, 205.

(23) J. Begg, op. cit., p.13.

(24) I. Levitt, C. Smout, *The state of the Scottish working-class in 1843*, (1979), passim.

(25) MS 13971, Tradesman's account book for 1831, Walkers of Bowland archives, National Library of Scotland.

(26) John Claudius Loudon, *Encyclopaedia of Gardening*, (1822), p.1249.

(27) T.C. Smout, op. cit., p.16.

(28) H. Eckford, *Curriculum Vitae*, undated, author's possession.

(29) Deni Brown, *Four Gardens in One*, (1992), p.8.

(30) 'Occasional Interviews', *The Gardening World*, (8-7-1905), p.543.

(31) A.M. Martin, *Let the trees be your guide*, (2002), p.10.

(32) D. Brown, op. cit., p.10.

(33) R.C. Mossman, 'The meteorology of Edinburgh, part 3', *Transactions of the Royal Society of Edinburgh*, Vol. 40, (1905), p. 481.

Chapter Two

In the snowstorms, Henry would have seen little of this new land, the Highlands. The last twenty-five miles or so into Inverness would have been dark, too, at these latitudes, as would the final thirteen miles on from Inverness to Beaufort. For garden writer and Lowland Scot John Claudius Loudon, writing in 1821, the Highlands consisted 'of little else than dreary mountains and some moors'.(1) Like the political agitator William Cobbett, pushing the lot of the rural poor, he hated unproductive land. Yet in the same breath almost, Loudon extolled Inverness-shire for its fertile valleys, and mountains 'covered with natural forests inhabited by the red and roe-deer, the alpine and common hare…'.(2) Loudon was the most influential horticultural writer of the first half of the nineteenth century, William Robinson perhaps the most influential writer of the second. Like Cobbett, these two were pushy and prickly agitators, falling over their own opinions and prejudices in pursuit of their view of the common good.

Residents on the Beaufort Castle estate on the June evening of the 1841 census included head gardener Joseph Bain and his wife Ann, seven apprentice gardeners of which Henry was one, and a female servant, Beth Grant. Four gardeners lived out in the parish. They presumably worked at Beaufort, otherwise it would have been one head gardener and seven apprentices! In the summer of 1835, a little before Henry arrived in the Highlands, two head gardeners were making garden visits. The gardens of Belladrum's, close to Beaufort's, were under the charge of a Mr. Westwood, who got wind of their visit and left. Joseph Bain it seems was more sociable, for he stayed. A report duly appeared in J.C. Loudon's *Gardener's Magazine* for November, 1835.(3)

They began at the large, oddly shaped and waved 'crinkle-crankle' walled garden, of about 250 metres by 125 metres. They were struck, on entering, by 'some long rows of hollyhocks at the back part of the flower borders, which run along the principal walks; the front of the borders being stocked with herbaceous plants, and the most showy kinds of annuals'. Mr. Bain told them he would have preferred a straight wall to this wavy one as he thought it caused air currents. On the wall were peach trees, apricots and a vine whose fruit 'covered with hand-glasses, were swelling freely'. They noticed 'two very handsome cedars of Lebanon', (which shows just how big those walled-gardens were) some large standard cherries, and other fruit trees in grass. 'In a dung-pit.... were the remains of an abundant crop of melons...'. They were also shown the bee-hive, there to ensure fruit-tree pollination.

They then visited the pleasure-grounds around the big house, where they 'peeped along the terrace front, and saw there was a rich display of dahlias, and other showy flowers; and in a flower-garden adjoining were clumps, or beds, of salvia[s] fulgens, involucrata, and splendens, Fuchsia[s] microphylla and globosa, Senecio elegans flore pleno, and several sorts of pelargoniums, in fine flower'. They 'peeped' only, to avoid meeting the owners, for the terrace or parterre gardens were planted close to houses where they could be enjoyed from the bedrooms above, as well as at ground-level outside. Although they thought the grounds well laid-out, a defect was noticed. 'This fine place, which contains sufficient space for every kind of ornamental tree and shrub... is entirely planted with a few species, most of them common'. They saw a summer-house, with a rosary (rose-garden), noting also some purple beech, 'Lombardy poplars and weeping birch. The whole of this beautiful place was in the highest order; scarcely a weed was to be seen and the grass was smooth as velvet'.

The visitors didn't mention a bowling-green by the house or a lake-side ice-house shown on the Ordnance Survey maps of 1872 and 1875. The bowling-green might have been there then; the ice-house would certainly have been, but this was a functional necessity and no more to be remarked on than would the wells and vegetables in the walled gardens be. Henry was an impressionable fifteen when he came here and you can see his influences: 'the most showy kinds of annuals' and 'a rich display of dahlias, and other showy flowers'. Ornamental trees and shrubs were of little regard; familiar species were grown mainly for effect, rather than their intrinsic interest. Standards were high. There were no weeds, the fruit was well-grown and the lawns scythed as 'smooth as velvet' or as smooth as you could get them before cylinder mowers, just developed, became commonplace. The only passions aroused by summer bedding in these early days were for the plants themselves; Salvias, Fuchsias, Senecios and Pelargoniums.

'An apprentice', advised Loudon, 'besides studying his art in the garden of his master, should, as often as may be, visit those of his neighbours, and observe what is going on there'.[4] At the gardens of Belladrum, the Beaufort apprentices would have seen that 'plants with showy flowers appear to be here the chief object; and in a flower-garden, walled round, there was a most beautiful display of dahlias, Fuchsia[s] microphylla, globosa, gracilis, &c.; Petunia phoenicea; Senecio elgans flòre pleno; Verbena[s] chamaedrifolia, venosa, &c.; and several very fine varieties of pelargonium'.

Joseph Bain lived, as head gardeners usually did, in a house just outside the main garden. About the only thing that Mr. Bain had in common with Lord Lovat was that they both had English wives. As had the Reverend Begg, incidentally. It might have been something of a status symbol for the aspirant class-conscious male. It hinted at wider horizons. Henry would have lived with the other apprentices in a lodge or *bothy* by the garden wall. There is a potting-shed and glasshouse area shown on those 1870s maps against the outer south-facing wall. If they were there in Henry's day, then hopefully the bothy was there too, and wasn't the small building on the outer north-facing wall. An article in the journal *The Scottish Gardener* for 1854 suggests, however, that it could have been.

'The gardener's bothy, in nine cases out of ten, is a portion of a back shade facing the north side of a high wall, into which no sun's rays ever enter, and is consequently the fertile abode of damp and ill vapours, most injurious to the health, and also to the habits of the inmates. Much has been written of late years on the improvement of all sorts of garden structures, but the gardeners themselves seem to be overlooked for the most part. We know of more than one instance in noble establishments, where even the head gardener's house, if it can be designated by the name, is worse situated, worse ventilated, and in all respects an inferior structure to the hen-house and dog-kennels, not to say anything of the byres and stables, these being in general comfortable structures, otherwise the fine cattle kept in them would not thrive. It is, however, to the assistant gardener's "bothies" or "lodges", as they are generally called, that special attention should be directed; these are now the very same in many fine places that they were forty and even sixty years ago, when habits of cleanliness, order, and comfort were very little cultivated by the common people of Scotland. It is due to the directors of the Royal Gardens at Kew, Frogmore, and the Edinburgh Botanical Gardens, to say that the Gardeners' Lodges at these places are really comfortable and commodious, with separate apartments for sleeping, sitting and cooking. The same cannot however, be said of several of the other public gardens, many of which expend annually large sums of money professedly for the advancement of horticulture, some of which possess also well-stored libraries, to which the men employed in the place have practically no access'.

The 'back of the wall', the writer concludes, is 'the proper place for mushrooms, worksheds and storerooms' only. (5)

This mood was maintained by George W. Johnson's *The Cottage Gardeners' Dictionary* of 1854: 'Bothy: the lodgings assigned to young gardeners in the northern part of the kingdom, and miserable hovels they often were, and, in some cases, still are'. (6)

Twenty-one years later, the standard of accommodation for gardeners seemed to have stagnated, or worsened. The Scottish magazine *The Gardener* ran the following article in September 1875:

> 'Bothies. The title of my paper, I doubt not, will offend the ears of some of your readers. I know that there are those who think that it should cease to be used, and a term more "refined" substituted; but I hold that while there is one of those miserable dwellings in existence, that it is a more suitable name for it than any other. Any orra loft or outhouse can be manufactured into one of those nondescript dwellings; and when a rude grate is stuck in the one end of it, and a small window in the other end, or a skylight in the roof, it is considered quite good enough as a habitation for gardeners.
>
> To the health of their inmates the situation of bothies is the most unfavourable that can be chosen about the premises. The majority of them are against the wall behind the hothouses, and, of course, facing to the north, and in the company of tool-sheds, stoke-holes, &c. It is needless to say that a dwelling with this aspect is seldom gladdened by the rays of the sun. In many instances, great trees and shrubs prevent even the few late rays of the long June day from penetrating so far as the dwelling; and a barrier is formed as well to the entrance of free air. Thus air and light – two elements essential to the welfare of plants – are denied to the growers of them. Verily, those whose lot it is to work outside all day have no need to lay in a stock of air and sunshine; but what is to be said of

many who are obliged to be at work all day in hothouses? Need we wonder that many of them are cut off prematurely?

The writer lived in a bothy attached to a garden of some consequence, the owner of which had the reputation of being a gentleman of liberality. This bothy was situated on the northern aspect of the hothouse-wall, and near to the main entrance gate to the garden. It was completely covered by ivy – its one window included; and a thick wood, bordered by an impervious shrubbery, came within a few feet of its front. It consisted of one apartment – the floor of which was roughly laid with rude flag-stones – and although great fires were kept burning in the large grate of the further corner dampness persisted in keeping possession. It was furnished with two close beds, a rickety table, three rheumatic chairs, and a cupboard which tradition said stood entire and fresh a century earlier. The cooking utensils were, as they too often are, very deficient in number and quality – one article having to serve the purposes of two or three. And this dwelling – shall we give it a dignified name, or dub it hovel? – was the playground of all the rats about the premises. In those circumstances three men lived, cooked their meals, and slept. I have often wondered if the proprietor knew for what the building was used. I rather think he believed it to have been got up solely as a picturesque ornament to the landscape, and that it never entered his mind that it was the lodging of his gardeners... I never knew a more miserable bothy... but many which I have known were only a slight degree better.

Another objectionable situation is in the stable-yard, and close beside, and in some instances in the same apartment with men who have different hours and pursuits from gardeners, and in the hearing of underworked, restless horses, which keep up an incessant knocking many a time all night long.

There has been a great deal said lately about the education and self-improvement of gardeners, but what opportunity have they for improvement in many bothies? In those of one apartment, with several men living in them, the studious man rarely has an opportunity of giving his whole attention to his studies... It is a very hard matter that a young man who wishes to improve himself should, every hour of his leisure, be obliged to associate with, and listen to the nonsense of any fool that his employer may choose to engage. The presence, simply, of the lolling, lazy, aimless man distracts the attention of the man of a studious turn. I have lived in bothies in which, when any of us wished for quietness, we would retire from the kitchen to another room, and not a sound sufficient to disturb our studies would reach our ears. This is as it ought to be...'

'...in justice to a few estimable noblemen and gentlemen... they have, with praiseworthy liberality, erected commodious and healthful dwellings for their gardeners, and some of them, as part furnishing of the increased accommodation, have given newspapers and journals of horticulture'.(7)

Surprisingly, the best Georgian estates could beat the run of Victorian ones when it came to gardeners' welfare. In 1836, a Scotsman in Australia recollected his youth: 'A gardener's Lodge, in a first-rate place, in Scotland, when I was a young man, was more like a school than a lodge. In the evening, the young gardeners would study (principally from books), by their own exertions, grammar, arithmetic, geometry, trigonometry, land-surveying, mapping, mensuration, horticulture, botany, garden architecture, and geography. It was customary for the head or principal gardener to attend in the lodge for an hour or two in the evening, to teach the apprentices and junior men. Thus, their time in the evenings was spent in study, until they obtained good situations'.(8)

Henry had perhaps gone to Inverness on the advice of William McNab, so it is likely that this last description would have fitted Henry's situation. Lord Lovat was 'much esteemed by his tenantry as a kind and considerate landlord' who 'by expending immense sums in improving and adorning his estate, provides employment for a great number of the peasantry, who, were this not the case, might be obliged to migrate elsewhere...' Lady Lovat was given to 'unostentatious charity, and active benevolence'.(9) So wrote the parish minister in 1841, Henry's last year at Beaufort. The local minister may of course have considered it wise to eulogise the local laird. He was hardly in a position to criticise him in public, should he have wished to. He could, however, have damned him with faint praise, or silence. A short perusal of other parish accounts show that ministers generally avoided the problem by ignoring the lairds altogether.

In later years, Henry spoke fondly of his early life. In his last years it was said that 'his mind often wandered to the Highlands where the fancies and frolics of youth were still remembered'.(10) Henry's later progress must surely have been built on the firm foundation laid for him here in Inverness-shire.

~~~

'The period of apprenticeship being finished, that of journeyman commences, and continues, or ought to continue till the man is at least twenty-five... During this period, he ought not to remain above one year in any one situation....'(11) Henry took this advice of Loudon's pretty much to heart on his return to the Lothians. Between now and 1847, he was to work at four different gardens. After a short stay at Stenhouse perhaps, he was off again, just seven miles this time to Kirkliston in Linlithgowshire, and the gardens of Newliston.

The house and gardens of Newliston face south-eastwards with the gentle lie of the land. At one time there was a grand approach to the house from the south. Just before Henry's arrival it was changed for something more subdued. To the visitor, the mansion now presents itself rather suddenly, half-apologetically. Imposing in a smaller setting, here it must always have looked out of scale. Henry was to work under the youthful head gardener Alexander Gibson. Mr. Gibson's profession was listed in full in the 1841 census, not as 'MS' for male servant as many of the other Newliston employees were. Henry's employer, James Maitland Hogg, was listed as 'of independent means'. The new Victorian age was making itself known in Newliston. Irish navvies were building the Edinburgh to Glasgow railway across an outer part of the estate. It was Irish labour that had built the Union Canal here, twenty years before, when a distillery was built at the local village of Kirkliston. J.C. Loudon was uncharacteristically impractical on the question of alcohol. To suppose that young gardeners 'would indulge in inebriety, or in alehouse society, is so entirely out of the question that we shall not enter on the subject',(12) he wrote.

Visualise a mansion surrounded, at least at the back, by an enormous garden. This garden or shrubbery was the 'pleasure ground' or 'the Policy' or 'Policies' in Scotland. Enormous walled kitchen gardens lie within or without it. Somewhere near the house, in a southern aspect, is a parterre. These grounds in their turn were surrounded by a much larger area of tree-studded parkland. Hardy fruit like apples may be growing outside
~~~

the walled garden in an orchard. Such was the theme and minor variations were played out on it.

The grounds here at Newliston were laid out to a formal design between 1722 and 1744. This was the time that the aristocrats' formal garden gave way to the equally contrived informality of the landscape garden. The garden grounds are not quite square, at 600 metres by 500 metres, and are out of alignment as if buckling under the stresses of the new ideas. The owner at the time had been ambassador at Versailles, which accounts for the unusual number of water features – ponds, cascades and canals, inspired by the famous garden there. Versailles-like, he also framed views to focal-points outside the grounds.

This same owner (the 2nd Lord Stair), having led an army during the eighteenth-century Austrian Wars of Succession, had avenues of lime trees planted in the south-east corner of the grounds in commemoration. They're formed into the pattern of the Union Jack, with a statue of Hercules or Atlas, still there, in the centre. The design that Henry saw would probably have been formed by pleaching the limes together – the large 'hedge on stilts'. They wouldn't have been pollarded as they were when they were all felled in 1964. Pollarded limes can't be formed into a design and anyway, why take a utilitarian practice into a pleasure-ground when you have this ideal, wholly ornamental substitute. J.C. Loudon too spoke of limes making good tree hedges.(13) Another unusual feature was the horse-shoe shaped enclosure, the present grassed area in front of the house. It is surrounded by a ha-ha and was used to school horses.

The fascinating eighteenth-century designs of the garden were all water off a duck's back to Henry, who never showed the slightest interest in this side of things. Henry was more fascinated by the new showy tender plants that were arriving in Britain from places like South America. And in this, he was of his age. These plants would have been grown, at least initially, in the walled garden. Even at 200 metres by 60 metres, the walled garden was smaller than Beaufort's and here lay within the south-west corner of the main garden, next to the bowling green. It had a very tall north-wall, providing an advantageously large south-face. There was, and still is, a delightfully strange eighteenth-century sun-dial in the middle of it.

Henry may have been here, in the walled garden, on the 3rd July, 1843, when Newliston lay in the path of a severe gale blowing eastwards to Edinburgh. It was reported that fruit-trees and bushes were stripped, and fast-ripening wall-fruit destroyed. 'So wild a tempest has not been experienced at this season for at least twenty years'.(14) Since the year Henry was born, in fact. What would he have made of it? A practical youth, he may have noted that a solid structure like a wall doesn't protect from wind. It merely causes it to eddy, creating more damage. Up to around 1970, there were 'fruit and rose gardens divided by hedges and herbaceous borders'[15] here, which may possibly have been how things were in Henry's day. It's a great pity that none of the scores of people who went to Staffordshire to write about Trentham Gardens came to Linlithgowshire, pushed open one of the four iron gates into the walled gardens at Newliston, and gave us a description. But nobody did!

The gardens and the gardeners have long gone and on a visit on a bright sunny day in late spring the still intact but crumbling walls enclose nothing but ghosts. Could such spectres materialise, the new things they would notice, among others, would be the clatter of trains on the main-line railway, a background traffic hum from the busy Edinburgh road and from Newliston's hemming in by the M8 and M9 motorways, and the roar of aircraft from being under the flight path of Edinburgh airport. As elsewhere – and you can understand the sentiment, if not the logic, of American tourists' complaints of our having built twelfth-century Windsor Castle so close to noisy Heathrow Airport – what we have created to enhance our lives has so often had the equally opposite effect of diminishing it.

~~~

About mid-way between Perth and Dundee lies Fingask Castle. Henry began work here on Tuesday 20th February 1844, under the head gardener Peter Loney. He stayed here a year, for he 'departed from Fingask 22 Feby. 1845'.[16] A visitor in 1828 described Fingask as 'beautifully situated on what are called the *Braes of the Carse,* commanding an extensive prospect over the rich and fertile vale of the *Carse of Gowrie,* which here, surrounded by cultivated hills, opens in one vast amphitheatre, with the river *Tay* rolling through it for upwards of fourteen miles, till it is lost to sight in the Bay of St. Andrews'. [17]

The mansion is sited on the western slope of a deep ravine formed by the Craig Burn. The design of the immediate landscape had been influenced by the attractive *picturesque* style at the turn of the century. This included planting up the sides of the burn with trees and shrubs, so becoming, as a writer from Perth enthused in *The Scottish Gardener,* 'more wildly romantic than it was before'. The young writer, David Gorrie, loved the flowers growing wild in this artificial landscape. 'The all-prevailing ivy, the wild Thyme and mountain Pink, the Primrose and the Cowslip, the
~~~

sweet-scented Violet... growing in all the profusion of nature...'.[(18)] Gorrie died three years after writing this, leaving us his youthful enthusiasm to share.

A later visitor, writing in 1864, was similarly entranced by this slightly artificial landscape, ignoring like the others the more artificial gardens and grounds. 'The view of the lower part of the dell is picturesquely terminated by the bridge across which the carriage avenue to the castle is carried, but the eye wanders at will far away up the hillside, and rests with pleasure on the undulating slopes which rise so gently and yet boldly too, from the mass of verdue at their base'. Such 'sylvan beauty' was 'alone worth the journey to Fingask to see'. From the house 'is obtained a splendid view of the Carse of Gowrie, which stretches away to the south and east level as the sea, till bounded by the river and the hills of Fife'.[(19)] No-one mentioned the large orchard, which lay in the south-east corner of the grounds across the ravine from the house. It was three times the area of the new walled garden, which when Henry had arrived, had just been built onto the front southern edge of the grounds.

The building of the kitchen garden, with a hothouse on the southern inner side of its north wall and a shed behind it on the other side, together with a little work elsewhere in the garden, cost the owner Sir Peter Murray Thriepland, Bart, £872 -9s -2d. The cost translates to twenty-eight years of Henry's wages of 12/- a week. This was a major job, and from the documentary evidence [(20)] it appears that outside contractors were involved. It might have been that workers from elsewhere on the estate were drafted in, but the wages, higher across the board than those of the regular employees, suggest irregular employment. Outsiders were also, it seems, employed for jobs like 'mole catching' and the regular mundane work of hedge-cutting and re-gravelling the paths. This work was on piece-rate and the workers wrote and presented their own bills. If they were from elsewhere on the estate, one would imagine that a claim on any garden's budget would have been made by their superiors.

The land-surveyor, James Young, took his first walk round, to look for a suitable site for the new garden, on 26th November 1838. A brickfield also had to be chosen. This was a field where clay could be removed to make bricks for the walls. They rented a field from a farmer from the nearby village of Rait. He was compensated by £18-12s-10d for the loss of a year's crops. The field was later reinstated with carted-in soil, at the cost of £15. Building began early in 1839. The bricklayers were on 4/- a day, a mason on 3/6, labourers on 2/-. Stones for areas like the gate surrounds

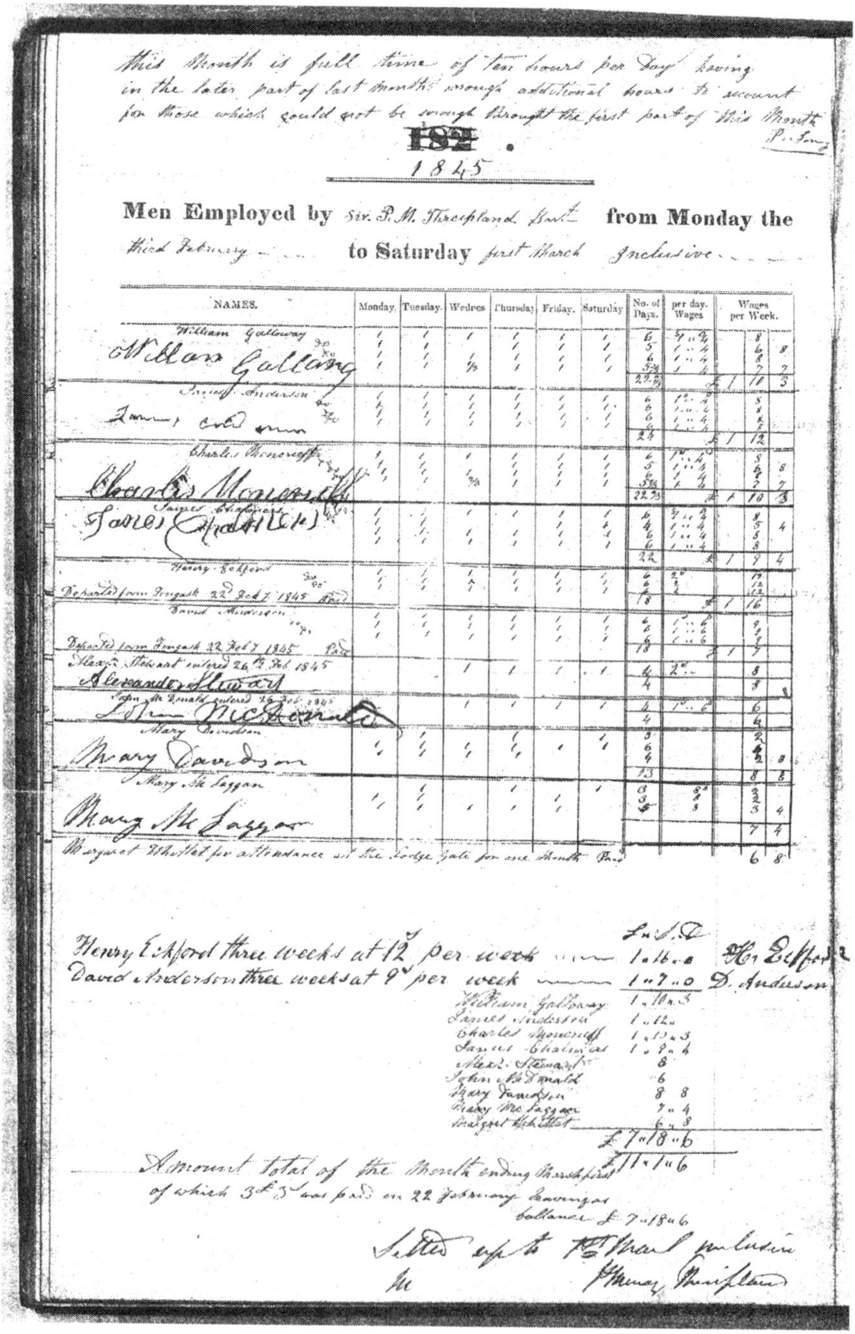

This Month is full time of ten hours per day having in the later part of last months wrought additional hours to account for those which could not be wrought throught the first part of this Month P. Fong

~~182~~ .
1845

Men Employed by Sir P. M. Threipland Bart. from Monday the third February to Saturday first March Inclusive

NAMES.	Monday.	Tuesday.	Wednes	Thursday	Friday.	Saturday	No. of Days.	per day. Wages	Wages per Week.		
William Galloway — William Galloway	/	/	/	/	/	/	6	1 .. 4		8	
	/	/		/	/	/	5	1 .. 4		6	8
	/	/	/	/	/	/	6	1 .. 4		8	
	/	/	⅔	/	/	/	5⅔	1 .. 4		7	7
							22⅔		£ 1	11	3
James Anderson — James Anderson	/	/	/	/	/	/	6	1 .. 4		8	
	/	/	/	/	/	/	6	1 .. 4		8	
	/	/	/	/	/	/	6	1 .. 4		8	
	/	/	/	/	/	/	6	1 .. 4		8	
							24		£ 1	12	
Charles Moncrieff — Charles Moncrieff	/	/	/	/	/	/	6	1 .. 4		8	
	/	/		/	/	/	5	1 .. 4		6	8
	/	/	/	/	/	/	6	1 .. 4		8	
	/	/	⅔	/	/	/	5⅔	1 .. 4		7	7
							22⅔		£ 1	10	3
James Chalmers — James Chalmers	/	/	/	/	/	/	6	1 .. 4		8	
	/	/			/	/	4	1 .. 4		5	4
	/	/	/	/	/	/	6	1 .. 4		8	
	/	/	/	/	/	/	6	1 .. 4		8	
							22		£ 1	9	4
Henry Eckford — Departed from Fingask 22d Feby 1845 Paid	/	/	/	/	/	/	6	2		12	
	/	/	/	/	/	/	6	2		12	
	/	/	/	/	/	/	6	2		12	
							18		£ 1	16	
David Anderson — Departed from Fingask 22 Feby 1845 Paid	/	/	/	/	/	/	6	1 .. 6		9	
	/	/	/	/	/	/	6	1 .. 6		9	
	/	/	/	/	/	/	6	1 .. 6		9	
							18		£ 1	7	
Alexr. Stewart entered 26th Feb 1845 — Alexander Stewart			/	/	/	/	4	2 ..		8	
							4			8	
John McDonald entered 26 Feb 1845 — John McDonald			/	/	/	/	4	1 .. 6		6	
							4			6	
Mary Davidson — Mary Davidson	/	/	/	/	/	/	3			2	
							6			4	
							4			2	8
							13			8	8
Mary McLaggan — Mary McLaggan	/	/	/	/	/	/	8	8		2	
							3	8		2	
							5	8		3	4
										7	4
Margaret Whittet for attendance at the Lodge Gate for one Month Paid										6	8

	£ s. D	
Henry Eckford three weeks at 12s per week	1 . 16 . 0	H. Eckford
David Anderson three weeks at 9s per week	1 . 7 . 0	D. Anderson
William Galloway	1 . 11 . 3	
James Anderson	1 . 12 .	
Charles Moncrieff	1 . 10 . 3	
James Chalmers	1 . 9 . 4	
Alexr. Stewart	8	
John McDonald	6	
Mary Davidson	8 8	
Mary McLaggan	7 . 4	
Margaret Whittet	6 . 8	
	£ 7 . 18 . 6	

Amount total of the Month ending March first £ 11 . 1 . 6
of which 3£ 3s was paid on 22 February leaving a
ballance £ 7 . 18 . 6

Settled up to 1st March inclusive
M
P. Murray Threipland

The Fingask 1845 Time-Sheet

were carted-in from the village of Kingoodie, seven miles distant, at a cost of £6-19s-10d. An earthquake occurred right in the middle of things, in 1840. The job was eventually completed however, and the final payment made, on the 13th January, 1841. All the work and the costs were beautifully written down in a seven-page invoice.

So, Henry had an almost brand-new kitchen garden to admire when he started work here three years later. It was, it must be admitted, a rather odd shape. Sir Peter had decided on an octagonal garden. There's nothing especially wrong with that. In theory, it should slightly increase your growing options by increasing the number of aspects. As well as the four points of the compass you also get north-east and north-west, south-east and south-west. At least, you would if the garden was so aligned to the compass. But it's not, quite. And the octagonal shape is unsymmetrical too. We could blame the earthquake but apparently 'designers' plans were rarely carried out exactly'(21) anyway. It didn't much matter. That's the beauty of gardening. It's rarely a tragedy when things go wrong. The sky won't fall in on you!

~~~

Victorian Britain was still a very unequal society. Henry's wages of £31-4s a year were slightly above the average for the 80% of Scots regarded as being of the 'manual labour class' in an 1867 report. The report also showed that one third of one per-cent of the Scottish population (just some 4,700 people) received on average £3,952 a year, or 25% of the national Scottish income. (22) From the evidence of the Fingask time-sheets, we can see that Henry was one of eight employees; five men, two part-time women and an apprentice. All of them were able to sign their names. The head gardener, Peter Loney, doesn't appear here. He would have been paid separately. Henry appears to be a foreman; at least, he is paid above the others. Everyone works a ten-hour, six day week, with Sundays off. During the short winter days from mid-November the hours were reduced to seven. The transition back to a ten-hour day occurred gradually between mid-January and mid-February. Almost no-one took any time off. Henry never did. There were no holidays, apparently, not even at Christmas or New Year. Any absence, say for illness or injury, even of less than half a day, meant no pay. Henry, on the highest wages of 2/- a day, is on 12/- a week, winter and summer. The four men are on 1/2d a day, 7/- a week in the winter and 1/4d a day, 8/- a week in the summer. In November 1843 the apprentice, Andrew Smith, began the second year of his apprenticeship
~~~

with a rise of 1/-, to 6/- a week and left, or died, before his second year was up. The two part-time women were on just 8d a day. They must have been supplementing other income as this was not a living wage. The 'Harvest Month' from the middle of August to mid-September 'gives additional wages to the Labourers'. Nearly everyone except Henry earned a higher rate then.

Year round, the ordinary men's wages averaged out at 8/- a week. This hardly compared to England. In 1837, even the agricultural rates there were averaging 10/3d.[23] Fingask's 8/- a week equals the very lowest English agricultural rates of the rural west-country. It was no wonder, surely, that Scotland had so many emigrants. Scots with 'get up and go' got up and went.

~~~

'The last thing we shall here recommend to the young gardener is', wrote J.C. Loudon, 'to keep a *pocket memorandum book*, for taking notes of every thing, whether professional or general, which strikes him at the time as interesting'.[24] Very few of these very useful gardeners' tools survive. None of Henry's have. They were an *aide mémoire,* a reminder of work to be done, of how success was achieved or mistakes made. We are fortunate here to have the *Diary of operations performed by P. Loney at Fingask* from 18th February 1839 to 1st November 1840.[25] Peter Loney was still head gardener here in the 1870s, so he had probably just taken charge now. Keeping a record of his first year or so of work was an excellent idea. The diary is perfectly literate, with few spelling mistakes. The dialect is occasionally revealed in the grammar. A slight criticism of him is that he only writes of what he himself was doing, not of the jobs he might have set others to do elsewhere, or in his absence. With one or two exceptions, the work described was the sort of work that Henry would have done too. Here are some extracts, beginning with April, 1839:

Sunday 5th	Watching about Fingask Grounds (rain wind North east)
Monday 6th	Cleaning the leaves &c out among the laurels between the Castle and Stables (dull day wind east)
Tuesday 7th	Cleaning the bank out the side of the approach road. At the low lodge for a parcel (Sun east wind)
Wednesday 8th	Making Cuttings and planting them into flower Pots (Sun east North)
~~~

Thursday 9th	Turning the hot bed to increase the heat in it and with the woman at the approach road cleaning (sun cold north east wind)
Friday 10th	Cleaning the coach road and the walk from the flower Garden to the upper fog house (Sun with cold east wind)
Saturday 11th	At Perth for a hamper of flowers from the farrier and having no other way to get them home had to wait for the coach to convey them and me to the lodge for which I had to pay 2/-. Planting and securing the plants that was in the hamper (cloudy Sun wind North west).

Spending his Sunday off looking around the grounds was a good way to get the 'feel' of a new place. The 'fog house' was a little garden summer-house. Mr. Loney had every right to moan about his stage-coach fare. Two shillings was about two-thirds of his day's wages. We can be sure he improved his travel arrangements the next time he visited Perth! Much of the following week was spent in preparing the ground for Dahlias, which were planted out on the 20th. There's still a danger from frost with planting now but we're near the sea, which moderates the climate. Another possibility is that Dahlias were hardier then; that later hybridists, neglectful of hardiness, allowed the plant to become more tender.

The entries for his work done on Christmas Eve and Christmas Day 1839 and New Year's day 1840 were rather gabbled. He obviously had better things to do with his free time then! Sundays were spent either at home, walking around the grounds, crossing the Firth of Tay to Fife, or going to church. Sometimes his wife went to church without him. The following March he had the nearest he got to a holiday with five days away in Edinburgh. This was to get cutting material. His itinerary would probably have included visits to the Botanic Gardens and the adjoining gardens of the Caledonian Horticultural Society at Inverleith. A month later he was getting cutting material locally:

7th Tuesday	At Megginch for Cuttings and at Rossie Priory for Cuttings and planting them into Pots. (Sun & cloudy wind north)
8th Wednesday	Making and Planting cuttings of greenhouse Plants and Potting dahlias (sun & cloudy wind)
9th Thursday	Lifting and Planting roses, hollyhocks, Canterberry bels [sic] Sweet williams &c and watering the flower

garden and gathering Sweet scented violets and other Flowers. (sun & cloudy)

In August he had another chance to socialise with his peers.

28th Friday At Perth Competition of Flowers & fruits vegetables roots [i.e. root crops], &c (Cloudy with sun)

These competitions were popular events. That was the third one he'd been to at Perth since May. In the contemporary *Rural Life of England,* William Howitt (who saw no incongruity in including Scotland in a book on England) wrote of a country excitement that had 'sprung up of late years, and have done much good – the floral and horticultural shows. These have been warmly patronised by the aristocracy; and it forms a striking feature in modern country life, to see carriages and pedestrians hastening... to... where different flowers and fruits, in their respective seasons, are displayed with great taste, and with brilliant effect. The place of meeting is sometimes at a country inn, where, on the bowling-green, tents are pitched, [!] in which the flowers or fruits are exhibited, and the whole scene is extremely gay'.(26)

Peter Loney also travelled to Perth to buy seed and all the other things he needed for the gardens. For seed only he went farther afield to Edinburgh, to 'Peter Lawson & Son, Seedsmen & Nurserymen, N°.1, George 4th Bridge'.(27) Knowing the city-centre street of George IV Bridge today, it's difficult to imagine the slower age when a garden nursery could thrive there. Twenty days before Henry arrived at Fingask, Mr. Loney had made the forty-five mile journey to Edinburgh to buy seed from Lawsons'. This would have been a ten-hour round-trip by coach, but Mr. Loney was buying large quantities of seed and would have wanted to see what he was buying. It was also a rare day out in the city.

Although primarily after flower seed, the invoice shows that he also bought thirty-one different vegetable varieties. These included one packet of Early Purple Artichoke at 6d, ½ lb. (250gm. – the seed weight) of Late Curled Leaved Broccoli at 9d, and one packet of New Curled Leaved Turnip Rooted Celery, at 3d. As Mr. Loney bought most of his vegetable seed locally, his intention here may have been to try out some new varieties.

The expenditure on vegetable seed alone came to 18/-, or nine days of his foreman Henry Eckford's wages. But the real money, £5-18-7 to be exact, was spent on flower seed. The invoice lists fifteen items. They included 178 varieties of annuals and biennials at £2-4-6; forty packets of perennial flower-seed at 10/-; Asters, wallflowers, larkspurs, lupins and Tagetes; ½ lb. of mignonette seed and eighty-two different kinds of stocks. Three shillings-worth of sweet pea seed concluded the list, to become the plants for Henry to look after here in 1844.

With a large 2/- box to put all the different seeds in, the bill totalled £6-18-7. It was paid on 25th May, with a discount of six shillings and seven pence. Although this was nearly four months after purchase, tradesmen expected to wait a year or more to be paid, so the discount was probably for this 'early' settlement. By May, too, the seed having been sown, Mr. Loney could see how much of it had been viable!

'Dickson & Turnbull, Nursery seedsmen and florists' were Mr. Loney's suppliers in Perth.(28) We have a good idea of what Henry was working with from the quite detailed invoices. They can be followed from a little before he started at Fingask to the day before he left. The seeds purchased show that a wide variety of vegetables were grown. Between September 1843 and February 1845 these were principally peas – twelve varieties, with names like 'Blue Sabre' and 'Improved Green Imperial' – and beans like 'Windsor' and 'Royal Dwarf'. Peas and beans were the favourites here as elsewhere in Victorian Britain.

'October 8th, 1843. 3 Pruning knives, 6/-'. A knife is a tool that professional gardeners buy themselves and use without sharing. Here they were bought for them, for they cost the same as a day or more's wages.

'September 27th, 1844. 1lb. Tobacco, 4/6'. Tobacco, containing nicotine, was used as an insecticide and could affect the users along with the insects. These were the days before health-warnings!

'February 16th, 1844. 8 Large mats, 12/-, 6 small mats, 4/6'.

'November 19th. 1844. 3 Nets, 6/-'. Nets and mats were used to protect the fruit and vegetables, although 'the best use' for mats was 'as a packing envelope; for, as a protection to wall-trees, they are inferior to netting... They are very serviceable, however, to place over beds of early spring radishes, &c, to prevent the night radiation.'(29) Nets were 'employed to prevent the radiation of heat from walls, and the rude access of wind to trees grown upon them, as well as to prevent the ravages of birds'.(30)

'November 3rd, 1843. 28lbs. Cast-iron nails, 5/10; 4lbs. Shreds, 2/8.' Shreds were strips of cloth used to support wall-trained fruit-trees. Tied around a branch, the two ends of the strip were doubled over, with the nail driven through four thicknesses of cloth into the wall. (31)

'August 16th, 1844. 24 flower-pots N°30, 2/-; 24 ditto N°24, 3/-; 24 do. N°18, 4/-; 24 do. N°12, 6/-; 12 do. N°8, 4/-; 6 do. N°6, 3/-; 6 do. N°4, 4/-'. The numbers related to the size of the flower-pot. It was the number of pots that could be baked at one time on a potters tray or cast, [32] so the smaller the number, the bigger the pot. In his 1822 encyclopaedia, Loudon sized them both by inches (the top of the pot's internal diameter) and by numbers.[33] By 1877, and probably earlier, numbers were written of as the 'old name'.[34] If this was so, they took a remarkably long time to die. When I began work as a gardener in the 1960s this sizing by numbers had actually been growing. For example, in the most popular size, 60 (3 ½"), there were now 'small sixties' (3") and 'large sixties' (3 ¾ "). Sizing by numbers only declined with the advent of plastic pots, and metrication in the 1970s finished it off. Since when, of course, we went back to sizing them in inches!

'May 2nd, 1844. ½ peck Paceys Perennial Rye Grass, 1/-; 2lbs. White Clover, 2/6'. Spring is the best time to sow grass-seed, and clover was commonly in the mix then. Because the only grass species is Perennial Rye, this probably went down in the orchard.

'July 12th, 1844. 6lbs. Crested Dogstail, 6/-; 2lbs. Poa trivialis, 2/-, 2lbs. Poa pratensis, 2/-; 4lbs. Hard fescue, 4/0; 4lbs. Paceys Rye Grass, 1/4. Bag, 1/-'. This is a general, hard-wearing grass-seed mixture. Buying in July, it won't keep till the spring, so Peter Loney must have had a summer or autumn sowing in mind. The relatively small quantity of seed suggests patching up, by over-sowing, and the presence of rye-grass (*Lolium perenne*) means it would be to the lawns, not the bowling-green. However, if the bowling-green got treated like the one William Howitt mentioned earlier, as a pitch for tents, it would need all the rye-grass it could get!

'April 19th, 1844. 1 ½ dozen scythe-stones, 9/-; 3 patent scythes, 36 in., 10/6; 1 scythe, 26in., 2/6'. Lawns were cut with scythes, and their blades were sharpened with scythe stones. 36" and 26" were the lengths of the blades. Sharpening stones, like blades, wear out over time. But the reason they're buying so many stones is this. You sharpen your blade and put the stone down in the grass, forgetting where. Even if you know, someone else won't, and comes along with a barrow and breaks it. Sharpening with a broken stone would put your hand too close to the blade, so you throw it away. Thus, eighteen scythe-stones.

'November 19th, 1844. 1 Ryder trained Peach, 10/-; 1 dwarf trained Nectarine, 7/-'. When the odd fruit tree needed replacing in the walled garden, it was more convenient to buy in from a nursery than to grow your own. Being 'trained' means that they'd been grafted onto a rootstock and grown on for two years.

'February 1st, 1844. 3 pecks Royal Dwarf potatoes, 4/-'. They don't seem to have bothered growing potatoes at Fingask. Here, they seem to be buying in a few out-of-season ones, perhaps new, having been forced at the nursery. They got most of their potatoes from local farmers, which they bought in in tons, or rather, by capacity in 'Bolls'.

Apart from '1 large garden line, 1/6' (when did you last see one of those?) there were many items such as gloves, a spade, grass edging shears, dutch hoes and draw hoes, that are familiar today. Shortly before Henry left Fingask, five thousand five hundred small larches and scots pines (called 'Scots Firs') arrived. They must have been for the park beyond the gardens and of no direct concern to Henry. However, they would have had to be 'heeled-in' immediately on arrival. They are an unpopular delivery if they arrive late in the day as everyone has to stay behind. One thousand five hundred of them were bought on Saturday 21st February 1845, Henry's last day working at Fingask. It's an interesting fact that Henry never seems to have developed an affinity with trees, forestry or ornamental!

There was a wide variety of different trees on the estate. In the park and in the woodland garden by the burn they were a mixture of evergreen and deciduous conifers, broadleaved ornamentals and fruit. Hawthorns were used for hedging, though occasionally allowed to grow naturally, as ornamental trees. Some fruit trees, the few that benefitted from special conditions, were grown in the walled garden. The majority of fruit trees though, grew in the large orchard on the other side of the burn. It was so large in fact that it grew more than the estate required, or at least more than Sir Peter required, and the apples and pears there were grown for income. Once a year they held a roup at Fingask, an auction where the apples and pears were sold to the one highest bidder.

In the summer of 1844, when Henry was there, it was held on Monday 5th August. They realised £101-5-11, minus £5-1s. for 'ready money'. The buyer, David Whittet, gathered the fruit as it ripened between 25th September and 21st October. In the buyer's zeal to keep outsiders away from 'his' property, he had to be reminded to allow estate workers access to the orchard! Mr. Whittet had acquired sixteen kinkins (small barrels) of apples and six kinkins of pears for his money. The apples included the varieties 'Strawberry', 'Red Apple', 'Golden Russett', 'Lemon Pippin', 'Irish Codlin' and 'Ribston'. A decent number from around one thousand five hundred named varieties in Great Britain then. The only pear mentioned was

'Galston'. Many pears then had French names, having been raised there. France and the Channel Islands are still the best places to grown them. 'No fruit ripens in England but a baked apple' was a French observation.[35]

On Saturday 22nd February, Henry worked a ten-hour day, as always. It may have been longer if those pines and larches had arrived late on the day they were bought. In the days before the retained employee apparently became the norm, foremen were hired by the year. Henry had been at Fingask for a twelve-month and on the following day he left. There was no overlapping with the man who came to replace him next week. Just as when Henry had started, his predecessor had left three days before. There were no wages wasted on the old foreman showing the new foreman the ropes. That was entirely the job of the head gardener. Nor, incidentally, could any little dodges or wrinkles that might undermine the head gardener's authority be passed on directly.

Henry would have been living in the bothy with the other unmarried gardeners, or in other tied accommodation provided by his employers. So part of his income would have been in terms of a cheap or free rent, an allowance of coal and oil for heating and lighting, of fruit and vegetables from the garden. And, hopefully, oatmeal, potatoes and milk. The indication of higher wages being paid to self-employed outsiders shows that such allowances were reflected in a lower wage packet. Being single and without dependants, Henry would have been wise to put some of his 12/- a week aside for when he was married. What would 12/- a week have bought in the 1840s? Still in the relatively affluent south-east but sixty miles to the south, one answer is provided from the account-book of a family from near Galashiels.[36] These particular people were well-off and had a town-house in Edinburgh, but some accounts from May to June 1845 appear to be from local Scottish border suppliers rather than expensive high-class Edinburgh establishments.

Looking at items that might not have been directly available to Henry from his employers, they seem mostly beyond the reach of a manual worker. A representative example is ½ lb. butter at 6d., Trout 7d., Crab 4d., 4 Flounders 8d., 6lbs. barley 1/6, six dozen eggs, 3/6, 6 Whitings 7 ½ d., three dozen Turkey eggs 3/-, brown loaf 3 ½ d., 5lbs. mutton chops, 2/6, 1 ½ lb. Salmon, 2/-. Bread, eggs and some of the cheaper fish like Whitings would have been possibilities. However, when Henry's daily 2/- wage could only buy 2lbs. of butter or forty-eight biscuits, with nothing to save for illness or old age, the 'going to the shops' or having goods sent from them still seems to have been an activity restricted to at least the relatively affluent. The Co-operative movement, begun at this time, was to bring down prices, as would, in a relative sense, increasing wealth. A trend away from markets and other street vending, in favour of shops, can be seen from the eighteenth century, if not earlier. However, it's easy to see how the traditional weekly or twice weekly market days, with the opportunity for isolated communities to socialise, remained an important part of ordinary people's lives throughout the nineteenth century and well on into the twentieth in many places too.

~~~

Take the borders road out of Liberton. After eight and a half miles, still in Midlothian, you will arrive at Penicuik. That delightful name comes from the Celtic, apparently, and means 'the hill of the cuckoo'. For some present-day inhabitants, the association made by those early settlers is now undignified, best ignored; pushed out of the Penicuik nest, in fact. Henry was like a bird. He'd flown off to Inverness and then come back over the nest to Newliston. Then he'd flown off Perth way and now, between the pull of home and the desire to make his own way in the world, he'd circled back.

In those days, Penicuik House [(37)] stood two full miles south-west from the then village of Penicuik, within an estate of some seven to eight thousand acres. The Pentland Hills lie to the north-west, while southwards the River North-Esk flows north-east on its way to the sea. Interestingly, on its journey from here the river is used to divide the parish of Cockpen where Henry's parents were married, from Lasswade where his mother was born. The park and the grounds (the policies) surrounding the house came to around a thousand acres. They were designed in the eighteenth century to the landscape ideals of the time. It included much tree planting, a subterranean passageway and three ponds, the largest being a lake with small islands. In 1730, a semi-circular walled garden was sited on the north bank of the river, facing south. By the 1820s it was already noted for the 'extensive range of hot-houses' it contained.[(38)] Vineries, stoves, conservatories and hot-houses seem to have been interchangeable names for any glasshouse, lacking the specific connotations they later acquired. With advances in glass manufacture in the 1840s, these 'hot-houses' were rebuilt.

There were two ranges of glasshouses in the walled garden and they appear to have been called the 'east vinery' and the 'west vinery'. They were sited in the traditional style on the inner south face of the north
~~~

wall. Behind the wall, on the north face, were the sheds and the bothy. In 1839, a C.H.J.Smith, landscape gardener, presented his bill. For plans and a specification for a new range of vineries with back sheds, of £20, plus travelling and 'postages' costs of 10/- and a day at Penicuik of £1-1-0, the bill totalled £21-11-00. It wasn't paid until 15th November, 1841. 'Messrs. Baxters, 18 Leith Street, Edinburgh, 5th November, 1842. For making, erecting, glazing and painting a range of vineries: £183-0-0.' This bill was paid by the estate's Factor, or manager, John Irving, a mere four months later on 13th April 1843. 'To James Thorburn, for building a vinery at Penicuik Garden, 1842: to building and finishing all the Mason work of Vineries and Back Sheds and Gardeners Room.... £110. Extra work in taking down the outer or back wall of the east vinery to the foundation and building it up again - £10', total bill £120. His bill was paid on 12th May 1843. 'To James Tod & Sons, Coach Spring makers and Smiths, 19th April 1842. To mounting for Ventilators of Vinery in two houses complete and put up at Penicuik Gardens... £12-8-10'. Bill presented 22nd October 1842, bill paid 22nd October 1843.[(39)]

Even the best employers thought nothing of not paying their bills for months or years. Eventually however, money had to be found. John Irving was in Edinburgh on the 30th June, 1844, and wrote from there to his employer, Sir George Clerk, in the following terms: 'The expense at Penicuik is great and constantly increasing. Your garden is a most unprofitable concern. I am certain that if you had had no garden you could have been £12,000 richer and could have given each of your younger children £2,000 more than otherwise they will get. The walks and shrubberies', he continued, painfully, 'are extending...'[(40)] Irving also complained of the unprofitability of the estate collieries and appears to have persuaded Clerk to lower the colliers' wages. All in all, Irving's surviving letters, and there are dozens of them, are a pretty depressing read. It takes the spring out of your stride, climbing up the steps of the General Register Office in Edinburgh's Princes Street, where these papers are housed. Another load of those Irving letters to get through! At least he will have died of something by now. Apoplexy probably.

One profitable source of income at this time was due to the building of the Edinburgh – Peebles railway. It cut through a far eastern corner of the estate, from which the valuation of damages due to loss of land came to £11,738-3-1.[(41)] Even this tidy sum was regarded with derision by Irving who wrote to Clerk, again from Edinburgh, on 7th April 1846: '...it will be a great nuisance as it is found scarcely possible for ladies to walk out where these Railway labourers are located in the neighbourhood. This is much complained of and is a loss not taken into account in estimating damages. The game and fences also suffer greatly.'[(42)] Irving seems more concerned to have missed a chance to claim compensation for sexual harassment, than for the harassment itself. Irving may have persuaded Clerk to lower wages, but as for cutting back on the gardens he may as well have talked to the bricks in the new garden wall. A new 'American' garden of acid-loving plants was formed in the 1850s and another walled garden was built later in the century.[(43)] Meanwhile, glasshouse rebuilding, and contractors' bills, continued apace.

11th November, 1844: To plastering fruit room and glasshouse by William Wilson, £6-19-1. 22nd October, 1845: To cash paid labourers of new vinery, £3-3-4. 3rd November, 1845: To glazing and plastering per William Wilson, Penicuik, £3-7-5. 10th August 1846: To painting, plastering and glazing by William Wilson, Penicuik, £9-5-1.[(44)] 27th February 1847: The Shotts Iron Company, Leith Walk, Edinburgh. For the expense of heating apparatus for the said new vinery at Penicuik House: £74-9-4. This included additions to the heating apparatus in the mushroom house, a building which would have been sited on the ideally dark and damp outer side of the north wall. Their work had been completed eleven months before their bill was paid. This 'new vinery', they explained, helpfully for us, 'was erected on the site of the West Vinery erected by the deceased Sir James Clerk of Penicuik, Baronet which had become ruinous and past repairing....'[(45)]

Henry's probable time here of 1845 was in the middle of exciting developments taking place at Penicuik. One was the large-scale tree planting, which was more on the forestry side of things and didn't involve Henry directly. The other was these new glasshouses, their modern technology raising new possibilities. Henry never showed much interest in trees and other hardy perennial plants, but besides fruit and vegetables he was drawn to those *arrivistes* the sub-tropical plants. Such plants needed to begin their lives under glass in Britain. On moving outdoors they gave rise to that nineteenth century innovation, the bedding scheme or 'bedding out'. Henry would have worked in the new glasshouses, and raised the new plants in them. Somewhere along the line, and it may have been here, Henry began to think of the development of these plants, and of his own future. It was later said that it was in the first half of the nineteenth century, very early in his career, that Henry had begun 'experimenting with flowers – fuchsias, geraniums and verbenas, notably in the direction of cross-breeding, and his successes in this unusual field of investigation were conspicuous.'[46]

Henry's head gardener at Penicuik was James Ramsay. He had arrived only recently, in the autumn of 1843. In the spring of that year, some evergreens had been ordered for 'close by the house' from the Leith Walk nursery in Edinburgh.[47] Penicuik may have been between head gardeners then, because they were requested by James Manson, who seems to have been in charge on the forestry side of the estate. The order included yews, Portugese laurel , box, *Rhododendron ponticum*, 'American Arborvite' (*Thuja occidentalis*), 'Chinese Arborvitae' (*Thuja orientalis*), *Aucuba japonica* and hollies 'smooth green' and 'striped'. To this list was added '2 double Ayrshire Roses' and '100 Irish-lvy Plants'. Evergreens supply winter colour but apart from splashes of cream and dull yellow, and red berries in winter on female Aucubas, this is almost unrelieved dark green. Nice and cool for Italy, say, but gloomy in Britain and in winter depressing. Its main practical purpose here would have been to keep the cold winds from the house in winter. The idea of restricting estate shrubberies to near the house was old hat even by the 1820s.[48] John Irving's complaint about the cost of extending the shrubberies was thus doubly futile, for it went against the fashion of the times.

So, these were the new plants around the house when Henry arrived a couple of years later. As far as we know, they held no interest whatever for him. On the list Manson sent to the nursery, he told the manager Mr. Shankley to 'put the plants carefully up so as they may be protected from the drought or frost. Let the cart as soon away with the plants as possible. The cart will come to the Leith Walk Nursery and take them...' Lifting bare-rooted, i.e. uncontainerised, evergreens is a skill. The soil must remain around the roots and the damp rootball tied firmly in sacking. However, the nursery would know the danger of allowing the roots to become dry or frosted. They were the well-known firm of Messers Dickson and Shankley and would hardly have been in business for seventy years if they hadn't. The reply this would have provoked has, perhaps fortunately, been lost. Manson might have been standing in for the head gardener and been a little over-zealous in consequence. At this time, the estate was buying hundreds of thousands of evergreen and deciduous trees from this nursery. It seems surprising that Penicuik didn't grow its own. They had plenty of room, and anyway the trees were only required small, at a year or two old. The convenience of being so close to the Edinburgh nursery must have tipped the balance in favour of the specialist growers.

James Ramsay, who Henry worked under here at Penicuik, had arrived from Dunbar with a very good reference. '...The only forcing he has had with me has been that of Mushrooms, which all Gardeners consider difficult and he has kept my Table constantly supplied with them. I believe he lived before he came to me in places where there were often these Hot Houses and I have no doubt he has perfect knowledge to manage them. With me his principal business has been the management of... Kitchen Gardens... From the little appearance I have had of him in the management of Flowers I should say he has rather a turn that way'.

'...I should not have parted with Ramsay if I had intended to have kept up my garden but I mean to remove from hence to Thirlstone...'.(49) Ramsay had only been at Dunbar for two years. Even as head gardener he was no more sure of a secure tenure than he would have been today. From the Reference, Ramsay's forte, or at least, his most recent experience, can be seen to have been with fruit and vegetables. It was thanks to gardeners like Ramsay, advancing Henry's abilities, that gave Henry security in his old age. Henry's later public success was to come from flowers, one in particular, but it was a vegetable, his production and hybridisation of the garden pea *Pisum sativum* that paid the bills.

At Fingask, Henry had been one of eight workers; five men, a boy apprentice and two weeding women. Here at Penicuik, the numbers of workers appear, from the evidence of the garden bills, to have been about the same. For a ten month period beginning in December 1846, one-hundred and forty-two items were 'Bought of Dicksons & Co., Nursery Seedsmen and florists'(50) of 1 Waterloo Place, Edinburgh. Here are a few of them:

10 December, 1846:

4 Dutch hoes	6/9
2 Draw hoes	3/6
3 Leather scythe stone bags	7/6

(A good idea – stops you losing – and breaking – them)

7 Pruning knives	14/-
1 gallon tobacco liquor	3/-

(An insecticide, also known as tobacco water. Diluted 5:1, water: tobacco water, it was still lethal enough to kill most things including, sometimes, the plants it was supposed to protect).

4 pair brooms	16/-
2 Edinburgh Bills	5/-

(One of a variety of billhooks used for hedge pruning).

Nearly one-hundred different seeds, mostly different kinds of herbs and vegetables, were bought. Mostly purchased in January, they were sown early, under glass or other suitable protection, for early cropping. Some fruit-trees and shrubs – cherries, gooseberries, apples and currants – were bought in February. They would have been bare-rooted, and planted in suitable weather before coming into leaf. Only a few flower-seeds were bought:

January 21st, 1847:	
12 Sorts German 10 week Stocks	4/-
½ lb. Sweet Peas (About 2500 seeds)	2/9
Prize Pansy	1/-
3 sorts new annuals	1/6
Prize Hollyhock	2/6
2ozs. Mignonette	1/6
10 sorts hardy annuals	2/6
12 sorts German asters	4/-
February 12th:	
Tropaeolum canariense	1/-
March 13th:	
1 Tropaeolum lobbianum	2/6

Sweet peas were one of the few flowers being bought regularly here. ('1lb. of Sweet Pease' were bought for 5/- in December 1855.)(51) And this is odd, because little mention was made of them in the contemporary garden literature. One exception to the rule was J.C. Loudon, who wrote in his *Gardener's Magazine* for October, 1837, that 'the Sweet Pea is esteemed by most lovers of the flower-garden for its rich profusion of flowers, and the delicate perfume which they put forth after a refreshing shower'. (52) The seed company James Carter, advertising in the same magazine during the 1838-1839 winter, listed seven kinds: black, Painted Lady (pink and white), purple, scarlet, striped, yellow and white. We can be pretty certain that the yellow wasn't yellow but cream and that the black would have been a dark purple or maroon.

Seven kinds, but only two or three of them generally grown. The familiarity of sweet peas and their lack of variety caused them to be overlooked and unregarded for another fifty years. At Penicuik, buying sweet pea seed in December indicated that they were sown under glass in January and February. Combined with later outdoor sowing in spring, it would have provided a long period of bloom with which to decorate the house. The idea of picking the flowers to prolong the flowering period doesn't seem to have been known about. Hardy annuals have a set flowering period, extended by successional sowing. The sweet pea is a hardy annual. Why expect it to be different, would have been their logical reasoning.

Although Henry was one of only a handful of gardeners here at Penicuik, Mr. Ramsay could count on help from elsewhere on the estate when it came to filling the ice houses. Ice houses were large larders, solid constructions in the garden in which items like meat were preserved through the summer in (usually) ice taken from the large garden ponds in the winter. It must have been one of the least popular winter jobs. Here is the entry from the estate's Farm Journal for Wednesday, 18th March, 1846: 'Began to fill the Ice House with Snow, there having been neither Snow nor frost previous to this date for fillin Ice house with – finished fillen Ice house Thursday 19 March'.(53) The work took fourteen men on the Wednesday and seventeen men on the Thursday. Ice houses must have been filled late in the winter, for the ice to last longer. Here, they appear to have been caught out by a mild March. There had actually been frost and snow in the second week of February and surely some ice went in then. It seems rather chancy otherwise!

Loudon, in his 1822 *Encyclopaedia of Gardening* wondered why gardeners were involved at all. 'However unsuitable or discordant it may appear, it has long been the custom, in country residences, to delegate to the gardener's care certain minor articles of culinary luxury, as ice, and the breeding and rearing of certain animals, as bees. In some cases also he has the care of the dove-house, fish-ponds, aviary, a menagerie of wild beasts, and places for snails, frogs, dormice, rabbits, &c....' (This is the nearest Loudon ever gets to humour!) 'We may observe, that it would be an improvement in rural architecture to constitute the ice-house one of the domestic appendages of the mansion, and to put its management, at least after being filled, under the house-steward, rather than under the gardener'. (54)

There was, however, one aspect of ice that everybody wanted to be involved with. 'October 28th, 1843. Manson, I have given permission as requested in the enclosed to the Caledonian Curling Club to play at Penicuik on the 23rd January or any day about that time when there is ice. I have agreed with Mr Cassels to give you a day's notice before they come, that you may have some of your men there to prevent any mischief being done by idle persons who may be attended there that day. It may be as well to give the Police Man notice that he may come round that day in the course of the duty. Your Ser. G. Clerk'.(55) Curling matches here were apparently competitions between those from the north and those from the south of the Forth. It was difficult to advertise a fixed date, due to the state of the weather, and only a few hours notice might be given. On the 9th January, 1844, the club's Henry Cassels wrote to James Manson, both to ask him to let him know the state of the ice just before the provisional date of the 23rd, and also to remind him that the Caledonian Curling Club was prefixed by the title 'The Royal Grand', Cassels helpfully writing it out carefully in large letters. (56)

Curling had been Scotland's most popular sport for about fifty years, and was played by all classes. Deep frost gave a slack period to the farming calendar, so an employer could allow his men some entertainment

The curling match, High Pond, Penicuik Estate, January 1847

without unduly interrupting work. You get a feel for its popularity from reading the estate's Farm Journal. Normally merely a functional account of what seed was sown in what field, it suddenly explodes into life. '15th January, 1847. The great national game of Curling... by the Caledonian Curling Club when forty-two Rinks started at about ½ past 12 o'clock at the signal of a gun being fired, and finished at about a quarter before 4 o'clock... The above 42 Rinks at 8 players per Rink making in all 336 Curlers. When the path round the pond was surrounded by about 650 people more to witness the Roaring Game'. (57)

How far a blind eye was turned, by how much Sir George Clerk's 'have some of your men there' meant all of them, is unknown. According to the journal, everyone worked, as they always did, Henry and everyone, six days a week, fifty-two weeks a year. Sir George Harvey's 1835 painting *The Curlers* in the National Gallery of Scotland depicts an all-male affair, where whisky and yellow and reddy-brown colours breathe fire into one short winter's day. Work was the canvas and the frame of your life, and you coloured it as you could.

~~~

As Henry seemed to be following Loudon's advice 'not to remain above one year in any one situation' until at least twenty-five, then the early spring of 1846 meant it was time for a move. He didn't go very far this time. At Penicuik he was only seven miles from Stenhouse, his home village. His next move over to Oxenfoord Castle, still in Midlothian, brought this down to six. As at Penicuik, Stenhouse was an easy cart-ride away. Oxenfoord Castle was the most grandly described of all Henry's situations. 'This beautiful country seat of the Earl of Stair is situated on the banks of a small river called the Tyne...'(58) This 'magnificent seat... on the west bank of the Tyne, the grounds around which are very picturesque...' (59)

'A magnificent edifice, with extensive and beautiful grounds...'(60)

Oxenfoord Castle lies in the parish of Cranston. The name was said to have derived from 'the Anglo-Saxon *Craenston,* signifying the crane's district, or resort'.(61) It is just north of the parish of Borthwick, where the father Henry never knew was born. In the 1840s, Cranston was said to contain 4778 acres, with Henry's new employer, the eighth Earl of Stair, owning most of them. As at Henry's other situations, much re-landscaping was taking place. Here, terraces and slopes were being cut into the lawns at the back of the house, on its south and east sides. Perhaps it was a desire to gain a distance from the here predominating styles of the 18th century and earlier; more Italianate than Italian now. They were similar to the terracing at Castle Kennedy, a property of the Earl on the west coast, in Wigtownshire.(62) Parkland was extended into the estate, and a Pinetum, an arboretum of pines and other conifers was created. A Dutch garden in front of the house, self-contained within its straight hedges of yew was kept, perhaps because of its formal design. Loudon visited here on 4th
~~~

August, 1841.[63] As we have come to expect, his admiration is qualified and his views are enjoyably opinionated. None of the usual anodyne eulogies from *him*.

'The castle is in a commanding situation, but has the common fault of being entered on the side that has the best views, and showing a stranger not only these, but the whole of the lawn, before he alights at the main entrance. The kitchen-garden is undergoing a thorough reform by Mr. Gardiner, a master in his art. A great many hollies are planted in the young woods, and the plants are protected from hares and rabbits by circular fences, 1 ½ ft. and 2 ft. 6 in. in diameter, formed entirely of the branches of young ash trees; their ends being stuck in the ground so as to form a circle round the plant, and their points woven into one another, as in the finishing of a common wicker-work hamper. There are a new church, new parsonage, handsome new factor's house, lodges, cottages, farm offices, all seen more or less from the public road, and all most substantially built of stone, and in good taste, at the earl's expense. The Edinburgh approach to the castle is excellent, but the other is less fortunate, showing only one side of the house, instead of coming up to it diagonally, so as to show two sides. Additional to the main door, there is a side or subordinate one, called the luggage door; a characteristic of Scotch mansions, arising, no doubt, from the hospitable habits of the country'.

Oxenfoord overlooked the grounds of the Preston Hall estate on the opposite bank of the river. Loudon went here after visiting Oxenfoord, and as he advised young gardeners like Henry to do likewise, so shall we. 'The park is crowded with magnificent trees... There are a large and very superiorly designed kitchen-garden, and an excellent gardener's house of three stories, large enough for a farmer; but, as we generally enquire into details we found this house, like many, we may say most, other gardeners' houses in Scotland, without a convenience essential to delicacy and cleanliness. The number of large and commodious gardeners' houses in Scotland which are altogether defective in this particular would not be credited in England.' Loudon seems to be complaining of the lack of an indoor lavatory. Presumably there was an outdoor privy. 'Forty different kinds of fig are cultivated in the garden here', continued Loudon, obliviously 'and, by the aid of glass and artificial heat, figs are sent to table from the middle of May till winter.' So, now we see the advisability of lavatories both indoors and out.

Henry was overwhelmingly unimpressed by all this. More to his taste would have been the half-a-mile square kitchen garden, north-west of the Oxenfoord House and itself surrounded by a woodland 'garden'.[64] The first Ordnance Survey map shows extensive glasshouse ranges in the kitchen-garden, at least some of which would have been there in the mid 1840s. Much later visitors, in 1889 and 1905 respectively, spoke of being struck by the 'quantity and quality of the grapes' [65]and of fruit-trees being grown within them. [66]

The head gardener that Henry worked under, Robert Gardiner, is shown in the estate papers to have been an important figure on the farm side, prior to reorganising the kitchen garden here. [67]A report in *The Gardener's Magazine* in 1839 shows him to have also been an innovative gardener. They printed a design of his for a trellis for fruit-trees[68] whilst a gardener at Kinnaird Castle. It seems likely that Gardiner was brought in to push through changes here. The large-scale frame of reference of the farmer, combined with some horticultural knowledge and, say, a forceful personality and organisational abilities, were the qualities required.

A report in *The Gardeners' Chronicle* in September 1848 [69] focuses on some of these changes. Gardiner was replacing the laurel shrubberies with Rhododendrons. He propagated them by seed, which is fine with species, which they were. He found that they'd germinate quicker on a hot-bed than in a cold-frame, the usual practise, and grow on happily afterwards. Most of the article is given over to Gardiner's descriptions of changes that had been going on in the kitchen garden since the winter of 1841. Principally it involved root-pruning and transplanting fruit-trees and Gardiner detailed the process. 'By such treatment', he wrote, 'fruit trees are kept dwarf, very prolific, and ripen their fruit much earlier, which is of consequence in a cold exposed situation'. Nowadays these effects are produced by grafting onto dwarfing rootstocks, but root pruning is still a valid technique for large fruit-trees. Another disadvantage of large fruit-trees is that gardeners sometimes fell from them. The occasional death or serious injury this occasioned was commented on in the columns of papers like *The Gardeners' Chronicle.*

To our minds, it looks as if Gardiner is describing new techniques of his own. Yet it was one described in Loudon's *Encyclopaedia* of 1822, [70] with the principle understood since at least the time of John Evelyn in the seventeenth century; possibly even as far back as the Romans. Many plants were new in the nineteenth century, and cultural information was needed. But even where is wasn't, gardeners still described their practice, through the columns of the gardening magazines, as if none of them believed what they read in books (which few of them owned). A point to bear in mind here is that although most people in Scotland could read English, it's possible that most people in England and Wales could not. The 1871 census found that most Scots (89% of men, 79% of women) getting married over the previous ten years could at least sign their name in the marriage register. English male illiteracy, by such definition, was twice that.(71)

In England, discounting its 'markedly inefficient' private elementary schools, an 1861 report found the philanthropic 'public' elementary schools were only being attended by some 13% of children, and theirs was an irregular attendance, over an average time of four years.(72) Of those Britons who could read, most could not afford to buy books or even up-to-date newspapers. Literate gardeners relied on their employers to buy the horticultural magazines for them. What might be happening in the horticultural press is that a tradition of verbal repetition, a method most likely to ensure retention of information by the recipient, was now being written down. It was now a time of transition from a principally verbal popular culture to a written one, although a cheap 'popular press' lay still over the horizon. Hastening this process was the arrival of the 'penny post' in 1840, soon latched on to by political agitators who could now combine the verbal oratory of the mass meeting with the posting of written propaganda. (73)

One imagines that what really rubbed off on Henry, here at Oxenfoord, was Robert Gardiner's enthusiasm; his belief, widely shared, of empirical practice, and that through such practice you might discover something new, something different. Henry's probable time here, 1846 to early 1847, was also the time of the great potato failure and famine in Ireland. It affected Scotland too, along its entire west coast. In contrast to Ireland it was to remain localised, causing few direct deaths, for Scot's farming was diverse, not dependant on potatoes as Ireland was. Nevertheless, there was great concern at the time. West of the walled garden at Oxenfoord, close by the Edinburgh to Lauder road, lay Cranston Cottage. On the 6th August, 1846, John Dodds, the tenant there wrote to Stair's factor George Guthrie:

'We have beautiful weather now, particularly for the Turnip crop... Our potatoes are looking well, and there is no appearance of the Disease as yet. We had, I may say, no disease last year, and I hope we may be as fortunate this... The farmers in the low part of East Lothian are getting alarmed at the non-appearance of the usual supply of Irish Shearers – I suspect the Railways are monopolising the most of them...'(74) A week later, on the 13th August, a William Broadfoot Esq., who must have been on Stair's western estate (he gives no address) wrote 'The potatoes I fear are all gone and will not pay for the expense of taking up, it will be a great National calamity and a serious loss to the Farmer...' (75) A tenant's letter of the previous autumn shows how bad 1845 had been, too:

Doonhill 19th November. 1845

George Guthrie Esqre

Sir,

I take the liberty of writing you to say that I would take it a very particular favour if you would allow my present half years Rent to stand over for a month or six weeks which if you would be so kind as do for the latter date I would without fail then remit it to you by a Bank order. My reason for not being able to pay you on the 27 instant is caused by different Calamities befallen me this year. For one I may mention that of my Potato Crop which I had calculated would have done me much good. I had laid out here about £30 for Manure which beside the expense otherwise will prove to me a very serious loss so that had it proved otherwise with Potato Crop I would have been able to have paid you up to this year, but if you are so condescending as grant me the favour asked I shall make good my Promise to a day when I will have got some other part of the Farm Produce brought to Market. I will feel very much obliged by your answer to this and remains

Sir Most Respectfully

Your obedient humble servant

Mary Keppie.(76)

Let us hope that Stair, through his Factor, took a charitable view. Not everyone did. 'When the last Scottish famine was raging in the Highlands in 1846-8', writes the historian T.C. Smout, 'the last Scottish grain riots were also taking place in the little Lowland ports from Thurso round to Peterhead: the inhabitants had discovered that meal was being purchased by relief agencies and others for export coastwise and they were obstructing its shipment in case the price rose too much in their own localities'.(77)

Henry Eckford as a young man, c. 1847

~~~

It was the spring of 1847. Henry was fast coming to the end of the Journeyman stage of his career. Where would he go now? Henry's first move from Mid-Lothian had coincided with a serious typhus outbreak in Edinburgh. The move he made now coincided with the potato famine in Ireland and the concern that potato-blight was causing to Scotland. His first move had been to Inverness. His next move was to be even farther away and more drastic, for he was to leave Scotland altogether. The railways were heralding in a new age of mobility. It was a mobility that facilitated the growing shift from countryside to town. Millions of Britons would migrate abroad on the new steamships over the next thirty years, and particularly around the year 1848. There was now a series of bad harvests (in England at least), a severe trade depression from 1839 to 1843, a bank crisis in 1847 (ending the railway boom) and a brief return of the depression in 1848. It all led to the decade being remembered as 'the hungry forties', [78]people
~~~

flocking city-wards in hope of relief. A third of a million people migrated to London between 1841 and 1851. [79]1848 was remembered too as the 'year of revolutions' elsewhere in Europe. The very air seemed agitated. Henry, over six feet tall and in his youthful prime, felt the changing winds blow about him. And the strongest of these winds blew south.

References to Chapter Two

(1) John Claudius Loudon, *Encyclopaedia of Gardening*, (1822), p.1249.
(2) Ibid, p.1257.
(3) 'Notes On Gardens in Inverness-shire', *The Gardener's Magazine*, Vol.10, (November 1835), pp. 553-555.
(4) J. C. Loudon, op.cit., p.1328.
(5) 'Gardeners' Lodges', *The Scottish Gardener*, Vol.3, (1854), pp.129-130.
(6) George W. Johnson (ed.), *The Gardeners' Dictionary,* (1877), p.129.
(7) 'Bothies', *The Gardener*, (9-1875), pp. 430-431. (No author but signed 'D', possibly for Richard Dean, *The Gardener's* assistant editor then).
(8) 'Scotland', *The Gardener's Magazine*, Vol.12, (11-1837), p.521.
(9) Rev. C.Fraser, Parish of Kiltarlity, *The New Statistical Account of Scotland, Vol.XIV,* (1845), p.497.
(10) 'The Late Mr Henry Eckford', *The Journal of Horticulture*, (14-12-1905), p.547.
(11) J. C. Loudon, op.cit., p.1199.
(12) Ibid, p.1334.
(13) Ibid, p.1146.
(14) Robert C. Mossman, 'The Meteorology of Edinburgh, part 2', *Transactions of the Royal Society* of *Edinburgh*, Vol.39, (1900), p.104.
(15) 'Newliston', *Inventory of Gardens and Designed Landscapes in Scotland*, Countryside Commission for Scotland, Vol.5, (1985), p.177.
(16) Thriepland of Fingask Papers, Perth and Kinross Council Archives, (uncatalogued).
(17) J.P.Neale, 'Fingask Castle, Perthshire', *Views of the Seats of Noblemen and Gentlemen in England, Wales, Scotland and Ireland*, 2nd Series, Vol.4, (1828), no paging.
(18) 'Features in Scottish Scenery,' *The Scottish Gardener*, (1854), p.11.
(19) 'A day at Fingask Castle and Grounds', *The Dundee Advertiser*, (25-10-1864), p.2.
(20) Thriepland Papers, op.cit.
(21) D.Lambert, P.Goodchild, J.Roberts, *Researching a Garden's History*, (1995), p.18.
(22) T.C.Smout, *A Century of the Scottish People, 1830-1950*, (1997), p.111.
(23) Pamela Horn, *Labouring Life in the Victorian Countryside*, (1976), p.259.
(24) J. C. Loudon, op.cit., p.1327.
(25) Thriepland Papers, op.cit.

(26) William Howitt, *Rural Life of England*, (1845), p.86.
(27) Thriepland Papers, op.cit.
(28) Ibid.
(29) G. W. Johnson, op.cit., p.105.
(30) Ibid, p.567.
(31) Ibid, p.735.
(32) A.G.L. Hellyer, 'The Encyclopaedia of Garden Work and Terms', *The Gardeners' Golden Treasury*, (1966), p.92.
(33) J. C. Loudon, op.cit., pp.327-328.
(34) G. W. Johnson, op.cit., p.352.
(35) Ralph Waldo Emerson, *English Traits*, 1856, (1902), p.55, (quoting the Comte de Lauraguais).
(36) Ms. 13980, (1845), Walker of Bowland Papers, National Library of Scotland.
(37) See: 'Penicuik. Estate and Mansion' , *Dalkeith Advertiser*, 9-8-1894, p.3; 'Penicuik' , *Inventory of Gardens* ... Vol 5, (1986), pp. 186-192; Colin McWilliam, 'Penicuik' , *The Buildings of Scotland: Lothian except Edinburgh*, (1978), pp. 358-388'; Frances H. Groome, (ed), 'Penicuik' , *Ordnance Gazetteer of Scotland*, Vol. V, (1884), pp. 174-176; Rev. W. Scott Moncrieff, Parish of Penicuik, *The Statistical Account of Scotland*, Vol. I, Edinburgh, (1845), pp. 29-43, (written in 1839); J.P. Neale, 'Penicuik, Mid Lothian' , *Views....*, 2nd Series, Vol. 2, (1845), no paging.
(38) J.C. Loudon, op.cit., p.1250.
(39) GD18/1719/24-27, (1841-1843), The Clerk of Penicuik Papers, National Records of Scotland.
(40) GD18/5789/1, (1844), The C. of P. papers, N.R.S.
(41) GD18/5789/3, (1846), The C. of P. papers, N.R.S.
(42) GD18/5789/1, (1846), The C. of P. papers, N.R.S.
(43) 'Penicuik', *Inventory of Gardens and Designed Landscapes*... Vol. 5, (1986), p.190.
(44) GD18/1741, (1844-1846), The C. of P. papers, N.R.S.
(45) GD18/1719/28, (1847), The C. of P. papers, N.R.S.
(46) 'The Cult of the Sweet Pea', (advertisement), *The Daily Mail*, (21-2-1911), p.10.
(47) GD18/5789/2, (1843), The C. of P. papers, N.R.S.
(48) J. C. Loudon,op.cit., p.190.
(49) GD18/5789/2, (1843), The C. of P. papers, N.R.S.
(50) GD18/1842/2, (1847), The C. of P. papers, N.R.S.

(51) GD18/1842, (1855), The C. of P. papers, N.R.S.

(52) 'A Mode of training the Sweet Pea in Flower-Gardens', *The Gardener's Magazine*, (10-1837), p.446.

(53) GD18/1737/4, (1846), The C. of P. papers, N.R.S.

(54) J. C. Loudon, op.cit., pp.381-382 and p.384.

(55) GD18/5789/2, (1843), The C. of P. papers, N.R.S.

(56) GD18/5789/2, (1844), The C. of P. papers, N.R.S.

(57) GD18/1737/5, (1847), The C. of P. papers, N.R.S.

(58) 'Oxenford Castle', *The Gardening World*, (3-8-1889), p.774.

(59) Rev. Alexander Welsh, Parish of Cranston, *The New Statistical Account of Scotland*, Vol. I, Edinburgh, (1845), (written in 10-1839), pp.193-194.

(60) 'Oxenfoord Castle', F. H. Groome, (ed.), *Ordnance Gazetteer of Scotland*, Vol.V, (1884), p.146.

(61) Rev. A.Welsh, op.cit., p.191.

(62) 'Oxenfoord Castle', *Inventory of Gardens*... Vol.5, (1986), p.183.

(63) 'Oxenford Castle', *The Gardener's Magazine,* (12-1842), pp. 581-582.

(64) 'Oxenfoord Castle', *Inventory of Gardens*... op.cit., p.182.

(65) 'Oxenford Castle', *The Gardening World*, op.cit., p.774.

(66) 'Oxenfoord Castle', *The Gardeners' Chronicle*, (9-9-1905), p.195.

(67) GD135, (1840s, passim), The Earl of Stair Papers, National Records of Scotland.

(68) 'Design for a Trellis for Fruit Trees...', *The Gardener's Magazine,* (1839), pp.599-600.

(69) 'Garden Memoranda', *The Gardeners' Chronicle,* (16-9-1848), pp.623-624.

(70) J. C. Loudon, op.cit., p.461.

(71) T.C. Smout, op.cit., p.216.

(72) S.J. Curtis, M.E.A.Boultwood, *An Introductory History of English Education Since 1800*, (1966), p.69.

(73) David Thomson, *England in the Nineteenth Century*, (1969), p.81.

(74) GD135/2393, (6-8-1846), The E. of S. papers, N.R.S.

(75) GD135/2393, (13-8-1846), The E. of S. papers, N.R.S.

(76) GD135/2390/42, (1845), The E. of S. papers, N.R.S.

(77) T.C. Smout, op.cit., p.13.

(78) David Thomson, op.cit., p.80.

(79) Simon Gunn in C.Williams, (ed.), *A Companion to Nineteenth-Century Britain*, (2004), p. 241.

Chapter Three

The Scots gardener had long been in England and long been favoured by employers there. He was also favoured in Ireland, if the following advertisement from 1907 is representative: 'Wanted. Leading Foreman (working) for an Irish Nursery, Scotchman preferred, skilled propagator, grafting, budding roses, fruit-trees...' [1] Robert Uvedale (1642 – 1722) of Enfield, Middlesex, said to be the first to acquire the sweet pea from Sicily, wrote: 'I never yet found any blue apron man (but one poor Scotchman who was my gardener) that had any relish for that part of gardening or indeed for any other, in comparison with their wages and perquisites'.[2] (One wonders what that unrelished part of gardening was…).

In the 1760s there was resentment in England, London in particular, at the arrival of numbers of Scots. There was fear of Scottish influence in English affairs – there was even a Scottish prime-minister, Lord Bute! Too grand to be a gardener, the politician became a botanist and first director of the Botanic Gardens at Kew in Surrey. Scottish gardeners were the Polish plumbers of their day; much sought after, they provided lively competition to the established trade. In 1822 Loudon wrote that 'about the middle of the last century, Lee, Gordon, Russel, and Malcolm, all Scotch gardeners, commenced their nurseries at Hammersmith, Mile-end, Lewisham , and Kennington' (all places now in inner-London). 'Their success excited the jealousy of the established commercial gardeners, who, between 1760 and 1770, held several meetings, and entered into resolutions not to employ young men from the north. These resolutions were not long adhered to…'. [3]

On the 19th June, 1822, William Cobbett was travelling north out of London. Observing harvesters in what are now built-over suburbs, he wrote: 'It is curious to observe how the different labours are divided as to the nations. The mowers are all English; the hay-makers all Irish.

Scotchmen toil hard enough in Scotland, but when they go from home it is not to work, if you please. They are found in gardens, and especially in gentlemen's gardens, tying up flowers, picking dead leaves off exotics, peeping into melon-frames, publishing the banns of marriage between the 'male' and 'female' blossoms, tap-tap-tapping against a wall with a hammer that weighs half an ounce. They have backs as straight and shoulders as square as heroes of Waterloo; and who can blame them? The digging, the mowing, the carrying of loads; all the break-back and sweat-extracting work they leave to be performed by those who have less prudence than they have. The great purpose of human art, the great end of human study, is to obtain ease, to throw the burden of labour from our own shoulders and fix it on those of others.' (4)

What might have brought Scots gardeners to England though; to London? A result of their education perhaps? William Howitt thought so. In 1838 he wrote that in the rural districts of Scotland, 'every child, by national provision, has a sound, plain education given him... The consequence is, that almost all grow up with.... a determined resolve of depending on their own exertions: and though no people are so national... yet, if they cannot find means of living at home without degradation, and, indeed, without bettering their condition, they soberly march off, and find some place where they can, though it be at the very ends of the earth'(5) The modern historian T.C.Smout gives a reason for Scots to march off, soberly or otherwise. He points out that 'in 1860, Scottish wages were often up to one-fifth below those for the same English trades'. (6) Smout partly contradicts Howitt by stating that in around the 1830s and 1840s the Scottish 'Highlanders contributed to their own problems by their passion for the land they occupied, preferring a life of deepening poverty in an increasingly overcrowded environment to the risk of seeking their fortunes permanently abroad, or in the Lowlands.' (7)

An English garden journalist of the eighteenth century, Richard Bradley, highlights a still common English attitude to manual labour: 'The country people generally pick out such of their children to employ in husbandry as they judge are not worthy of good education, and whom they suppose have so little genius that they are only fit to drudge in hard labour...' (8) It is an attitude that seems to have a Europe-wide pedigree. Back in seventeenth-century Spain, Cervantes has Don Quixote exclaim 'Away with anyone who gives letters the preference over arms... the argument that such people usually adduce and depend upon is that brain-work is superior to physical work... as if the warrior who is in charge of an army or the defence of a besieged city did not labour with his mind as much as with his body.'[(9)]

In *A History of Gardening in Scotland*, in 1935, E.H.M. Cox devotes a chapter to the causes of emigration by Scottish horticulturists. The author contends that farmers and gardeners prefer an equitable climate, where plants grow better. An employer will prefer a man trained in an unkindly climate, where more effort is required to obtain good results. There was more chance of advancement by going south, the author considering the Scottish working man of the eighteenth and early nineteenth centuries more ambitious than his English counterparts. Apprenticeships were harder and longer, north of the Border. Also, Scottish gardeners who 'left the country to work elsewhere have always been nationally minded. We are a clannish people, and wherever they may have found themselves they have always been willing and eager to help horticulture in Scotland.'[(10)] And, as we will find, to help fellow Scots, for Henry sought out Scottish head gardeners and employers to work for in England.

Many of the foregoing points were considered and rejected a century earlier by a writer in Loudon's *Gardener's Magazine* in May, 1840.

On the Preference for Scotch Gardeners. By J. Wighton.

Scotch gardeners are often preferred in England, as if they had a better knowledge of their profession than English gardeners. As gardening is certainly as well understood in England as it is in Scotland, it may be worth while to enquire into the cause of this preference; and also the reason why so few young men in England, after serving a regular apprenticeship as gardeners, ever arrive at the head of their profession. They remain only a step above a common labourer, and seldom remove from the place of their birth; while most young men who learn gardening in Scotland become in time head gardeners, either at home or abroad. Various reasons are assigned why the preference is given in England to Scotch gardeners: one is, that they are usually better educated; another, that the greater coldness and changeableness of the climate in Scotland obliges the gardener to take greater care and pay more attention, which renders him more skilful in his business; another cause assigned is, that Scotchmen are generally a more steady and calculating race. However well founded these reasons may appear, they are not sufficient to account for the decided preference given to the Scotch gardener. Education can do but little where there is deficiency of natural abilities; and, though Scotland is colder than England, the English gardener has quite enough of coldness and variableness of climate to call forth his energies. If the Scotch are more cool and calculating, they must acquire those habits by early training.

It is probable that the Poor Law system in England has had the greatest share in producing the superiority of the Scotch over the English gardener. It mainly depends on the difference of training, when acquiring the knowledge of his business. In England, the mansion and gardens of the wealthy are more frequently situated adjoining a populous village; the proprietor, in consequence, often finds his property burdened by too many labourers. When his gardener wants an apprentice, his employer obliges him to take one who belongs to the parish; as he cannot think of employing strangers. The young man chosen begins with the honest intention of becoming a gardener, and has at first, probably, an anxious wish to learn. But this too often cools, from associating with others of his native place, who are not gardeners.

His attention is much taken off by such connexions, and he is less disposed to give his mind wholly to gardening, if his parents are, as it usually happens, of the agricultural class; because he shares in their ideas and feelings, and especially in the notion that he must be employed, because he belongs to the parish. If, however, he escapes these evils, when the time of his apprenticeship is expired, he finds it difficult to procure a situation as under gardener, on the same principle that caused him to be chosen for an apprentice, namely, his not belonging to the new parish where he makes application. This forms a serious obstacle to the advancement of young gardeners, and is the greatest cause why so many never remove from their native place. Seeing the difficulty of procuring a situation elsewhere, they grow indifferent about advancement, and give up all thoughts of becoming master gardeners; after a time they marry, and settle down for life in the place where they were apprenticed.

In Scotland, the residences of the wealthy are less frequently situated near a populous village; and the proprietor does not find his property overburthened with labourers. He often leaves his gardener to choose his own apprentices; as it matters nothing whether he employs a native or a stranger, there being no Law of Settlement to interfere with the labourer's independence. There is a lodge at the garden; and the apprentice is at once placed there, to live with the under gardener. It is curious that these lodges have got the Gaelic name of Bothies, there being so few Gaelic words in the Scotch dialect. Being thus thrown from the first upon his own resources, which are slender enough, he learns to think and manage for himself. He has thus every opportunity of learning his business from the gardeners, with whom he constantly lives, and has no village companions to divert his attention. At the expiration of his apprenticeship, he knows that he must seek a situation as a journeyman gardener elsewhere, as it is unusual to remain in the same place. This cuts the taproot of his connexions. By serving in various places for some years as an under gardener, he acquires sufficient knowledge to take the situation of head gardener whenever it offers; and, from being often transplanted, he readily takes root in any clime, though he always retains the love of his own country.

Though these advantages are peculiar to the Scotch gardener, it is not denied that there are many good English gardeners. The greater number of gardeners in England, however, remain little above the common labourer, in consequence, no doubt, of the evil operation of the Law of Settlement; and, though this part of the law has been lately abolished, the alteration will not soon produce an effect in those parts which are thickly inhabited. It may be observed, in conclusion, as a proof that there is no want of ability in English gardeners, that those young English apprentices, who are training under Scotch gardeners in England, are no way different in their habits and fortunes from those who are apprenticed under English head gardeners.

Cossey Hall Gardens, January 21, 1839. (11)

Weighton was one of many contemporary critics of the laws that had hindered mobility in England and Wales, and there was much widespread anger against the Poor Laws. His themes may have been well-worn but they are well expressed, an unintended tribute to his, presumably Scottish, education. Gardeners throughout Britain gradually became able to compete on equal terms. Amendment of the Act of Settlement and Removals, the later arrival of free and compulsory education for all the nation's children, and the rise of English horticultural colleges, beginning at Swanley in Kent (1889) and Studley in Worcestershire (1898) eventually saw off the dominance of the Scottish gardener. But by then, too, the era that saw the dominance of the professional head gardener was also ending. And – but we have got a little ahead of ourselves.

~~~

It was the late eighteen-forties. Just a stone's throw from Stenhouse – well, a few stone throws, half-a-mile – lay St Catherines, former home of Sir William Rae, one of the senior lawyers and politicians of the age. The novelist Sir Walter Scott used to stay over at St Catherines on his journeys to Edinburgh from the borders. Scott's journal shows that there were usually some grand guests at the dinner table. It's possible that Henry's father James, being locally important, made Scott's occasional acquaintance here. So Scott may even have met Henry, who would have been nine when Scott died. Walter Scott was a keen gardener who was developing the ground around his new home at Abbotsford and designing, among other things, ways of moving semi-mature trees.

William Rae had been Lord Advocate, a post similar to the later Scottish Secretary, during Sir Robert Peel's minority administration of 1834-35 and had briefly served again during the early part of Peel's second ministry of 1841-46. He died in 1842. Representative of 'old money' his kind were at odds with the classes being formed from the wealth of industry and commercial enterprise. As an indication of such change, St Catherines was now owned by a Mr. Wright, a seedsman from Edinburgh.[12] That there was money to be made in running a seed nursery was a lesson Henry learnt well, but one he felt unable to put into effect for another forty-six years. A more immediate concern was how he was actually to get to London. Mindful of his stage-coach journey in 1838, he may have been waiting for the new railway to link up, capital to capital. The threat of potato blight may have advanced such plans, for he left Scotland before the North British Railway's Edinburgh to Berwick line, opened in 1846, met the Newcastle
~~~

and Berwick's railway line coming north. (That eventually occurred with the completion of Robert Stephenson's Royal Border Bridge in 1850).[13]

Another factor influencing Henry now may have been the weather. The 1820s was the end of a prolonged period of very cold winters in Edinburgh stretching back to 1782, the *black auchty-twa*[14] of not so fond memory. The winters were now volatile, swinging from very cold to very mild and back again. The winters of 1822-23 and 1837-38 were to be the sixth and fourth coldest winters respectively for the entire period of 1764-1963.[15] That 1837-38 winter's snowfall was the greatest since the *black auchty-twa's,* and it went on snowing to the 17th May. Its effects were felt nationwide, drawing frequent responses from the horticultural press as elsewhere. On 20th January the thermometer in the Oxford Botanic Gardens recorded 1°F (-17°C). By contrast, the winters of 1843-44, and of 1845-46 when no snow fell in January, were to be the ninth and eighth warmest respectively for this same period.

Henry left for London early in the year of 1847. Part at least of the journey would have been by coach. Travel was expensive. As we saw from Peter Loney's diary at Fingask, a ten-mile coach trip cost 2/-, the day's wages of a foreman gardener. For the railways, an example from the September 1853 tariff of the Great Northern Railway Company's London to Perth route reveals fares of £1-10-0 for 3rd class, £2-18-0 for 2nd class and £4-7-0 for 1st class.[16] Removing the journey from Perth to Edinburgh, even the cheapest ticket to London cost almost two and a half weeks wages. Second thoughts were thus a luxury that Henry couldn't afford. He had every incentive to stay and succeed in the south of England.

If Henry wasn't ready to start his own nursery he could at least experience working in one. With that commendation from the curator of Edinburgh Botanic Gardens in his pocket, he arrived at 'Mr Low's Nursery' in Upper Clapton. Formerly a village, Upper Clapton now lay within a northern metropolitan borough of London, though still in Middlesex. 'The area', wrote a contemporary in 1846, 'has long been admired for its very rural character considering its immediate vicinity to the Metropolis'. [17]Printed on a contemporary map, it already has a hint of nostalgia about it. The 1841 census showed a population increase of a third over the previous ten years. The first Ordnance Survey map of 1868 shows increasing suburbia, whose open spaces are mostly brickfields.

Mr. Hugh Low was a Scotsman in his early fifties. He had arrived at a then new nursery in 1823, acquired the ownership in 1831 [18] and turned it into one of London's leading nurseries. By 1851 he was employing thirty-one men. Two of his sons were destined to find fame with plants. Hugh (later Sir Hugh) Low, junior, became one of the famous Victorian plant collectors, living mostly abroad. Stuart Henry Low, by contrast, stayed with the family firm to become 'one of the most financially successful nurserymen of all time'.(19) On his father's death in 1863, he took on the business, worth a respectable £2,000, and by his own death twenty-seven years later had turned it into one worth £123,885.

Henry worked with two of the elder Hugh Low's sons, this Stuart Henry Low and his brother James. Hugh Low junior had gone off plant collecting abroad but had returned, for a while, by October 1847.(20) Henry, at the nursery in 1847 but for only a short time, may have met the young Hugh briefly. Whether he did or not, the desire to go plant hunting in the mountains of Borneo never seems to have affected him! Henry would have known Stuart better. There were of similar ages, being only three years apart, and were both to become successful nurserymen. Fame went Henry's way, with sweet peas and fortune Stuart's, with orchids. They must have made a lasting friendship, for twenty-one years later, when Henry was hybridising Verbenas, he named a mainly white one 'Mr Stuart Low'. The flower was described by *The Gardener's Chronicle* as being of 'good shape' and having a 'greenish eye, with an almost indistinct rose ring round it'. Another described as 'orange cerise, shaded dark round a white eye' he named 'Hugh Low'. (21) Without the prefix 'Mr.', this would have been named after the seven year old son of the second Hugh Low. Confusing, isn't it! Three Hugh Lows. Henry was fond of naming his plants after the babies and children of his friends and acquaintances.

In Henry's time here the nursery was noted for a broad range of hardy and glasshouse plants. If it specialised in anything it was Camellias, judging by its advertisement in the first, 1841 edition of *The Gardeners' Chronicle.* *(22)* In 1847 Mr. Low senior, 'a successful importer of plants from almost every quarter' (23) was raising a stock of the just introduced Chilean Incense Cedar (*Austrocedrus chilensis*). Thujas and *Thuja*-like conifers like this were popular. *Thuja* species had been planted at Penicuik just before Henry had arrived there. Henry didn't stay long at the Lows' nursery but with a reference from Low senior in his pocket it may have been an amicable parting. The change of country; his first commercial nursery; the change of pace here as one of two and a half million, the greatest concentration of people in the western world; it was all a bit much, one suspects.

Cities aren't a natural magnet for a gardener either. Industrial cities like London were foul and becoming more so. The American Ralph Waldo

Emerson had arrived in England in the autumn of that year, 1847. He wrote of 'the darkness of its sky.'

> The night and day are too nearly of a colour. It strains the eyes to read and to write. Add the coal smoke. In the manufacturing towns, the fire soot or *blacks* darken the day, give white sheep the colour of black sheep, discolour the human saliva, contaminate the air, poison many plants, and corrode the monuments and buildings.
>
> The London fog aggravates the distempers of the sky, and sometimes justifies the epigram on the climate by an English wit, "in a fine day, looking up a chimney, in a foul day, looking down one"'. (24)

So, Henry went away. So did the Lows, eventually, to the suburbs, when they and their plants, the orchids especially, could no longer cope with the smog. (25)A short trip to the city centre followed by an eighty-three mile journey south-westwards brought Henry to Salisbury in Wiltshire. He might, this time, have gone the whole way by train. 1847 was the year the railways arrived in Salisbury, as a branch-line from Southampton.(26) It probably wasn't any more comfortable on the South-Western Railway than going by coach, but at least it was over sooner. Funnily enough, Emerson took the same London to Salisbury train the following year and failed to remark on the journey,(27) so it can't have been too bad! It's likely that Hugh Low knew the Scotsman William Dodds, gardener to Colonel Edward Baker in St. Ann's Street, Salisbury, and Henry's new position. William Dodds was a *Dahlia* specialist, who also grew a lot of Camellias and was noted for his grapes, pineapples and other greenhouse plants(28) which he occasionally sold. An article on a visit to Mr. Low's nursery in *The Gardener's Chronicle*, quoted from above, was tantalisingly signed 'Dodman', which could be Dodds. Contributors often used pen-names.

Colonel Baker's garden was nothing like the estates that Henry was used to. However, at 120 metres wide by 145 metres long, it was large enough to feature on maps of the town. In 1820 it had been the formally laid-out grounds of a Friary. It was taken over soon afterwards by Baker, who swapped formality for informality, creating a large pond with an island in the process. They were though long referred to as *The Friary Gardens*. A large kitchen-garden, 107 metres by 34 metres in size was apparently added

Col. Baker's House

to the east side of the property in the 1850s. [29] Baker's grand but terraced house gave little indication of the garden behind, as a nurseryman, visiting for Loudon's *Gardener's Magazine* had discovered in 1831:

'July 5th... I called at Colonel Baker's, in Salisbury; a place which, before entering it, one might suppose, could present nothing attractive, from its being apparently situated in the centre of the city: but the contrary is the case; for you are ushered into a large piece of ground, laid out with great taste as a pleasure-garden, and so arranged as that all the surrounding objects, except the beautiful spire of the cathedral, are completely hid, so that one may at once fancy oneself in the country. This place was then under the able direction of the late Mr Shennan, formerly gardener at Gunnersbury House, a well-known pine [apple] - grower, and a man of very superior abilities. His pines were looking extremely well when I saw them; and we had some very handsome fruit in various stages of growth. Several specimens of hot-house climbers were in high perfection; such as Combretum purpureum, [?probably *C. coccineum*] Passiflora Bonapartea [*P. x Buonapartea*], &c. Some specimens of Ixora coccinea were also well worthy of notice. The grounds were bespangled with ornamental and rare plants; but the kitchen-garden department was confined to a small space, and evidently looked on as of minor importance. The crops, &c., were, however, by no means deficient, and the whole place was characterised by neat and orderly keeping.

I remain, Sir, &c.

Wm. Sanders.

Laurence Hill Nursery, Bristol, Dec. 15. 1832.' [30]

William Dodds arrived at this time. They were growing the latest tropical plants, for Combretums were only just arriving in Great Britain. Dodds may have worked under Mr. Shennan, for he began as an under gardener, soon becoming head gardener, Mr. Shennan perhaps having died, or retired. Dodds stayed for thirty years. When Col. Baker died in 1862 he moved away to Bristol, the same city as the visitor of thirty years before. Henry probably lived-in here, at 40-42 St Ann's Street. Another young Scots under-gardener is indicated to be living-in on the 1841 census. Thirty-seven years old William Dodds got married round-about the time Henry arrived here. Jane Dodds was thirty-eight and had probably been a widow, so if Henry shared the house it would have been with their eight-year-old daughter Fanny and baby Margaret. (Col. Baker was away much of the time at his London residence). Jane came originally from the Wiltshire village of West Knoyle, twenty miles away near the Dorset and Somerset borders. She may have made an impression on the young Henry, and a liking by him for the ladies around there, for he was to marry one.

It seems odd from our perspective that some of the plants grown by a head gardener in a private garden were sold commercially, but this seems to have been the case. Henry was to do the same with sweet peas, thirty years later. John Harvey, in his *Early Nurserymen* writes of 'one Dodds of Salisbury who, in 1836, supplied 27 bulbs of hybrid Amaryllis for £1.2.0d., for the gardens at Nynehead near Wellington, Somerset; and in the following year pineapples... He was very likely William Dodds... who as gardener to Colonel Baker was winning prizes for Camellias at a meeting of the Wiltshire Horticultural Society in 1833'. (31)

When William Dodds became noted for his Dahlias, Col. Baker, now retired, became as enthusiastic as his gardener. (32) Much of this enthusiasm rubbed off on Henry too, for Dahlias were among Henry's first successes, and he and William became life-long friends. Many of William Dodds' new *Dahlia* varieties were marketed through his friend John Keynes's Castle Street Nurseries in Salisbury. The *Dahlia* had only recently arrived in Great Britain, (in 1814, for all practical purposes) yet by the time of Henry's birth in 1823 it had became, Loudon tells us, 'the most fashionable flower in this country'. (33) The attraction of the bright and the bold was reflected in clothing too. A contemporary noted 'the shirts embroidered with dahlias, death's heads, race-horses, sunflowers and ballet-girls....' of young London clerks commuting to work in 1862.(34) Victorians were either not all staid and respectable or hadn't yet become so, it seems....

The exotic ostentatiousness of Dahlias gained them popularity but, as now, not a wholesale acceptance. They are one of those plants like Primulas and Chrysanthemums whose natural attractions are lost on the hybridist. Perhaps they're best seen as a bold group in a mixed border, or individually, near the front, rather than grown separately *en masse*. 'Hurrah!... it is a frost!' cries a character from the novel *Handley Cross*, written in 1843. 'The Dahlias are all dead!'(35) William Robinson in *The English Flower Garden* spoke of the over-heavy blossoms of the early Victorian 'show' and 'fancy' Dahlias, 'hence the greater popularity [in 1883] of the many lovely Cactus varieties'.(36) A century later, cactus Dahlias were being described as looking 'like miniature hedgehogs which have curled up in a ball and been blasted with an electric shock that makes them go pink'.[37] Poor old Dahlias!

When Col. Baker died his property and contents were auctioned, which fortunately gives us an inventory[38] and a complete list of his pot plants; a rare snapshot. Although 1862 was thirteen years after Henry had left, Dodds was still head gardener and had been for thirty years. By then, perhaps, he was settled enough in his ways to be growing the same sort of plants that were growing here between 1847 and 1849, in Henry's time. There were 82 Camellias, all named, including *C.* 'Jubilie' a variety raised by Hugh Low. 104 heaths, 1,105 strawberries (in pots), 48 orchids, 238 pineapples, 60 ferns, 153 other greenhouse plants and 321 other stove-house plants were also listed, most of them named. There were also 1,881 empty flower-pots. Assuming that a few plants 'went missing' before the auction, this number of plants is work for two full-time gardeners, especially in the summer with the extra watering. As there was a garden to look after as well, it's likely that one or two more gardeners were employed in the summer. So, Henry would have been one of three or four gardeners then, which was more sociable then being one of two! I wonder if one of them was Henry's future wife, Charlotte Stainer?

Probably not. The furniture listed from the 'Man Servant's Room' was or was similar to that which Henry would have been provided with. There was a 3ft. 3in. painted French bedstead; a feather bed and bolster; three blankets and a coloured quilt; a large deal board and two trestles; three deal forms and two stools. Hard wooden seats; and a board to go onto trestles for a table to write letters on, inconvenient but space-saving in a small room. These (now second-hand) items realised £2-17-0.

It is odd, considering the store set by practical trial and error, that Camellias were grown in heated glasshouses. Some of them needed to be, but the many *C.japonica* varieties were perfectly hardy in Wiltshire. J.C. Loudon, in 1822, knew that they could be grown outdoors, with the red-flowered forms the hardiest. Nevertheless he advised that they should be protected with mats in the winter.[39] The reason for growing a flowering plant was for the employer's enjoyment. *Camellia* flowers, arriving early in the year are, like early-flowering Rhododendrons, easily spoilt, especially if planted with an easterly aspect. Glasshouse protection prevents this. Pot culture means that Camellias can be grown even if the land is chalky. An employer, often away, would wish to see plants in flower during his residence. The option of outdoors and heated indoor growing conditions allow pot-grown Camellias to have their flowering period held back or advanced, as required, by the gardener.

Camellias can be a problem outdoors in very cold and exposed conditions. These can occur throughout Scotland, away from the west coast, with even the hardiest forms 'a challenge' anywhere 'north of Aberdeen away from the coast'. (40) Most *Camellia* varieties cannot be grown successfully in Scotland, needing higher night-time summer temperatures for flower development. While there may have been an element of 'playing safe' in glasshouse cultivation, the form of *C.reticulata* that they had then wasn't the true, more robust form introduced later, in 1924, but the tender plant, now called *C.r.* 'Captain Rawes', introduced by Robert Fortune. (41)

Scottish gardeners like Henry and William Dodds must have revelled in the climate of the south of England. It was now possible to achieve an endless number of hybrids impossible to cultivate in Scotland. They had so much success, in fact, that Camellias became the most popular glasshouse shrub in Victorian Britain.(42) *The Salisbury Journal* called them 'the queen of flowers' in 1862.(43) Nor was its success as a status-symbol limited to Britain. In *Brigitta,* a novel set in Hungary by the then contemporary Austrian writer Adalbert Stifter, the plants were seen in the gardens of another (fictitious) military man: 'We stopped to watch a number of women who were engaged in cleaning the leaves of Camellia plants. In those days the Camellia was quite a rare and therefore an expensive flower. The Major examined the plants that had already been wiped and made a comment or two…'. (44)

In 1864, most of Baker's Georgian house was pulled down and a museum of natural history created, both in its place and in the garden behind. This would be of passing interest except that one of the museum's founders was the Earl of Radnor. Henry's longest period of employment was as head gardener to the Earl, though this was a few years away as yet. At the time, too, there was a pub, almost opposite Baker's house, called *The Radnor Arms(45)*. The Radnors were a dominant family in this area and as such would have known Col. Baker, a politician and military man. Baker would have been closer to his gardeners that the large Scottish estate owners could have been, and it seems a reasonable supposition that Henry was first introduced to the Earl here, at Baker's house.

If you walk up St Ann Street (as it's known today) you will see a pub, *The Bakers Arms*, next to where Col. Baker's house once stood. But it's a new name and is, it seems, just a coincidence.

ST. ANN'S STREET.

~~~

At the age of thirty-one, Henry found that his paths had led him to the gates of the Earl of Radnor's estate at Coleshill, on the (then) Berkshire – Wiltshire border. Before that, however, they had two other places to show him: Trentham Hall, in Staffordshire and Caen Wood, or Kenwood as it is now, back near London. Trentham, for the mid-nineteenth century's horticultural press, was the most important garden of the age. Under the head gardener, George Fleming, everything seemed to be going on there and every plant seemed to be grown there. Beds of sub-tropical plants – bedding – were the fashion, now fixed by Trentham. Even the old-fashioned flowers were creeping back around the edges of the pleasure-ground there. Their description in a contemporary guide was as flowery as the plants themselves.

'Looking towards the wood the eye crosses an astonishing variety of plants, shrubs and trees – the effect is singularly beautiful and harmonious – but it is produced by materials whose character is not strained or divided by anything foreign to their nature. Plants and flowers familiar to our childhood and associated with the glowing dreams of our youth, here seem as much at home as in the sunny glades of the bird-singing wood, the
~~~

cottage garden, or the ripening corn-fields. Ferns, white and yellow Broom, Furze, Paeonies, and a host of others, contribute to the general effect.'[46] One or two such plants even found their way into the gardens proper, where they were used to line the 'walks' or paths; plants such as musk (*Mimulus moschatus*), hollyhocks (*Althea* sp.) and, interestingly, sweet peas. In the general neglect of hardy annuals and hardy herbaceous perennials, these large estates were much alike. At least flowers in general were no longer relegated to the out of sight kitchen-garden, as they had been in some of the largest and most fashionable eighteenth-century estates.

Trentham lies three and a half miles south-east of Newcastle-under-Lyme, in the valley of the River Trent. There's enough of Trentham left now to get a feel for it, for how it might have been, especially with recent restoration work. The framework of the Italian garden survives. How inelegant and un-Italian the Italianate style was! Wiltshire, helped by Harold Peto, tries to draw it into itself, as at Iford Manor, near Bradford-on-Avon. Here at Trentham it's open, loud, brassy. Were it human it would be warm too, for it suits Staffordshire and Staffordshire people, or has become so suited over the years. The estate was owned by the Dukes of Sutherland, who owned much more here and elsewhere in Britain then. The south side of the mansion looked out over a lake to Tittensor Hill beyond. Atop the hill, an obelisk and statue provide a focal point. The gardens were terraced between the mansion and the lake and included, by the lakeside, the ten-acre Italian garden. The lake, two miles around, covered eighty-three acres and contained wooded islands of one acre and four acres in size. This lake was surrounded by a fifty-acre pleasure-ground, to the south beyond which lay the park of some four-hundred acres. From the highest point in the park was obtained 'an excellent view over a bare and somewhat flat country, embracing however... the pottery towns of North Staffordshire, the site of each being well marked by huge volumes of fire and smoke'. [47]

Aside from the stone ornamentation, the horticultural interest lay in an Orangery; a 200 yard (183 metre) long trellis-covered walk; a 'Rosery of the best specimens, in circular beds with ivy borders';[48] a shrubbery in the pleasure-ground; the kitchen garden, rather small at only five acres and thus surrounded by 'slips', extra land used for stock-plants, cut flower and vegetables like potatoes; an 89' (27m) long, 60' (18.3 m) wide and 14' (4.27 m) high conservatory (one of four), and flowerbeds. An 'American' garden, being planned in Henry's time here and formed just afterwards, was reached across the 90' (27.4 m) iron bridge over the River Trent which ran through the grounds. From this bridge 'onwards to a considerable distance the water is fringed on either side by Water Lilies and other flowering aquatics, and from here one observes the gigantic operations which has

Glasshouses at Trentham
Taken from 'The Gardeners' Chronicle and Agricultural Gazette', 1872

been resorted to in changing the course of the river from the bed of the lake, and which formerly used to silt it up in all directions; that beautiful sheet of water is now fed exclusively from a rivulet on the opposite side'[49] This redirection and draining of some ten to twelve acres round-about wasn't started until 1853, but preliminary work may have begun a year or two earlier during Henry's time here.

The scale and variety of planting and an emphasis on innovation, experimentation and teaching was the draw for Henry and many like him. The inner south and east facing walls of the kitchen garden, for example, were covered by a narrow glasshouse, 6'-7' (2m) wide, 10' (3m) high and all of 600 metres long. All the front glass ran on castors and slid sideways, in sections, for ventilation and access. A slate path ran down its centre – 'so neat and clean' – wrote a visitor from Lanarkshire. 'On the side of the path next to the... wall are bedding-out materials in pots – I should not like to suggest how many thousands.' (Trentham used one hundred thousand bedding plants annually). 'Peaches and Nectarines are the principal occupants of this wall on the south side; Cherries, Pears, and Apples on the east side'. On the side between the path and the glass front were 'fruit trees in pots by the thousand, in endless variation, comprising Apricots, Peaches, Nectarines, Plums, Apples and Pears... chiefly in thirteen-inch [33 cm] pots placed rim to rim' ('pot thick' we say nowadays). [50]

Mid-May would have been the start of a busy period. Whatever plants had been bedded out for the winter and spring were lifted and removed. These were likely to have been hardy evergreen perennials which, disliking root disturbance, may well have stayed in their pots. Soon after Henry's arrival at Trentham he discovered that Fleming was already experimenting with providing winter interest in the form of flower, colour, height, form and texture. He used heathers, the winter flowering ones presumably, young (and therefore small) plants such as conifers and hollies, with varigated ivy as an edging. [51] It was an alternative to 'spring', sometimes known then as 'winter' bedding when it was referred to at all. Planted out in the autumn, it sat dully through the winter before flowering in the spring. Some of the most innovative designers today are, in effect, returning to these practices pioneered in the 1840s.

Once the beds were prepared, the one hundred thousand summer bedding plants were planted and watered. Although gardening was a male profession, (as it mainly still is), these places had 'weeding women'. Demarcation lines can dissolve under pressure though, so perhaps the women joined the men for a week or two, if only to fetch and carry plants between the standing-out grounds and the flower beds. Into this busy time, imagine a pause. Too many plants, perhaps, have been arriving for the planters, and the fetchers and carriers, young men and women mostly, take a short break. There has been a shower and now the sun, with the first real warmth of the year, is making itself felt. The break from effort, the sudden blue sky, thousands of flowers, smiling faces, one or two of them, unusually, women's... Gardening is the only, the best job that anyone can have and he, Henry realises, is the luckiest of men, and feels sorrow for all those unfortunates indoors or down mines. And then the moment closes over, and passes.

Bedding begins life under glass, and also under glass flowers and vegetables were being 'forced', brought forward into early growth. Such 'forcing-houses' were being built there between November 1851 and February 1852, so Henry would have been aware of the latest technological advances. Here at Trentham they were to be used principally to bring on cherries and vines. There was experimentation with new vine-pruning ideas. George Fleming invented a 'salting machine', an idea to keep the vast acres of gravel paths weed-free. Exhibited inside the new *Crystal Palace*,

George Fleming portrait, 'Cottage Gardener', 1854, (RHS Lindley Library)

set up in London's Hyde Park for the Great Exhibition of 1851,[52] it was a boiler on wheels that sprayed steaming salt water, (the salt mixed at 2lbs. per gallon, if you want to try it out) onto the ground. 'At Trentham this contrivance is found to be very effective', said *Beeton's All About Gardening* 'and much more economical than hand-weeding the paths, besides giving a freshness and brilliancy to the gravel.' [53]

George Fleming had his critics. His weeding machine 'threw people out of work', the sliding glasshouse panes and other ventilation projects 'already patented.' But Fleming also had a lot of support. Leading figures always attract admirers and detractors and, thus, controversy. Out of the public eye when he died in 1876, he didn't even rate an obituary in the gardening press. A line or two in *The Gardeners' Chronicle* – of late years Mr. Fleming's attention has been given over to agriculture rather than horticulture – was all there was.[54] Fleming was a pioneer of many things, especially of the bedding-out system in Great Britain. Bedding-out, en masse, was part of the brilliance and splendour of the great estates. In their decline, the bedding-out system declined also. New ideas would be needed to fit and shape the changing times. Henry, we can imagine, buried the thought away in the back of his mind.

∼∼∼

In the 1850s, 1852 seems the likely year, Henry left the north midlands for London. He was heading for Kenwood, the then Middlesex, now north London, residence of the Earl of Mansfield. Henry didn't put Kenwood down on his *Curriculum Vitae*, thought some of his obituarists did. Perhaps he never went there. It's more likely that he did and forgot, or wanted to forget. Why might that be? Bullying by someone in authority perhaps; an unhappy love affair. It is just possible, in view of a London marriage, that he met his first wife, Charlotte Stainer, here. She was to die in childbirth and Henry, in his old age, may have been unable to face the memory of it. Henry's later admirer, William Robinson, wrote in Henry's lifetime that he was at Kenwood at this time, and Henry didn't contradict him.[55]

One attraction of Kenwood (or Caenwood as it was sometimes known then) is that it reunites us with our old friend J.C.Loudon. But it's almost the last time, for he died, suddenly, in 1843. 'As an example of peaceful sylvan beauty' Loudon wrote, 'nothing can surpass Kenwood'.

It was 'beyond all question the finest county residence in the suburbs of London, in point of natural beauty of the ground and wood, and in point also of the main features of art. The park may be said to consist of an amphitheatre of hills; the house being situated on one side, backed by natural oak woods rising behind it, and looking across a valley, in which there is a piece of water, to other natural woods, also chiefly of oak, which clothe the opposite hills…'. These oak woods, 'probably the oldest about London, are remarkable for being composed almost entirely of Quercus sessiliflora' now *Q.petraea*, the sessile oak, and gave Caenwood its name, derived from 'acorn wood'.

'A stranger walking round the park' Loudon continued, 'would never discover that he was between Hampstead and Highgate, or even suppose that he was so near London. It is, indeed, difficult to imagine a more retired or more romantic spot, and yet of such extent, so near a great metropolis'.

One William Keane, writing nearer to Henry's time here in 1850 also found 'the scenes of Kenwood romantic. The trees cover the tops of the hills, and hang on the steeps of others; in one place they front, in another they rise above, in another they sink below, the point of view…'. A pathway 'enters the bosom of a dark wood, under an arbour, formed of the living trunks and large branches of rhododendrons like forest trees…'[56] Neither Keane nor Loudon commented, let alone expressed concern, on the unusual fact that the Rhododendrons, *R.ponticum* that would have been, had been seen to be seeding themselves at Kenwood since the 1830s.[57]

A large, seven-acre kitchen garden, now ornamental, lay a little eastwards of the mansion, beyond the stables. It ran, further than it does today, along the south side of the Hampstead – Highgate lane.[58] Keane described them as sloping 'beautifully to the south... diversified by natural ridges, with east and west aspects, that are useful for a succession of fruits and of vegetables. The horticultural buildings, in the several divisions, are two greenhouses, each forty feet long, a stove [hot-house] twenty feet, a vinery ninety, and a pit [a low, sunken and heated greenhouse] sixty-six feet long. The soil is loam in some parts, peat in others, and rather sandy in other places…'. Immediately to the west of the mansion lay a one acre flower garden. Having the kitchen and flower gardens on each side of the mansion would have involved Henry and the other gardeners in much criss-crossing between the two. Presumably the sight of the workforce from the mansion didn't offend the occupants' sensibilities, the usual case elsewhere. [59]

Keane lyrically described the flower garden. 'The large beds are disposed on grass, some are planted with choice shrubs, and others are adorned with the most beautiful flowers, blended and grouped in the best manner to give a pleasing effect to the whole. The fine orange trees also contribute to heighten the effect in the summer season, and to give out their delicious odours in the "incense-breathing morn", or in the dewy twilight hour. Here we saw a beautiful Araucaria imbricata [now *A. auracana*, the monkey-puzzle tree] fifteen feet six inches high, and a large Ligustrum lucidum [Victorians called this the *shining privet*[60]] in full bloom. The standard roses are also worthy of particular notice'.[61] Perhaps this flower garden had improved since Loudon, more discerning, described it in 1838 as 'the only defective part of the place. It is naturally shaded and confined by a lofty lime avenue on the one hand, and by a rising hill of oak wood on the other; and the area of the garden contains by far too many small trees and shrubs among the flowers: in consequence of this, the turf is almost always damp on the surface; and the flowers come up with slender and etiolated stems, and pale colours. Most of the flower-beds, also, are too large; and they do not combine so as to form a whole. Were it ours, we should clear the whole area…'.

This flower garden was once the kitchen garden for which the 'rising hill of oak wood' on its north side provided natural shelter. The landscaper Humphrey Repton had had it moved in around 1800,[62] when he was reintroducing flower gardens to the large estates. Re-siting the kitchen garden further from the house on the other side had the effect of placing decorum above convenience, as the head cook would, unavailingly, have complained. The site of the new kitchen garden was more exposed and Repton, disliking walls, was obliged to build one on its north side. It's still there today. Look closely and you'll see all the little holes from the nails that generations of gardeners, including Henry, drove in to tie in and otherwise protect the wall trees that once blossomed and fruited here.

Loudon also noticed that there was no great variety of trees and shrubs at Kenwood. In this respect, Henry's many gardens followed a pattern. They were not plantsmen's gardens and as a result Henry was never inspired to be one. There were, however 'a few very fine specimens of foreign trees and shrubs... such as a cedar of Lebanon... a larch... and a Robinia Pseùd-Acàcia' now *R.pseudoacacia*, here. The presence of the *Robinia*, the 'false acacia', is interesting. In the same family as the sweet pea, Leguminosae, its white flowers share the same distinctive vanilla-like fragrance; perceptible sometimes from quite a distance. Henry could hardly have failed to notice it. Loudon deprecated the fact that Kenwood, (unlike Trentham) 'being at no season of the year open to strangers' [63] could be enjoyed by so few. Public access would only arrive with public ownership in the 20th century.

On the 22nd March, 1854, almost the whole household bar the gardeners arrived at Hampstead from Scone Palace, the Mansfields' Scottish seat. Their annual six-month stay, taking in the 'London Season', saw them return to Scone on the 22nd September.[64] George Cockburn, the head gardener here, had to ensure that the gardens looked their best between April and August. For Henry as for the other gardeners, spring and early summer remained the busiest time of the year. Especially so, in the season of balls and parties, and visitors coming to stay. The gardens were being shown off and had to look perfect.

It is often assumed that it was aristocrats, living in garden-less London addresses for the 'Season' that led to the comparative neglect of hardy spring – flowering plants and to attempts at brightening the flower garden in the autumn, when the wealthy returned to their country estates (and gardens). David Thomson was saying as much in his *Handy Book of the Flower Garden* in 1887. He saw that hardy spring flowers had been overlooked but, thinking out loud 'it would be difficult to say that any absolute reason exists why this should be so... There can be no doubt that the fact of the most opulent and fashionable families being, in the majority of cases, away from their country-seats in the spring and early summer, has been the chief course of directing the efforts and attention of gardeners to the crowding of as many flowers into the autumnal months as possible.' [65]

Did most of the wealthy families go to London? The Mansfields, that did, went to Kenwood, (no more central London address being known) an estate. During 'the season' at Kenwood the gardens would have been a focus of attention, as were the gardens of those not at London or visiting for short periods only. Think of Trentham, and their massive summer bedding schemes. If plants were being encouraged into autumn flowering, then, it could be argued, this was because it was an important era in the history of plant hybridisation. Extending flowering periods was one of the hybridists' objectives *per se*. The opposite argument from that usually proposed could be made, namely that the flower garden had for many to be at its best from April to August. Tender summer bedding can't be extended back into spring, so the reliance, in spring bedding terms, was as it is now, on a limited range of hardy flowering plants. You get an echo of the times in Parks bedding today. Spring bedding is at its best when it is removed, in late May, to make way for the summer flowers. This summer bedding, often already in flower, is quick to establish, and looks good until around late September, when it starts getting blowsy.

As far as wages and conditions and items bought for the garden go, Kenwood was similar to the estates that Henry had worked at in Scotland. The garden accounts for the years 1852-1854[(66)] show that the gardeners, numbering from between ten and more commonly thirteen, all men and unnamed, earned on average 12/6 a week. Two boys earned 8/- a week. Everyone's hours and wages were reduced slightly during the shorter winter days. Compared to the rest of the estate staff, Henry and the other gardeners were on an average wage. They got about the same as Henry Byrne, the baker, on £30 a year, and a lot more than Harriet Lambert the scullery maid, on £12. No-one was near those like the cook, the butler George Skeffington on £84 a year, or Henry's boss George Cockburn the head gardener, at £90 a year. Although things like rents, food and heating might have been cheap or free, even these three could not have considered themselves to be among the middle-classes of the time. Lower middle-class perhaps – just. [(67)]

Henry may have accompanied Cockburn to the summer show of the (not yet Royal) Horticultural Society in their gardens at Chiswick, beginning a tradition of meeting up afterwards with friends in nearby Gunnersbury. It was later recounted of Henry that 'in company with a number of others of kindred spirit he would betake himself in the evenings of a show week to a little hostel in Gunnersbury, and there each would show his triumphs, and the talk would be of things floral. Evenings such as these yielded him far greater pleasure than... public praise…'.[(68)] Public praise would come later. The summer of 1853 was a washout, as was 1854's, and if Henry went then, many did not and the Society floundered as a result.[(69)] Here at Kenwood, as on Henry's other estates, specialist workmen were sometimes brought in from outside. In 1853, J.Thompson was paid £1-12-0 for renewing the iron bands and screws that supported the elderly false acacia. Seven guineas (£7-7-0) was paid for '15 dozen Iron rods for tree Rhododendrons and gates for drains etc.'

'Repairing Iron hurdles, fences and 18 braces to arbours and tree umbrellas' cost £5-2-6. The 'romantic scenes of Kenwood' were maintained in part with unromantic iron bands, rods and braces. The following selection comes from a further browse through the 1852-53 accounts:[(70)]

		£. s. d.
April 17th	Repairing garden Mens beds.	0 2 0
May 1st	6 dozen garden brooms.	0 18 0
May 29th	Dodds for 2 ½ cwts. of manure.	0 11 3
July 24th	6 dozen garden brooms .	0 18 0
Oct 9th	Refreshment to men removing Orange trees.	0 5 4
Oct 23rd	Collecting 24 bushels of acorns.	1 4 0
Nov 20th	6 dozen garden brooms.	0 18 0
Dec 25th	Christmas Boxes to Clerk and pew-opener Highgate Church, Postmen and Chimney Sweep.	0 17 6
Feb 12th	6 dozen garden brooms.	0 18 0

The 'Mens beds' being repaired would be those of Henry and the other bachelors. They lived in the bothy, which from the maps of the time, appears to have been in the kitchen garden. A T.A.Evans was paid two guineas 'for sheeting &c for garden mens beds', and a T.Fernee was paid for 'whitewashing [the] Garden Bothy'. The Mansfields looked after their men. At least, they were as well cared for as the animals would have been, which was not always the case on these estates. Two hundred and eighty-eight brooms a year for fifteen employees shows that the Kenwood paths were swept on a daily or almost daily basis. The broom heads, being made of birch twigs or heather on an ash handle, were quickly worn. I can remember, as a parks department gardener in the 1960s, sweeping paths with birch brooms in a semi-circular sweep. Eventually all the brooms would be worn to inflexible and unusable stumps. The arrival of new brooms was, as it would have been in Henry's time, something of a red-letter day! There would have been some elbowing by the Kenwood men to grab the best, avoiding the springy ones that had to be pushed down hard and tiringly to make an effect.

'Refreshments to [the] men' would have been 'small' or weak beer, safer to drink than water. The orange trees, having been displayed in the flower garden for the summer, were now being moved back into the warmth of the greenhouses for the winter. The 'pew-opener' is the verger, though quite what those Christmas Boxes are doing in the garden accounts is a puzzle. The answer is perhaps mixed. Cockburn saw them going without and may also, by keeping them sweet with Mansfield's money keep them amenable to any little favours he might require of them. Such deference was worth shaving a slice off the gardens' accounts for, and objections by Mansfield would have been viewed as parsimony.

~~~

Late winter, 1854. What a depressing summer we had last year, Henry mused, surveying the gloomy evergreens out in the park. What was he doing, flitting about still? He was thirty now. This journeyman thing had run its course. Society was changing too. Someone like him should be in retained employment. Married too. Head gardener? It was about time. What happened to the contacts he'd made in Salisbury? The Radnors?

Early in 1854, Henry was appointed head gardener to the Earl of Radnor at his estate at Coleshill, in Berkshire.[71] He was to stay there for the next twenty-one years.
~~~

References to Chapter Three

(1) Advertisement, *The Horticultural Advertiser*, (25-9-1907), p.2.

(2) J.G.L. Burnby, A.E. Robinson, *And They Blew Exceeding Fine*, (1976), p.8.

(3) J.C. Loudon, *Encyclopaedia of Gardening*, (1822), pp. 1319-1320.

(4) A.M.D. Hughes, *Cobbett*, (1946), p.117.

(5) William Howitt, *Rural Life of England*, (1845), pp. 204-205.

(6) T.C. Smout, *A Century of the Scottish People 1830-1950*, (1997), p.112.

(7) Ibid, p.13.

(8) Richard Bradley, The Monthly Register, 1724, quoted in M. Leapman, *The Ingenious Mr Fairchild*, (2000), p.86.

(9) Miguel de Cervantes Saavedra, *Don Quixote*, 1605-1606, (2003), the John Rutherford translation, p.354.

(10) E.H.M. Cox, *A History of Gardening in Scotland*, (1935), pp.202-204.

(11) 'On the Preference for Scotch Gardeners', *The Gardener's Magazine*, (5-1840), pp.244-246.

(12) George Good, *Liberton in Ancient and Modern Times*, (1893), p.135.

(13) John and Julia Keay, eds., 'Railways', *Collins Encyclopaedia of Scotland*, (2000), pp. 835-836.

(14) Robert C. Mossman, 'The Meteorology of Edinburgh, part 2', *Transactions of the Royal Society of Edinburgh*, Vol. 39, (1900), p.80.

(15) A.B. Thomson, *Mean Winter Temperatures in Edinburgh 1764-1963*, (1971), p.15.

(16) TD 95/75, Vol. 211, (Sept. 1853), The Mansfield Papers, Scone Palace via National Records of Scotland.

(17) 'Plan of the parish of Saint Mary, Stoke Newington, Middlesex', (1846), Archives dept., London Borough of Hackney.

(18) *Journal of Society for Bibliography of Natural History*, (1968), p.328.

(19) J.L.G. Burnby, A.E. Robinson, *Now turned into fair garden plots*, (1983), p.39.

(20) Ibid, p.40.

(21) 'Societies', *The Gardeners' Chronicle*, (19-9-1868), p.994.

(22) Advertisement, *The Gardeners' Chronicle*, (2-1-1841), p.2.

(23) 'Home Correspondence', *The Gardeners' Chronicle*, (27-4-1850), p.261.

(24) Ralph Waldo Emerson, *English Traits*, 1856, (1902), p.22.

(25) J.L.G.Burnby, A.E.Robinson, *Now turned into fair garden plots*, op. cit., p.40.

(26) John Chandler, *Salisbury: History and Guide*, (1992), p.73.

(27) R.W.Emerson, op.cit., p.159.

(28) Henry Eckford, *Curriculum Vitae*, (undated), author's possession.

(29) G23/701/31, 'Plan of Col. Baker's leasehold kitchen garden', (12-8-1856), Salisbury book of plans, c. 1845-1900, Wiltshire Record Office.

(30) 'Observations on several gardens in England', *The Gardener's Magazine*, (2-1833), pp.16-17.

(31) John Harvey, *Early Nurserymen*, (1974), pp. 101-102.

(32) See William Dodds' obituaries: *The Gardeners' Chronicle*, (1-9-1900), p.179 and (5-1-1901), p.15; *The Garden*, (5-1-1901), p.16; *The Gardeners' Magazine*, (5-1-1901), p.16.

(33) John Claudius Loudon, *Encyclopaedia of Gardening*, (1822), p.962.

(34) George A. Sala, 'Twice Round the Clock', quoted in M. Paterson, *Life in Victorian Britain*, (2008), p.224.

(35) Alice M. Coats, *Flowers and their Histories*, (1956), p.66; quote.

(36) William Robinson, *The English Flower Garden*, 1883, (1905), p.529.

(37) James Bartholomew, *Yew and Non-Yew*, (1998), p.49.

(38) 451/334, Sale of Contents Catalogue, (23-4-1862), Wilts. R.O.

(39) J.C. Loudon, op.cit., pp. 1038-1039.

(40) K.N.E. Cox, R. Curtis-Machin, *Garden Plants for Scotland*, (2008), p.32.

(41) *Hilliers' Manual of Trees and Shrubs*, (1975), p.60.

(42) E.H.M. Cox, op.cit., pp.131-132.

(43) 'Horticultural Notes', *The Salisbury Journal*, (1-3-1862), p.3, (quoting Bell's Weekly Messenger).

(44) Adalbert Stifter, 'Brigitta', in *Famous German Novellas of the 19th Century*, (2005), the Edward Fitzgerald translation, p.112.

(45) G23/1/165PC, 'Plan of the City of Salisbury,' (1835), Wilts. R.O.

(46) 'Notices of Books', *The Gardeners' Chronicle*, (22-8-1857), p.583.

(47) 'Trentham', *The Scottish Gardener*, Vol.8, (1859), p.82.

(48) 'Notices of Books', *The Gardeners' Chronicle*, (22-8-1857), p.583.

(49) 'Trentham', *The Scottish Gardener*, Vol.8, (1859), p.82.

(50) 'James Anderson: Jottings from my note-book, in a visit to the south', *The Scottish Gardener*, (1859), p.362.

(51) Joan Morgan, Alison Richards, *A Paradise out of a Common Field*, (1990), p.37.

(52) David Stuart, *The Garden Triumphant*, (1988), pp.16-17.

(53) *Beeton's All About Gardening*, (c.1890), pp.129-130.

(54) 'Obituary', *The Gardeners' Chronicle*, (5-8-1876), p.180.

(55) See, for example, 'Mr Henry Eckford', *The Garden*, (2-1-1897), frontispiece; 'Mr Henry Eckford', *The Garden*, (28-7-1900), p.70.

(56) William Keane, 'Kenwood, Highgate', *The Beauties of Middlesex*, (1850), p.63.

(57) J.Morgan, A.Richards, op.cit., p.175.

(58) Parish map of St. Pancras, (1849), Camden Local Studies and Archives Centre.

(59) J.Morgan, A.Richards, op.cit., p.179.

(60) George W. Johnson, *The Gardeners' Dictionary*, (1877), p.491.

(61) W. Keane, op.cit., pp.60-65 passim.

(62) Julius Bryant, Carol Colson, *The Landscape of Kenwood*, (1990), p.8.

(63) J.C. Loudon, *The Suburban Gardener and Villa Companion*, (1838), pp. 661-674 passim.

(64) TD95/74, Vol.211, The Mansfield Papers, Scone Palace.

(65) David Thomson, *Handy Book of the Flower Garden*, (1893), p.177.

(66) TD95/74, Vols. 210 & 211, (1852-1854), The Mansfield Papers, Scone Palace.

(67) R.D. Baxter, The taxation of the U.K., 1869, quoted in G. Best, *Mid-Victorian Britain 1851-75*, (1971), p.90.

(68) 'The Cult of the Sweet Pea', (advertisement), *The Daily Mail*, (21-2-1911), p.10.

(69) Miles Hadfield, *A History of British Gardening*, (1969), pp. 318-319.

(70) TD95/74, Vol.210, The Mansfield Papers, Scone Palace.

(71) H. Eckford, *C.V.*, op.cit.

Chapter Four

Coleshill is a Cotswold village that, through no effort on its part, has over the years managed to wander in and out of three counties; Wiltshire, Berkshire and Oxfordshire. At present, it straddles Oxfordshire and Wiltshire. In Henry's day, it lay firmly in Berkshire (or Wiltshire!). Coleshill is as beautiful now as it was said to be then, with a slightly grand and overbearing air about it. I remember picture-books and posters from my infants' school in Beckenham, in the nineteen-fifties, in which gentle habitationless hills and vales rolled about under big, blue, white-flecked skies. They were an ideal of a southern-England, which, in the years since, has remained disappointingly elusive. At the top, eastern part of Coleshill, looking south across the busy Oxford to Swindon road, you certainly don't see it. Halfway down the hill, looking south-west from the church, the sprawl of Swindon safely over the horizon, you can.

> 'Coleshill, Highworth, Wiltshire, the seat of Earl Radnor', begins an 1870 account in *The Gardeners' Chronicle.* 'This fine country seat, set down in the midst of some rich woodland, from whence fine views can be obtained of the surrounding country, lies midway between the Faringdon and Shrivenham stations of the Great Western Railway, being distant about four and a-half miles from each'.
>
> 'Limes [lime trees] are grand at Coleshill, and they are supposed to be some of the finest old Limes in England, and near the mansion is a pleasant walk under a short avenue of them – most deliciously cool and inviting on a sweltering summer's day'. (1)

> 'Coleshill is a parish', begins the 1869 Post Office Directory for Berkshire, less lyrically; '... court district of Faringdon, rural deanery of Vale of White Horse, archdeaconry of Berkshire, and diocese of Oxford, 3 ¾ miles west-south-west from Faringdon, and 71 ¾ from London'.
>
> 'Several cottages built of stone, and of ornamental design, with gardens sloping and neatly laid out, are scattered over the village... The soil is clay and loamy... The population in 1861 was 464: the area is 2,301 acres'. (2)

That population of 464 had risen slowly from just 324 in 1821, for Coleshill had always been a small village, and the area rural. As late as 1988, a description of the area begins: 'The Vale of White Horse, one of the most remote areas of southern England...' (3) It may have been here that Henry's later expressed wish for 'somewhere away from it all', referring to Wem, in Shropshire, first developed.

To us, Henry seems a little young, at thirty-one, to become a head gardener. But this wasn't so. The British mainland population was expanding rapidly, making it a predominantly youthful one. According to the 1851 census, half the population was under twenty. It's easy to imagine the Victorians as elderly. All those men in top-hats sporting mutton-chop whiskers, walrus moustaches or great long beards like Henry's; the women, hiding their faces behind bonnets and other headdress as fashion dictated. But the hair, hats and bonnets hid youthful faces.

Henry was to marry (twice) and have all his children here, in this little corner of England. It makes a long local connection with nursery-rhymes seem appropriate. Henry was to order most of the week by week necessities for the garden from the nearby towns of Highworth and Faringdon. Faringdon had once been the home of Henry Pye, one of the worst-ever poets laureate. Walter Scott described him as 'eminently respectable in everything but his poetry'. An ode on King George III's birthday was so crammed with allusions to feathered songsters that it provoked the now well-known nursery-rhyme:

And when the Pye was opened,
The birds began to sing;
And wasn't that a dainty dish
To set before the King? (4)

According to local tradition 'Coleshill is the hill of King Cole of nursery-rhyme fame, who is supposed to have lived at Cole's Pits'. [(5)] Let us hope it is true. Perhaps Coleshill's name is also associated with the Berkshire man Thomas Cole, a successful tradesman in the thirteenth century – the 'rich clothier of Reading' [(6)] he was called, and an ideal example for Henry who was to partly emulate him. Coleshill couldn't be named for being on a hill above the River Cole, the village's western boundary. That's stretching things too far.

What Henry did not do at Coleshill is as interesting as what he did do. He ignored plans, mostly or wholly of Loudon's, to completely redesign the grounds around Coleshill House. Victorian gardens mirrored Victorian house interiors. Most people's house interiors had been sparse, but cheaper factory products and rising incomes were soon to lead to a clutter of ornamentation. The new ideas for the gardens at Coleshill called for a 'Botanic' flower garden, a Pinetum, a Winter garden, an American garden, a Thornery, 'Spiraes' (a Spirea collection ?), a new Rosery and a new kitchen garden, to be joined to new pleasure grounds by a tunnel.[(7)] There were other ideas too, an undated proposal calling for a new hot-house or vinery, a conservatory and an Italian garden in front of the house[(8)].

To elucidate, a pinetum is an ornamental conifer collection which might or might not include pine-trees. A winter garden now is an outdoor collection of plants of particular attraction in winter. Then it probably referred to a conservatory, with their year-round floral displays, particularly cherished in winter. An American garden, by that time, was a collection of acid-loving plants, not necessarily American natives. A thornery would have been a garden of thorny plants like hawthorn (and never proposed by gardeners who have to work in them). A rosery, by then, was a formal garden for roses (it hadn't always been for roses). 'Sixteen acres would have been cleared for the construction of the pleasure ground and terrace gardens, with fountains' and water basins together with 'sixty-two vases on pedestals and nearly 4,000 yds [3657 m] of new parapets and stone edging'[(9)] was proposed. All the plans were characterised by nice clean straight lines as though Kent, Brown, Repton and the entire eighteenth - century landscape movement had never existed.

Changes had been wanted, plans painstakingly produced; why were they ignored? Well, the Earl may have wondered why he should be involved in such middle-class fashions. Mr. Cowie was head gardener in the 1840s and a James Cowie, gardener, was sacked for drunkenness in late 1847 or 1848. [(10)] There was to be at least one more head gardener between then and Henry's arrival in 1854, for Henry took over from a Mr. Hugh Fraser[11]. Henry, although he undoubtedly would have known of the plans, had as we know little interest in such things. When the head gardeners are inebriate, short tenured or indifferent; when the client is in two minds and the designer dead (Loudon died in December 1843) then such plans will gather dust.

To find out what Henry did do at Coleshill, it is instructive to see what the grounds probably looked like when he arrived. In May, 1847, it was stated that 'The garden and grounds here are extensive, and of late have been much improved. In the kitchen garden are two fine houses, built last summer, for the cultivation of Grapes, each 40 feet by 16 ½ feet; the Vines were planted in autumn, and are now starting away very strongly... In front of these houses is a range of pits, 65 feet by 8 feet, which were erected at the same time.' These were being used to produce out-of-season winter strawberries and cucumbers for the mansion. 'In front of these are cold pits for wintering half hardy plants, succeeded by dung frames, which will with difficulty ever be dispensed with where a large supply of early vegetables must be produced. In one of the pits, planted out in rows, was a fine crop of French Beans, producing as abundantly as if they had been out of doors. Contiguous to these is the Mushroom house, a good erection for the purpose, 30 feet long. In the middle of the house is a path with three rows of shelves on each side, all of Welsh slate; several beds were bearing abundantly... The houses and principal pits are all heated by hot water pipes, which in the pits have moveable covers, a point of importance where it is desirable to keep up a moist atmosphere. The principal Vineries have an air drain along the front, with apertures immediately over the pipes; the air thus gets heated before it enters the house'. [12] The head gardener, Mr. Cowie, was also forcing roses in 'gentle heat' to ornament the conservatory.

So, there was already some area of glass, which would have been heated by coal-fired boilers. When Henry took over, the bills show coal being bought 'For the Gardens' from R.Goddard, Coal, Salt and Slate Merchant, at Shrivenham Station. The bills are almost complete for between 1854 and 1863. [13] Coal cost between 15/- and £1 a ton but was generally 18/- a ton in winter and 17/- a ton in summer. Coal was delivered during most weeks of the year and paid for six months in arrears. There was a garden cart, but deliveries could be a ton or more at a time. Faced with a journey ending in that long uphill trek from the river, they probably got Goddard to deliver it, the garden cart a standby if they ran short.

It is said that before Henry's time (and the time of the third Earl) the river was crossed by a ford. 'Even after the bridge was made the carters continued to use the ford. The horses were accustomed to wade through the river and to take a drink of water, while the carters liked to wash the wheels of the wagons or to soak them if the weather was hot and dry', to tighten the joints. [14] It's likely that the horses still stopped for a drink but took the dry way across now, especially in winter. Henry does seem to have tried to buy more coal at summer prices, but he probably only had so much room to store it. Up until 1861, the amounts averaged out at 22 tons of coal a year. An exception was 1860, when it leapt to 32 ¾ tons. This was the summer of 'no sunshine, but continuous rain' as Shirley Hibberd recalled it in his book *The Amateur's Flower Garden* in 1871. [15] It was followed by 'the severe winter of 1860'-61, remembered by William Robinson in his book *The English Flower Garden* in 1883. [16] The foul weather continued into spring. [17] From 1861, coke was occasionally bought too, but it was expensive then, at 22/- a ton.

In 1862 coal consumption suddenly leapt up to 46 tons, with 48 ¾ tons in 1863. Fourteen glasshouse structures can be identified in the first O.S. map of Coleshill in 1876. [18] The 1850s and '60s was when glasshouse construction really took off in Britain, and 1862 looks to be the year it occurred at Coleshill. There are three simple reasons. 'In 1845 and 1850', writes the garden historian Christopher Thacker, 'the cost of materials was lowered through the abolition of the glass tax and the brick tax, and in 1847... an improved method for producing sheet glass, far clearer and larger in size than the panes previously available' had been patented. [19] The second reason is that half-hardy plants like Dahlias and, soon, Verbenas and Pelargoniums, were more Henry's interest, and they needed winter protection. Henry also bought tropical and sub-tropical plants, many of which require high minimum temperatures and light levels to survive. The third reason is that coal, the fuel that heated the greenhouses, could now be delivered economically and in quantity by the railways, which had arrived at Shrivenham and Faringdon in 1840. Some of the coal may have been for Henry's house and a bothy but most of it would have heated the greenhouses.

Henry also bought salt from Goddard, it being listed on the same bills. As with the coal, it was ordered on a regular basis but less often. It was mostly purchased in winter, sometimes in spring and never in summer. It was bought as so many lumps, a few at a time, Henry spending between 1/- and 5/3 per purchase. Being deliquescent and not easily stored, the salt would have been used soon after it was bought. Therefore its use as a weed killer in something along the lines of George Fleming's Salting Machine seems unlikely, it being purchased out of season. Because of its winter use, it was almost certainly a de-icer for slippery steps and paths.

Knowing the effect of salt on ice seems to contradict Jennifer Davies' view in *The Victorian Kitchen Garden* that 'at one time salt was added to the ice' in the Victorian ice house 'with the thought that, like salted beef, it would keep longer'.[20] Not so, apparently. In 1854 *The Cottage Gardener* was reporting that some gardeners were recommending 'salt and salted water to be poured on the ice, from time to time' in the ice house 'as the ice was being packed. That idea must have originated from not understanding the reason why salt is so largely used with ice by the confectioners for producing an intense degree of cold for freezing their [ice-cream] mixtures, but that very end is obtained at the expense of the ice, which melts much faster with salt than without it. I have known one or two gardeners who used salt in packing their ice, and kept a good supply notwithstanding; but that cannot affect the question…'. [21] One or two pretty stupid gardeners, it has to be said!

Salt has, over the years, been used as a supposed fertilizer on things like sea kale and Asparagus. Perhaps it was done to remind the sea kale of the sea-side and feel at home. *The Cottage Gardener* was recommending 'liquid manure and a little salt' for sea-kale in 1854, [22] and gardeners followed the advice. 'Sprinkled salt on surface of Asparagus beds and watered with liquid manure' was the diary entry of one south-London gardener in May, 1873. [23] Salt has also been used as an insecticide for caterpillars on cabbages and other brassicas. [24] William Cobbett, in *The English Gardener* of 1829 [25] suggested mixing it, in small amounts, into composted manure.

Cobbett, incidentally, had been a friend of Earl Radnor and had once persuaded him, in one of Cobbett's more mad-cap schemes, to plant 13,600 *Robinia pseudoacacia* as timber trees. Cobbett called it the 'locust' tree, its American name, although it was already known here as the common or bastard Acacia. [26] In his *Rural Rides* Cobbett described their appearance here in 1826 as 'the most beautiful clumps of trees that I ever saw in my life... If men want woods, beautiful woods, and in a hurry, let them go and see the clumps at Coleshill'. [27] By Henry's day they were reaching maturity, but when the first trees were cut the trunks proved too crooked for the saw-mills and thus useless as timber. [28] Grown in plantations, they had little ornamental value either and were perhaps used as firewood. Their scent in flower would have reminded Henry of the sweet peas he grew in Scotland.

As we know, there was at least one large 'locust' tree in the grounds at Kenwood. The scent, one could say, was following Henry around.

Aside from the glasshouses, the only other alterations here during Henry's time appear to have been improvements to the flower and kitchen gardens. Apart from a few flowers around Coleshill House, the immediate grounds retained their park-like appearance of mixed deciduous and evergreen trees in grassland. Some landscape work did take place in the grounds here under Henry. There was, for example, tree planting but the intensive production of fruit, vegetables and flowers together with the volume of glasshouses gives the place the look and feel of a nursery, more than anything.

Among the first items that Henry bought for the garden were mats from local suppliers John Fidel and Sons in October 1854; '71 best Mats, £8-5-8' which were paid for the following March. Buying in the autumn, Henry was looking to protect his plants in the winter. Mats didn't seem to change a great deal throughout the nineteenth century. In his *Encyclopaedia* of 1822, Loudon was describing them as woven or matted from bast, which was usually the fibrous inner bark of the lime tree. Manufactured in inland Poland and Russia, they 'are used in gardening for a great variety of purposes; for protecting wall-trees, by being hung before them, and removed in mild weather; for protecting espaliers and standards, by being thrown over them; for protecting more delicate shrubs' and plants. 'They are used to cover hot-beds, hot-houses, hand-glasses, and every sort of glass-case…'. (29)

One present-day use is as frost protection for cold frames. This was one of the uses that *Beeton's All About Gardening* was to recommend to its readership of 'middle-class families' in the 1890s. (30) 'As a substitute for the Russian garden-mats, which are expensive', obtain and cut 'reeds into lengths of four and a half feet for the width of the mat, work them in bunches about one and a half inch thick, the bunches to be tied together with strong cord, in three places, each with a single tie: the mat will thus present a succession of rolls of reeds strongly tied together, forming a strong warm covering for frames and pits'. (31) Henry would have had no such time to waste. He had his boss's money and local suppliers to help him spend it. You don't keep dogs and bark yourself!

The world of local suppliers, as for the rural poor in general, was still circumscribed by the distance a horse and cart could reach and return from in a day. The language you heard was that spoken in the dialect of your village and its surrounding area. Henry's original home, over three hundred miles away to the north, meant that he and the villagers would have struggled under a degree of mutual incomprehension. The words 'Henry' and 'Eckford' can run together to sound like one word, especially to a southern English ear. Where Henry is from, the cadence of 'Eckford' falls slightly at the end, in a dismissive brooks-no-argument way. When Henry spoke his name, its sound, combined with his status, would have ensured that the name the locals thought they heard was the one that they did. Thus, when William Spindloe presented his bill for half-a-dozen ash rakes, 4/-, on 29th November 1854, it was to Mr. Hickford! And for Mr. Spindloe so for William Scarcebrook, the coach and harness maker from Faringdon, for the saw-sharpener Joseph Woodward, and no doubt others too, Henry was 'Mr. Hickford' for years after.

The garden's cart took a lot of use and abuse, but probably no more than a gardener's van would today. In 1854, it was round at Scarcebrook's on May 9th, and back in again on 1st September: 'To repairing roof of garden cart and covering roof and sides with Patent Linen...' March 1st, 1855; 'to repairing garden cart with a strong plate of iron…'. You imagine William biting his lip; if *that* doesn't do it...! June 17th, 1856; 'to repairing garden cart with splicing on two new front ends to bottom sides... £1-17-6'. January 10th, 1860; a new drawing bar to the garden cart fixed 'with a strong leather brace…'. With bigger firms, Henry had accounts that were settled in arrears, usually yearly; sometimes longer, sometimes less. Mr. Spindloe and Mr. Scarcebrook couldn't be treated thus, living like most people from hand to mouth, and they were paid more promptly. That 29th November bill for ash rakes was settled on the 30th.

Old William Spindloe was, attractively, not of his time. Whether he was of an earlier or a later time is difficult to say. His bills weren't headed with elaborate drawings of ash-rakes entwined into a large coat-of-arms but merely the items and their cost neatly written on slips of paper. Mr. Spindloe was unusual, too, in still using the old long form of the letter 's'. Used in Secretary hand, it had been losing currency for a hundred years but it was his touch of gravitas. I'm sorry to report that eventually even Mr. Spindloe threw in his lot with the fashions, with bill headings hardly leaving room for the bill beneath; he was, finally, fighting it out with the best of 'em.

On 25th November, 1854, Henry bought a number of plants from Garaway, Mayes & Company, Durdham Down Nurseries, Bristol. The bill was headed by an attractive drawing of the firm's headquarters, with smoke puffing happily from their chimneys – a selling point, showing how industrious they were. As the bill attracted a 5% interest if unpaid after a year, it was duly settled eleven months later, on 27th October 1855. The bill was:

18 Fancy Geraniums (followed by eighteen named varieties).

1 dwarf trained Moorpark apricot.

1 dwarf trained Gallande peach.

2 dwarf trained Noblesse peach.

2 dwarf trained Royal George peach.

2 dwarf trained Coes Golden Drop plum.

1 dwarf trained May Duke cherry.

1 pyramidal Marie Louise pear.

1 pyramidal Winter Nelis pear.

12 Rose Devoniensis.

12 Rose La Marque.

4 Lobelia caerulea alba.

Basket & Mats.

'Moorpark. The first Apricot in the kingdom, taken altogether', (32) said *Johnson's Gardeners' Dictionary*, reflecting a widely held view. 'Noblesse' and 'Royal George' were considered good for forcing. 'Golden Drop' was a superior plum. Keeping the birds and wasps away, the 'May Duke' cherry fruited from early June to mid-August. 'Marie Louise' was an early dessert, wall-grown pear, as was 'Winter Nelis', fruiting slightly later. 'Devoniensis' was considered one of the best of the tea roses, and tea roses were good for forcing. Henry's buying strategy was to follow a pattern. He picked the best of early, mid and late season cookers and eaters, trees to widen the fruit's uses and availability. To extend the season further, he chose plants that were good for forcing, here meaning plants that responded well to being 'forced' to flower earlier under glass. Some roses too would have been grown in pots under glass for the same reason, for most rose cultivars still only gave one brief flush of flower. It's interesting to observe that, as at Henry's other workplaces, plants were often acquired ready-trained. Victorian gardeners didn't always grow everything 'from start to finish', as might be supposed.

Another interesting observation is the way certain concerns reappear under different guises. With doubts beginning to surface about the wisdom, in human and environmental terms, of the year-round importation of fresh fruit, flowers and vegetables, consider this in David Thomson's *Book of the Flower Garden* written in 1869: 'Would the charm of our summer and autumn galaxy of bloom not cease to produce those pleasant emotions with which it is contemplated, were it possible to sustain its *sameness* all the year? To the attentive eye, each change in the seasons brings its own peculiar beauty and charm. If, instead of change, we had one continuity of song, leafy woodland, and flowery garden, would it not become monotonous, and cease to be a source of exquisite pleasure to the mind? Lovers of flowers ought to be thankful that the year and the human heart have room for changes.' [33]

~~~

As the century wore on, no self-respecting business referred to itself merely as a foundry, an ironmongers, a seed merchant. The larger the letter-heading, the more you could fit into it (preferably with a nice drawing of your company premises) the better. Coats of Arms were popular. This led to absurd lengths, where a bill-heading was half or more the size of the bills which arrived at Coleshill House gardens. You deserved to be noticed and you flaunted your wares. No-one hid their light under a bushel, or rather their Colza oil-powered Moderator Lamp beneath the very latest in high quality bushel boxes. Faringdon's Thomas Anns & Son 'furnishing ironmongers, braziers, bell-hangers, locksmiths, etc.', were no more or less guilty of this than the others. Their headings seemed small for their bills were large, Henry's purchases averaging out at around one a fortnight. The items were totted up and paid for yearly. Here are some items from Thomas Anns's bill for 31st March, 1854. Bear in mind that Henry had only just arrived here from Hampstead.

> 8 good Mortice door locks with Brass Bolts & Wards 7/6, & 1 key to each lock for Garden Doors: £3.
>
> 4 Mortice door locks to match the above with... bolts [and] keys 8/6: £1-14-0.
>
> Fitting & adjusting the keys... £3.
>
> Fitting 2 locks on Iron Gates & altering & adjusting the lock [and other items]: £5. [34]
~~~

Henry was keen to make his mark. That the garden gate and door locks were found to be useless or non-existent was a disgrace and needed Henry's immediate attention. That attitudes at Kenwood, four and a half miles from the centre of the world's largest city, might of necessity be different to those in a village between Fatherthem and Shrivelthem (or something) was a notion that irritated Henry and he brushed it aside. He obviously didn't think much of the tools they had either. On the 16th April he was buying '6 best Garden Scythes' at 4/3d each, total £1-5-6; on the 9th June six garden spades at 3/3d each, total 19/6, and five days later six garden hoes for 7/-, which makes them 1/2d each. The records show Anns' bills continuing regularly to 1863, and no doubt well beyond. They show the ordinary everyday needs of the gardens. 28lbs. of wall nails, for example; a pruning saw, a flue brush (for the boilers), a lantern, a pair of vine scissors, a budding knife, a pruning knife, three balls of tarred cord.

One distinctive difference between then and now is that it paid them to repair and repaint rather than to throw things away. Take watering-cans, for example: '1 Green painted Water pot with large spout & 2 fine Roses' at 5/-; '2 large strong Water Pots', 7/6 each, and '1 Red painted Water pot with 2 Roses & long spout', at 5/9 were bought on August 9th. On 21st September Thomas Anns were 'mending 4 Water Pots, 2 new bottoms, new roses, spout etc., and painting' them: 8/-. Yes, let's hope they weren't the four Henry had just bought!

Only a few documents written in Henry's hand have so far surfaced. In all spheres, it seems, Henry was a practical man and a genuinely modest one. The few articles he was persuaded to write for the horticultural press were mostly practical in nature. Not until he was a 'grand old man' did he consent to give interviews when the journalists descended. And by then it was probably too much effort to object. One of the documents is Henry's *Curriculum Vitae*, collected by the late Bernard Jones and known about. The others turned up in this archival research. The first of them is 'Filling [the] Ice House' at 'Coleshill Gardens, March 14th, 1855'. (35) It took fifteen men and two boys four days to break ice from ponds, cart it into the pleasure grounds and pack it into the ice house there. Although winters were colder on average in the nineteenth century than they've been since, it still seems a gamble to leave this work until March. Henry had seen them all caught out at Penicuik nine years before when they had had to use snow. And, he was working in the warmer south now. On the other

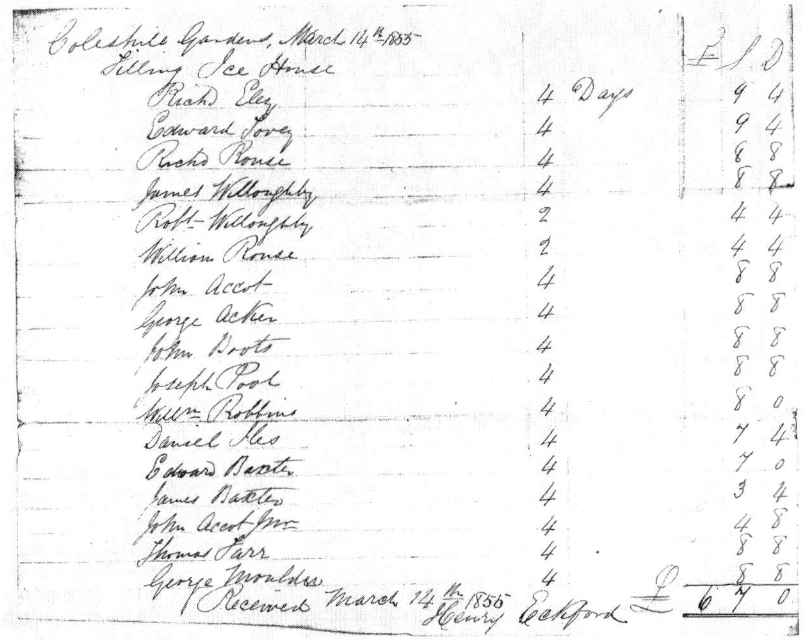

Coleshill Gardens, March 14th 1855
Filling Ice House

		£	s	d
Richd Eley	4 Days		9	4
Edward Tovey	4		9	4
Richd Rouse	4		8	8
James Willoughby	4		8	8
Robt Willoughby	2		4	4
William Rouse	2		4	4
John Accot	4		8	8
George Acker	4		8	8
John Broots	4		8	8
Joseph Pool	4		8	8
Willm Robbins	4		8	0
Daniel Iles	4		7	4
Edward Baxter	4		7	0
James Baxter	4		3	4
John Accot Jnr	4		4	8
Thomas Parr	4		8	8
George Moulder	4		8	8
		£ 6	7	0

Received March 14th 1855 Henry Eckford

hand, Henry had to store the ice as late as possible to ensure that it lasted into the summer.

Henry wrote down the workmen's names in a long neat list, stating the days worked (nearly all worked the full four days here) and the wages earned. It shows that the wage was 2/2 a day for most of the men, meaning they must have been on 13/- a week. The surnames Willoughby, Acot, Rouse and Baxter come up twice and indicate relatives such as brothers, or when one earned half that of the other, father and son. Not nepotism exactly, but one of the rare ways that ordinary people could feel in control of affairs outside the home. It remains routine on some remote estates today. The sixty-four days labour cost £6-7-0, which Henry received straight away. It seems that although the gardeners did the work, this job wasn't within the gardens budget and was paid for out of general estate funds. Perhaps Loudon's argument alluded to earlier, that the ice-house wasn't within the horticultural remit, was having some effect.

As we've seen, the emphasis of Henry's fruit-tree purchases were on widening their season. A good reliable mix of different fruits had to be available as often as possible. Therefore he didn't choose new varieties, necessarily. Old, tried and trusted would do just as well. The 'Moorpark' apricot, for example, has been around since 1700. (36) Henry's vegetable purchases were similar, with a wide range of different types and varieties. With ornamental flowering plants however, it was a different picture. His purchases were almost wholly hardy annuals and biennials, half-hardy perennials treated as annuals and greenhouse perennials. With just an exception or two, they were all recent introductions, in fact the more recent the better. Other flowering ornamental plants were generally ignored, unless pot-grown for forcing. There were sure to be some hardy perennial plants about, though. If Henry had wanted more he might have divided up what herbaceous plants there were in the kitchen garden, providing cut flower for the house. He could have taken cuttings from any flowering shrubs. But he seems mostly to have ignored them. It is easy to understand the later belief that the Victorians neglected hardy flowering plants, both the old-fashioned flowers and the new introductions. (37) To generalise from the particular estates that Henry worked at, one could say that their neglect by such places led not to their unpopularity necessarily, but to being, for a while, unfashionable. (The popular and the fashionable not being always the same thing!)

Henry probably had no choice but to value few plants for their intrinsic worth. Those he did were mostly the new introductions, displayed for their novelty. Plants were functional, like fruit trees or vegetables. With ornamental trees and shrubs, plants like box, yew, lavender and hawthorn were bought by the yard, by the half-mile even, as hedging. The shrubberies were dominated by evergreens with the purchase of plants like laurel, *Mahonia* and *Garrya elliptica*. On a dull day or at twilight, when the bright bedding deferred to the dark shrubbery, the Victorian garden became a gloomy, even eerie place. Frightening too, for a child, especially during the later craze for ferns and their general need for shade and damp.

'But as the last rays of the sunset withered away behind the tallest chimney stack, he would hear the warning voices telling him it was time for him to go, and he knew that somewhere in the approaching shadows were the ugly night-tenants of the garden. Then he would close the door of the summer house very slowly and carefully, and walk up the garden path until he reached the three stone steps that led down to the scullery. These he would take at one leap, and run quickly into the house with all the devils of darkness at his heels'. (38)

The fashion of the times in specimen trees too was for the new and the evergreen. Just think of the exotic monkey puzzle, or the great deodar cedar (Henry bought twenty-seven, including two cedar of Lebanon – imagine having the room to do that!) and the magnificent Californian redwoods like Sequoiadendron, the largest living organism on earth. It's tempting to see this fascination with the biggest, the newest, the brightest and strangest as the immature response of a youthful populace. But if we saw, for the first time, such living things from places we never knew existed till now, wouldn't we be the same?

The next invoice from Garaway, Mayes & Co. on 10th November 1855 (39) included, unusually, two hardy shrubs, a *Skimmia* and a flowering cherry. Six 'Skimmia elegans' (2/6 each), being evergreen, probably joined the others out in the grounds. There seems never to have been a *Skimmia elegans*. The nursery is probably referring to *S. japonica*. According to 'R.F.' i.e. Robert Fortune, in his letter to *The Gardeners' Chronicle* in November 1852, (40) he discovered it in a Shanghi nursery in 1848 and introduced it to England in 1849. Some nurserymen evidently took the view that they were as entitled as the botanists were to name a new plant. Confusing! Skimmias were new faces in Britain. Some of the flowering cherries, on the other hand, were becoming familiar. 'Prunus sinensis flore-pleno [now *P.glandulosa* 'Sinensis'], 3/6' had been here since at least 1774. Its bright pink double flowers now destined it to become one of the most popular of all Victorian and Edwardian garden plants. Being 'excellent for forcing' (41) however, Henry would have potted it up to flower indoors, in the conservatory. It must have been a success, for five years later he was to buy two more, at the same price. (42)

Garraways weren't Henry's only suppliers. There was the 'Royal Berkshire Seed Establishment' of Sutton and Sons, Seed Growers and Merchants. (43) Only the attractive bills remain, the invoices being lost. They feature drawings of the premises at 7 & 8 Market Street, Reading, and an idealised Victorian garden 'laid out with taste and economy', later replaced with the Royal Coat of Arms. Two London suppliers were William Bull of the Kings Road, Chelsea and James Carter & Co. of 237 & 238 High Holborn. 'Seedsmen to the Royal Gardens and the principal Botanic Societies of the British Empire, &c. &c.' ran the print around another coat of arms. Their invoice of the 14th September 1860 (44) shows Henry buying '10 doz. Hyacinths for early forcing, £2' and two dozen each of six *Cyclamen* species, including *C.persicum*, at 14/4. For Christmas flowering, one imagines. A fragrant breath of the Mediterranean for wintry middle England. Henry had an account with the local suppliers too, the Tuckers; cheaper than the rest and to whom a pear tree was a pear tree, a daisy a daisy, and they said so.[45] But Garraways & Co. seem to have been the main suppliers. Looking further down their November list, greenhouse perennials catch the eye, especially an Australian shrub called *Genetyllis tulipifera*, [46] and thereby hangs a tale.

On Saturday 20th May, 1854, an article on an inside page of the weekly *Gardeners' Chronicle* appears to have caught Henry's eye. It wouldn't have been the four lines of Latin description – few gardeners of those that could read could have coped with that, as now – but the heading: 'New Plants. Hedaroma tulipiferum. Genetyllis tulipifera of gardens. A bush of this remarkable plant was exhibited at the last meeting of the Horticultural Society by Messrs. Garraway and Co., of Bristol, who received a silver medal for it... The flowers of the specimen exhibited were... dull in colour... but we incline to believe that in the hands of some of our great cultivators the white would become clear, and the dull red brilliant, in which case this would really deserve the name of the "Tulip - bearer".' [47]

For Henry, soon to draw attention as a hybridist, it would have sounded as a command. Horticulture teaches patience, for the plant, being newly introduced, would need to be propagated from before it became more available. It would have been a seed or two or perhaps a barely rooted cutting or seedling that arrived at Coleshill the following year. As a new plant, a description and illustration duly appeared in the pages of Curtis's *Botanical Magazine* for 1st July, 1855. Following twenty-six lines of impressive Latin description, in which they managed to miss-spell one of the two names it laboured under, it lapses into English. 'During the distant excursions so frequently made by the venerable Drummond in Western Australia to the interior of the Swan River Settlement, he spoke with rapture of two species of *Genetyllis*, as among the most interesting of his discoveries...' In April we received beautiful samples 'flowering in great perfection, from the nursery of Messrs. Garraway, Mayes, and Co., of the Bristol Nursery... That now before us, *G. tulipifera*, though only two feet ten inches high, has from 150 to 200 heads of flowers upon it... It is a hardy greenhouse plant'. [48]

Well now, any plant is hardy in a greenhouse, with the right temperature. The 'venerable Drummond' was seventy-year-old James Drummond from Henry's own county of Midlothian. The plant had two names then because botanists weren't yet sure what to call it. In 1867 having acquired another name it acquired a fourth, its present one of *Darwinia macrostegia*[49]. Trying, it hardly needs saying, in having to learn four names, three now obsolete, for one plant. A common occurrence it was too, and led to gardeners sticking with the original one. *Johnson's Dictionary* was still confusingly calling our *Darwinia* 'Genetyllis tulipifera' in 1877. (50) This one case is a good example of the way a new age of science was beginning to emerge through strands of fog. The illustrations in Curtis's magazine make up for all the verbiage. They're wonderful.

The fog took a long time to clear. Twenty-five years later the situation was, if anything, worse. 'The nomenclature of garden plants... is admitted on all hands to be in a very unsatisfactory condition... Misstatements, accidental or wilful, add to the disorder, while the application of names, often absurd or cumbrous, by incompetent persons, botanical or horticultural, is the most prolific cause of perplexity…'.(51)

If Garraways were at the more esoteric end of horticulture, the Tuckers were at the prosaic end. The six-monthly account Henry had with them for January to June, 1857 (52) (and which, on the face of it, overcharges by eleven shillings) was:

		£.	s.	d.
January 26th	1 quart Mustard		1	
	1 quart Cress		1	
	1 pint Syon House French Beans			6
January 27th	7 std. Pear Trees		17	6
	10 std. Apple Trees		15	
	2 Cedrus Libanii		7	
	2 Cedrus Deodara		5	6
	100 yds. Box. 4d per yd.	1	13	4
	12 Yew Trees		12	
March 28th	½ lb Sea cale (seakale) seed		1	6
April 29th	½ lb Early Horn Carrot		1	9
	1 pint Scarlet Runners (beans)			6
June 5th	Pkt. Nemophila insignis			4
June 15th	2 quarts Champion (of) England Peas	2		
	1 quart Knights Tall (marrow) Peas	1		
	1 quart Early Emperor	1		
	£	5	11	11

As most of these delivery dates match their conventional planting and sowing times, the ones that don't, stand out. The 'Horn' carrot was the name for the short early carrot, so growing them now looks like trying for a quick late-season crop. The volume of carrot and seakale seeds looks enormous, but Henry had to allow for the possibility of low germination and adulteration. The seeds of the hardy annual flower *Nemophila insigis* (now *N.menziesii*) had arrived both too early and too late. Henry might have sown it now, in the open, for some late colour. However, he probably saved it for sowing in pots in September. A pretty blue and white display ('insignis' means 'showy') for the conservatory in the spring.

Of the peas, June is getting late. Henry might have sown them now, to extend the season with a late crop. Alternatively, like the *Nemophila*, he could have sown in the autumn to have a head start in the spring. Henry is choosing well. Twenty years later, *Johnson's Gardeners' Dictionary* was still describing the 4' to 5' high 'Champion of England' as 'a first-rate pea' and 'Knights Tall Marrow' (7' – 8' tall) as another 'good variety'. (53) The seed Henry has is sufficient for rows totalling around eight hundred feet (250 metres) in length. By the 1880s, over a hundred pea varieties could be listed, (54) a testament to its pre-eminence. *Johnson's Dictionary* however complained of 'a great sameness about those of the early kinds' (55) and thought one good variety enough for a small garden.

Nemophila was one of the very few hardy flowering plants to interest Henry. Five others, all from Tuckers, were sweet peas (bought on 2nd May, 1859), *Alyssum* (18th October 1859), thirty *Laburnum* trees (12th January 1860), and primroses and 'daisies' (probably *Bellis perennis*), bought on 1st and 4th April, 1861. In contrast to Henry's lack of interest in hardy ornamentals, every different type of fruit and vegetable was welcome at Coleshill. Flick through the rest of the plant invoices and to the foregoing you can add currants, grapes, thirty-six named varieties of gooseberries, mushroom spores, potatoes, raspberries, gherkins, borecole (kale), turnips, cabbages and onions. One plant that doesn't quite fit in anywhere is sphagnum moss, a three-shilling bushel load of which arrived from Garaways in March 1858. Here, its use would have been in composts. With all the new plant arrivals, gardeners were feeling their way a bit, finding what composts suited what plants. Standardized composts were not to arrive until the 1930s. Sphagnum moss is particularly associated with orchid composts, so it is likely that Henry grew them, though none appear in the invoices.

The nursery accounts confirm the impression that Henry is always widening the growing season, increasing the availability of flowers, fruit and vegetables whenever and wherever he can. Peas emerge as the staple vegetable at Coleshill, reflecting its popularity elsewhere. In those days, when people with gardens had vegetables growing in them, it's interesting to note that the flowers most often seen, yet overlooked as flowers, were those of the pea and bean family, Leguminosae. Part of the reason, for bean flowers at least, was that a fear of its scent was said, in 1902, to have 'lasted in rural England down to quite recent times'. [56] So it's possible that superstition, if only subconsciously, prevented hybridists from engaging earlier with the similarly flowered sweet pea.

~~~

In 1855 a new boiler system was purchased at a price of £61 8s 4d. [57] Its installation, between September and December, may have been a necessity; alternatively, one of greater capacity may have been acquired in anticipation of all the new greenhouses Henry wanted. The company chosen for the work was Burbage and Healy of 130 Fleet Street, London, 'Manufacturers of Hot Water Apparatus, for Warming Conservatories, Green and Hot Houses, Pine & Melon Pits, Churches…'. Firms with London addresses often had London manufactories too. This one's was in Dorset Street, Salisbury Square. The invoices list the cost of each boiler part, plus labour and expenses. The basic boiler cost £8-10-0. Labour, 'to Engineers time Altering Hot Water Apparatus in Vinery, 6 days' came to £2-8-0. That's 8/- a day, far more than even Henry would have been on, so they must have sent more than one engineer. A different but familiar hand wrote 'correct, H.Eckford' at the bottom.

In that the sum added up as stated, Henry was right. Whether Henry knew or cared about trade prices and percentage mark-ups is unknown. He would have had some idea of travel costs. It may have passed through his mind to compare the wages of horticulturists with those of engineers. For peace of mind, he would have let the thought pass out again. You start going down that road and you'll start thinking of other disparities, of that between the few rich and the many poor. But that was how things were and as far as Henry and others like him knew, always had been. These were days of easy migration. If you didn't like it or you couldn't manage you could move. He had. And in his own field, he'd done well. Better than being a soldier out in the Crimea, dying of disease. Though even the incompetent lot running that show were finally getting organised, so he'd read... The world would always have its troubles. He had other, more immediate concerns. Practical ones.

One of the regular accounts Henry had during his stay here was with 'Charles Deacon, Senr., Saddles, Harness & Collar Maker. Horses neatly fitted. Stable tools of every description'. The everyday welfare of, at a guess, a horse and a pony would have been the responsibility of the
~~~

garden staff. Tackle for the tack room was a regular, almost monthly (and sometimes weekly) expense. Examples from 1855 are: September 5th, Extra large Patent Leather Collar 15/-, Flocking 3 cart Collars & Saddles, 6/-; November 29th, 1 set Horses Boots lined through lace holes, £1. (58) This indicates that a pony was used to pull a roller or one of the new cylinder mowers over the lawns. Boots were worn by the animals to lessen the damage to the turf. An item from September 10th, 1861 seems to confirm this: New Backband for Mowing Machine, 8/-.(59) Henry corrected the 1858 bill of £10-12-9 to £10-12-3, (60) saving sixpence. He had to be careful. Lord Radnor checked things too, and made comments. More examples: July 10th 1856: Belt driving Reins 8/6, Best Water Brush 5/-, Cart Grease 2/-, Chamois Leather 1/9, sponge 5/6, (61) and so on and so on; horse transport as it always had been and always would be. For the next ninety years, certainly, but the rarest of sights after that.

An item from Thomas Anns & Son's receipt for 1855 catches the eye: March 24th – 1 pit Grate for Under Gardener's room at Lodge, 4/-. (62) It's always helpful when these list-makers elucidate! Here we can see that one at least of the three lodges sited at the main entrances were lived in by one or more of the gardeners. They may have been doing the place up a bit for a new tenant, for a receipt for 'things had for the Lodge' (63) comes in with a November 13th date. The bill, mostly of china and kitchen items, comes to £1-4s-3d. No workman could spend almost two weeks wages at once, so it would have been provided by the estate, ultimately by the Radnors, who may also have had a say in its composition. Two wine glasses at 1/4d, therefore, would either have gone on a shelf, for decoration, or more likely to Faringdon or Highworth on market days to be sold. The bill's heading, a quietly decorative and discreet 'E.Smith', hides a fact unique in our invoices. This Highworth 'dealer in glass, china and earthenware, glass carefully matched' was a woman, Elizabeth Smith.

A vignette. The following year, 1856, Henry had three outside men in, 'thinning and clearing young Beeches' (64) from woodland. Over a week and a half they coppiced 1 acre, 87 ½ poles, or about half a hectare. From this they fashioned thousands of faggots and hundreds of poles which they sold back to the estate. Whether it was his job or not, the workmen were likely to have been illiterate and Henry was obliged to make out the bill. Apart from the actual sums, it's written in a rather slapdash manner. Henry had better things to do! The bill went to Radnor who wrote 'Should be accepted, March 21st. 1856, R.'. So Henry, still half-connected to the affair, wrote 'settled, March 25th 1856 Hy. Eckford' and sent it back. Radnor sent it flying back with Henry's words crossed out. 'I meant', he wrote, 'by one of the men who was paid. R'. Henry was still learning that you attend to your work in all its aspects, be it interesting or mundane. Don't neglect anything. It's all part of the whole. Cross all boundaries, including, Henry may have realised, those you construct in your mind. Not a bad lesson for the plant hybridist to be, to learn!

The regular garden cycle continued to turn. Requirements for the gardens continued to arrive. Garden mats and deal laths from Fidels, those long bills from Anns and Son; a new nursery suppliers, Arthur Henderson and Co., with a suitable address in Pine Apple Place, Edgware Road, London. Alas, as with Sutton and Sons, no information beyond 'to goods as per invoice'. [65] A rare note in Henry's own hand records receiving three tons of carrots on 25th October, 1856. [66] They are from a Mr. Cooper, farming the lands of the recently retired vicar of Coleshill (Edward Pleydell Bouverie, related to the Radnors). Either Henry's crop had failed, or the gardens couldn't provide them on the scale required. Interestingly, Henry signed himself 'Harry Eckford' here. So he was either informally known as Harry and perhaps always had been, (Henrys are Harrys in Scotland) or the Coleshill people called him Harry – Harry Hickford! – and he'd gone along with it. 'Harry Hickford', too, would have been a way that local people, by giving Henry an identity of their own, were better able to acknowledge his presence among them.

There were some unusual items on Anns' yearly bill for 1856. [67] '14 Nov. 1 Gallon of oil for Leather pipes of Fire Engine', 5/6, a can (1/8), two oil brushes (6d.) and '1 Bottle best oil for the Engine', 1/3. Rich and poor alike lived with the ever-present destructive threat of fire. The gardeners were responsible for the maintenance of one engine, and living 'on site' would have been responsible for manning it, day and night. The 'leather pipes' made up a flexible leather hose which would have fitted onto a tank and hand pump on wheels. If they could have got a shaft-horse harnessed up quickly enough it would no doubt have been effective on a small fire somewhere. It could have done next to nothing for Coleshill House. Even the expertise available in 1952 wouldn't prevent a fire from completely destroying it.[68]

References to Chapter Four

(1) 'Garden Memoranda', *The Gardeners' Chronicle*, (31-12-1870), pp.1734-1735.

(2) E.R.Kelley, ed., 'Coleshill', *The Post Office Directory of... Berkshire...*, (1869), p.648.

(3) Marilyn Yurdan, *Oxfordshire and Oxford*, (1988), p.5.

(4) R.P.Beckinsale, *Companion into Berkshire*, (1972), p.19.

(5) R.Gathorne-Hardy, ed., *The Berkshire Book*, (1951), p.91.

(6) E.A.G. Lamborn, *History of Berkshire*, (1909), p.152.

(7) D/EPb P20, (1843), The Pleydell-Bouverie archives, Berkshire Record Office.

(8) D/EPb P14, (1840s), The P.-B. archives, B.R.O.

(9) J.A.B. Heslop, unpublished MSS. on the history of the park and gardens of the former Coleshill House, (c1990), Berkshire National Trust, p.15.

(10) D/EPb, C28, (c1848), The P.-B. archives, B.R.O.

(11) See: Martin Billings, ed., *Directory & Gazetteer of Berkshire and Oxfordshire*, (1854), p.95; D/EPb, A28/25, The P.-B. archives, B.R.O.

(12) 'Garden Memoranda', *The Gardeners' Chronicle*, (18-5-1847), p.307.

(13) D/EPb, A28, bdls. 11 & 12, (1854–1863), The P.-B. archives, B.R.O.

(14) Alfred Williams, *Round about the Upper Thames*, (1922), p.79.

(15) Shirley Hibberd, *The Amateur's Flower Garden*, (1871), p.62.

(16) William Robinson, *The English Flower Garden*, 1883, (1905), p.444.

(17) Miles Hadfield, *A History of British Gardening*, (1960), pp. 347-348.

(18) O.S.25" 1st. edition, surveyed 1876, Berkshire sheet VII, 15.

(19) Christopher Thacker, *The Genius of Gardening*, (1994), p.254.

(20) Jennifer Davies, *The Victorian Kitchen Garden*, (1987), p.82.

(21) 'Preserving and Serving Ice', *The Cottage Gardener*, (12-1-1854), p.278.

(22) 'Sea-Kale', *The Cottage Gardener*, (19-1-1854), p.302.

(23) *Diary of a Victorian Gardener: William Cresswell and Audley End*, (2006), p.44.

(24) A.G.L. Hellyer, 'The Encyclopaedia of Garden Work and Terms', *The Gardeners' Golden Treasury*, (1966), p.211.

(25) William Cobbett, *The English Gardener*, 1829, (1996), pp.16-17.

(26) George W.Johnson, *The Gardeners' Dictionary*, (1877), p.697.

(27) William Cobbett, *Rural Rides*, 1830, (1985), pp.355-356.

(28) B.K.Boom, H.Kleijn, *The Glory of the Tree*, (1966), p.88.

(29) J.C.Loudon, *An Encyclopaedia of Gardening*, (1822), p.340.

(30) *Beeton's All About Gardening*, (c.1890), frontispiece.

(31) Ibid, p.231.

(32) G.W. Johnson, op.cit., p.56.

(33) David Thomson, *Handy Book of the Flower Garden*, (1893), p.5.

(34) D/EPb, A28, (1854), The P.-B. archives, B.R.O.

(35) D/EPb, A29, (1855), The P.-B. archives, B.R.O.

(36) J.C. Loudon, op.cit., p.807.

(37) See, for example, E.H.M.Cox, *A History of Gardening in Scotland*, (1935), pp.143-144.

(38) Dylan M.Thomas, 'In the Garden', from *Dylan Thomas, The Collected Stories*, (1983), p.352.

(39) D/EPb, A28, bdl.11, (1855), The P.-B. archives, B.R.O.

(40) 'Skimmia japonica', *The Gardeners' Chronicle*, (20-11-1852), pp.739-740. See too
R.Fortune's further letter to *The Gardeners' Chronicle*, (11-12-1852), p.789.

(41) *Hilliers' Manual of Trees & Shrubs*, (1975 edn.), p.237.

(42) D/EPb, A28, bdl.12, (1859), The P.-B. archives, B.R.O.

(43) D/EPb, A28/25 (numerous examples between 1856-1861), The P.-B. archives, B.R.O.

(44) D/EPb, A28, bdl.5, (1860), The P.-B. archives, B.R.O.

(45) D/EPb, A28/25 (numerous examples between 1856-1861), The P.-B. archives, B.R.O.

(46) D/EPb, A28, bdl.5, (1860), The P.-B. archives, B.R.O.

(47) 'New Plants', *The Gardeners' Chronicle*, (20-5-1854), p.323.

(48) 'Genetyllis tulipifera', *Curtis's Botanical Magazine*, (1-7-1855), no paging.

(49) *Journal of the Linnean Society*, Botany IX, (1867), pp.176-179.

(50) G.W.Johnson, op.cit., p.373.

(51) 'On the nomenclature of garden plants', *Journal of the Royal Horticultural Society*, (1879), p.126.

(52) D/EPb, A28, (1857), The P.-B. archives, B.R.O.

(53) G.W.Johnson, op.cit., p.611.

(54) J.Davies, op.cit., p.71.

(55) G.W.Johnson, op.cit., p.611.

(56) 'Beans: Their History and Cookery', *The Gardeners' Magazine*, (5-4-1902), p.217.

(57) D/EPb, A28, bdl.13, (1855-1856), The P.-B. archives, B.R.O.

(58) D/EPb, A28, bdl.6, (1855), The P.-B. archives, B.R.O.

(59) D/EPb, A28, bdl.7, (1861), The P.-B. archives, B.R.O.

(60) D/EPb, A28, bdl.7, (1858), The P.-B. archives, B.R.O.

(61) D/EPb, A28, bdl.6, (1856), The P.-B. archives, B.R.O.

(62) D/EPb, A28, (1855), The P.-B. archives, B.R.O.

(63) D/EPb, A29, (1855), The P.-B. archives, B.R.O.

(64) D/EPb, A28, bdl.9, (1856), The P.-B. archives, B.R.O.

(65) D/EPb, A28, bdl.13, (1856), The P.-B. archives, B.R.O.

(66) D/EPb, A28, bdl.3, (1856), The P.-B. archives, B.R.O.

(67) D/EPb, A28, (1856), The P.-B. archives, B.R.O.

(68) Gemma Fox, ed., *Coleshill 2000*, (2000), pp.112-117, passim.

Chapter Five

Henry Eckford, thirty-three years of age, and Charlotte Stainer, twenty-six, were married at the Islington parish church of St.Mary's, Middlesex, on Monday 6th April, 1857. As far as it's possible to tell, those are the facts. How they both met, however, and how they got to be getting married in north London is somewhat of a mystery. Still, one can speculate. The first and most obvious connection is through the Mildmay family. Charlotte Stainer came from the village of Queen Camel in Somerset, where the St.John Mildmays were Lords of the Manor. Henry's employer, William, the third Earl of Radnor's late second wife had been Judith Anne St.John Mildmay of this family. (1) The most likely scenario therefore is that Anne had had Charlotte working for her at Coleshill, and Charlotte had met Henry when he also came to work there in 1854. (Anne had actually died three years earlier). (2)

The Mildmay family also owned the large Mildmay House in Islington. Therefore Charlotte may also have spent some time 'in service' there during the aristocrats' London Season. Or, she may have left Coleshill in 1851, when Anne died, to work permanently for the Mildmays at Islington. In that case, marriage now would provide financial security for her as Mildmay House was sold off shortly after Henry and Charlotte's wedding(3). If Charlotte had only ever worked at Mildmay House, she could have met Henry back in 1847, when he was working at nearby Upper Clapton and living, presumably, in cheap lodgings in Islington. In any London scenario though, it's more likely that she'd met him later, when Henry was working at Kenwood House in Highgate, from where he went straight to Coleshill.

As well as the Radnor and/or Mildmay connection to account for Henry and Charlotte's meeting, there are three other, less likely, scenarios. Anne Mildmay's father Sir John was Member of Parliament for Winchester and, later, Hampshire. Henry's Salisbury employer, Colonel Baker, was M.P.

for Salisbury (in Hampshire). It's just possible that Sir Henry got Colonel Baker to give Charlotte a job with the Colonel. Or, more likely, a job to Charlotte's father Job Stainer, who was a gardener. Job and Henry could have worked together at Salisbury. Another possibility opens up through Jane Dodds, [4] the wife of William Dodds, Henry's head gardener at Salisbury. Jane' s home village of West Knoyle, though in Wiltshire, is only some twenty miles from Queen Camel in Somerset. It is just possible that Jane and Charlotte could have known each other. A third possibility lies through Henry's old Newliston employer James Hogg, whose daughters owned a property in Islington. Henry could conceivably have rented a small part of it while working in London. There was a *Pelargonium* called 'Amy Hogg' too, in the 1870s, [5] and possibly one of Henry's, for he hybridised them.

However, the Mildmay–Radnor–Queen Camel–Stainer connections are the strongest. However they met, it does seem that Charlotte was living in Islington, if only to account for her being able to marry in St.Mary's there. Let us hope, a rather remote hope, admittedly, that the Mildmays opened part of the Mildmay House and grounds for the wedding reception. Years later, when Henry was a well-known hybridist, he named a *Verbena* 'Miss Charlotte Mildmay', the flower described by *The Gardeners' Chronicle* as having 'a large pale flesh-crimson eye…'.[6] Much like the baby it was named after! Henry was to often name the flowers of his crosses after the wives or children of former employers. (As well as his children, friends, relations, colleagues and Victorian notables). No longer working for them, he had no need to ingratiate himself with them and these little honours seem purely affection. Although, observing the age, it's difficult to tell genuine affection from conditioned respect.

A London wedding was a return to old haunts for Henry, with a few days taking in the sights. The idea of "holidays", as opposed to fairs and other festivities, was just beginning to appear (in some places, to reappear) in the national consciousness. Until now, people couldn't afford to take much time off work, stopping only when it was forced on them by circumstance, and to travel away was prohibitively expensive for most. With the coming of the railways, millions had flocked to the Great Exhibition of 1851, in Hyde Park's 'Crystal Palace'. Rising incomes, cheap fares and the speed of distant travel tempted many. 'After 1851's demonstration of how easy it was', wrote Geoffrey Best, 'trips to London were offered on every special occasion... on Easter, Whitsun and August weekends, and, by the [eighteen] seventies, on weekends throughout the summer. Excursions to London... demanded the spending of at least one night away from home, and since many of those advertised... were of from two to eight days' duration, one wonders where people stayed; did many of them visit relations, or was there already a flourishing cheap hotel and lodgings business for the tripper to London...?' (7)

One answer to that is provided by *The London Conductor*, a best-selling guide to London rushed out for visitors to that great 'industrial exhibition' of 1851. One wonders if the book's purchasers included Henry and Charlotte. 'On entering London', the author advises, 'the visitor's first duty will be to seek lodgings, for which purpose he will do well, if he wishes to live cheaply, to go either to Islington, the East End, or to some of the pretty villages south of the Thames...'. (8) At the time of the wedding, Henry and Charlotte were lodging separately but near each other in roads just off the east side of Islington's Essex Road. Henry stayed at the home of John Williamson, a coal dealer at 14, Windsor Street. (9) Charlotte's Pickering Street was three roads farther north. Both relatively new roads, (Windsor Street 1822, Pickering Street 1847)(10) they were nevertheless now in the poor east end of the borough. The middle-classes were deserting them in their outward flight from over-urbanisation.

St Mary's was the 'mother church' of the parish. From Pickering Street and Windsor Street it was a short walk away across the Essex Road. (The church was destroyed in the 1940 London Blitz). The Ordnance Survey map of 1871 shows that you walked up Church Lane, past the soup-kitchen on your right and down a tree-lined path through the graveyard to the church(11). Built in 1751, it was 'a plain heavy structure', or so it was described by another of those 1851 London guide-books. (12) The attractive mediaeval church at Queen Camel doesn't seem to have attracted our two. That was St.Barnabas's, and Barnabas was a saint associated with hay-making. Henry was to work on the land in rural England for the rest of his life, but the movement of the times was the other way. People were moving off the land and out of agriculture. Henry and Charlotte were caught in its swell and were, for a time, carried along by it.

Like Henry, Charlotte Stainer was one of eight children. Baptised on 28th August, 1831, she was the fourth of eight, possibly ten daughters born to Sarah (née Brooke) and Job Stainer of the Somerset parish of Queen Camel. She would have been named Charlotte in honour of the present (or recently late) Lady of the Manor Charlotte St.John Mildmay. This lady was formerly Charlotte Pleydell-Bouverie and so of the family of Henry's, and almost certainly her, employer at Coleshill – Lord Radnor.

At the time of Charlotte Stainer's parents' marriage in 1832, and at her birth, Job was described as a labourer. Neither parent could write. One of Charlotte's younger sisters, Ann, was a witness at the village wedding of Charles Reeves to Ann Cheeseman. The bride, who lived to be a hundred, was a contemporary of Charlotte too, being some five years younger. On her one-hundredth birthday in 1936 she recalled the days of her and Charlotte's youth for the local paper. Cary Fair flourished, as did Club-Day, 'the great holiday of the year, with its junketing and merrymaking. She remembered when the only means of getting to Yeovil [seven miles away] was by walking. She used to get a ride back on the 'lime-cart' when it returned to [Queen] Camel empty, and always gave the driver twopence.' She 'remembered with delight old-time circuses and fairs. Her memory went back to the time when she was taken to church at the age of two by her mother for the special service which commemorated the Coronation of Queen Victoria.'(13)

This was in 1837. Charlotte would have been five or six then. At a time when most people's lives were still a struggle in rural poverty, those 'old-time circuses and fairs' came as blessed relief. The 'club' of Club Day was a Friendly Society into which people contributed to provide financial security in their old age. As a rule, the day began with a business meeting, followed by attendance at church. A good dinner with ale for all was followed by a fair on the grass. Entertainment included merry-go-rounds, rifle galleries and bulls-eye tents. Writing in 1904, one of the new, nervous Edwardian middle-classes who'd gone up in the world but lost something in the process called it 'every form of amusement which can be divised for men, women and children of the more primeval and less tutored class.' (14)

Earlier writers were more sympathetic. William Howitt described fairs in general in his 1845 *Rural Life of England*. Society, it has been said – historians citing the Norman invasion, the Reformation and the Industrial Revolution as exacerbating factors – has become fragmented and divided by class. Howitt, in response, chose to emphasise the fairs' inclusiveness. 'Fairs are great sources of pleasure to all classes of country people. The farmers, and their wives and daughters; the villagers of all descriptions; the cottagers from the most secluded retreats; the squire and his family from the hall – all flock to the fair of their county town, and find some business to be transacted, and a world of pleasure to be enjoyed. There are cheese, cattle, horses, poultry, geese, and a hundred other things, to be sold; and multitudes of household articles, clothing, and trinkets to be bought; and, besides all this, a vast of seeing and being seen to be done.' (15)

The time spent at fairs, festivals and the like may have contained an edge of desperation too. Born in 1836, Charlotte's younger sister Mary died before her fifth birthday, for another Mary was born in 1841. Surrounded by death, life was made the most of, when it could be.

It seems most likely that Charlotte had been living and working in London. The economic forces that had taken her there now took her to the Cotswolds, the wife of an up-and-coming head gardener. No doubt her parents were pleased, for Coleshill was a lot nearer to 'Camel' than London was. It seems likely, too, that one or two of the places Charlotte had worked at had had running water, for 'Mr Anns Bill for water-pipes to Garden Cottage', £9-16-5, [16] appears shortly after her arrival at Coleshill, on 24th May. It was at least one way she could make her mark on events. Another change to occur coincidentally with her arrival was that old William Spindloe stopped calling Henry 'Harry Hickford'. Spindloe's bills would, from now on, be addressed to 'Mr Eckford', beginning with the one for half-a-dozen (home-made) ash rakes, 4/-, on 25th November[17]. Another November item to catch the eye comes from Thomas Anns' yearly bill. 'Nov. 3rd. 51 yards of very stout rope for pulling down trees. £1-4-0'. [18] A nice warm job for the winter, that!

The following year saw the birth of Henry and Charlotte's first child. Henry James (James after Henry senior's father) was baptised on 10th April, 1858. The annual village fête was held at the end of June and we hope that young Harry, as he was known, enjoyed his first social occasion and the attention he would have received. Radnor's importance earned the fête more than a mention in the Agricultural Gazette section of *The Gardeners' Chronicle*: 'During the past week the annual *fête* at Coleshill... has been held, and competitions of skill in various kinds of farm work and in the growth of garden produce have been decided'. It 'presents as good a model as any we know of the kind of country gathering and holiday which one hopes to see in fashion everywhere – a happy substitute for the yearly parish "revel" described by "Tom Brown", as prevalent in his school days, and still recurring annually in a degenerate form in most west country parishes – and at the same time a more satisfactory and independent, because striven for and earned enjoyment for all whom it attracts than is afforded by those annual occasions on which local societies meet with, as we believe, mistaken benevolence to praise long servitude and reward morality'.[19]

'Most gardens had a row of beehives' remembered Flora Thompson in *Lark Rise to Candleford*, recalling her youth in rural Oxfordshire in the 1880s. [20] Bees were necessary to ensure fruit-tree pollination and for

Henry, growing more fruit than most, they were essential. 'To E.Haines – to 2 swarms of bees, £2-2-0' came in the bill to the Garden Account on 24th August, 1858. (21) On April 20th, 1859, six beehives from Thomas Anns, at 1/6 each, (22) completes the picture.

Lawns at these places were field-sized, seeded from a mixture of grasses and clover. On light sandy soils Shirley Hibberd recommended including 'Avena [now *Trisetum*] flavescens, the yellowish oat grass' and two plants that would probably have arrived there anyway, birdsfoot trefoil and yarrow. (23) Gardening books often recommended laying good-quality turf but in view of the scale here one suspects such advice was ignored. Henry's immediate predecessor at Coleshill, Hugh Fraser, had. In 1853 he'd ordered the following from Suttons, enough seed for a four-acre lawn. This is what Henry inherited:

	£.	**s.**	**d.**
2 ½ bus., (bushels) 50lbs.			
Finest lawn grass seeds @ 21/- (per bushel)	2	12	6
10 lbs. Poa nemoralis @ 1/3 (per lb.)	0	12	6
6 lbs Cynosurus cristatus	0	6	0
20 lbs. White Dutch Clover	0	15	0
3 bus. Pacey's Perennial Rye Grass	1	1	0
2 bus. Italian Perennial Rye Grass	0	14	0
10 lbs. Festuca ovina @ 1/3 (per lb.)	0	12	6
1 sack, 1/6 & 6 bags, 4/6	0	6	0
	6	19	6 (24)

The idea of grass-only lawns seems to have crept in with the small suburban lawns at the turn of the century.

A 'machine for mowing lawns' had been patented in 1830. (25) Henry, usually the first with the latest novelty, waited until now, 1859, to purchase one. No doubt there had been problems adapting a cylinder that trimmed cloth in a factory to one that trimmed plants out-of-doors. The price was still high – well beyond any gardener's pocket for example – though of course that was no problem for Henry. A new plant and Henry wanted it. With technology he was – one hesitates to combine the words canny and Scot – circumspect, shall we say. Following the time, in his impressionable youth, of almost freezing to death on the Edinburgh – Inverness stagecoach, you can't help noticing that everywhere he went after that had to have a railway-

SAMUELSON'S

SILENT-WORKING, SELF-CLEANING, LAWN MOWING, COLLECTING, & ROLLING MACHINES,

WITH MANWARING'S AND BOYD'S PATENT IMPROVEMENTS.

SOLE MANUFACTURER, B. SAMUELSON, BRITANNIA WORKS, BANBURY

These Standard Machines are now constructed with a simple and durable Silent Driving Gear.

(Copies of testimonials from all parts of the country as to the saving of labor, and improvement of Lawns, may be had, Post Free, on application to Mr. Samuelson's Works, or to any of his Agents

PRICES FOR 1861.

Including Packing Case and Carriage Paid to any Railway Station in the Kingdom.

		Boyd's and Manwaring's Patent.	Manwaring's Patent only (Silent Gear.)
Manual Power..	12 inches wide	£5 0 0	£4 12 0
	16 " "	6 15 0	6 7 0
	19 " "	7 10 0	7 0 0
	22 " "	8 0 0	7 10 0
Pony Power....	22 " "	9 5 0	8 15 0
	25 " "	12 0 0	11 10 0
	30 " "	14 14 0	14 0 0

ORDERS WILL BE EXECUTED ACCORDING TO PRIORITY OF RECEIPT.

May be procured through any Ironmonger or Seedsman in the country

Samuelson's mower advertisement.

line going to it. Although various bicycles had been invented throughout the nineteenth-century, it wasn't until the introduction of the pneumatic tyre in 1888, guaranteeing a comfortable ride, that Henry thought of having one himself – in his late sixties! Anyway, on the 12th July 1859 Henry purchased a thirty-inch wide, pony-powered lawn mowing machine from the Britannia Works of Bernhard Samuelson at Banbury. Thirteen guineas

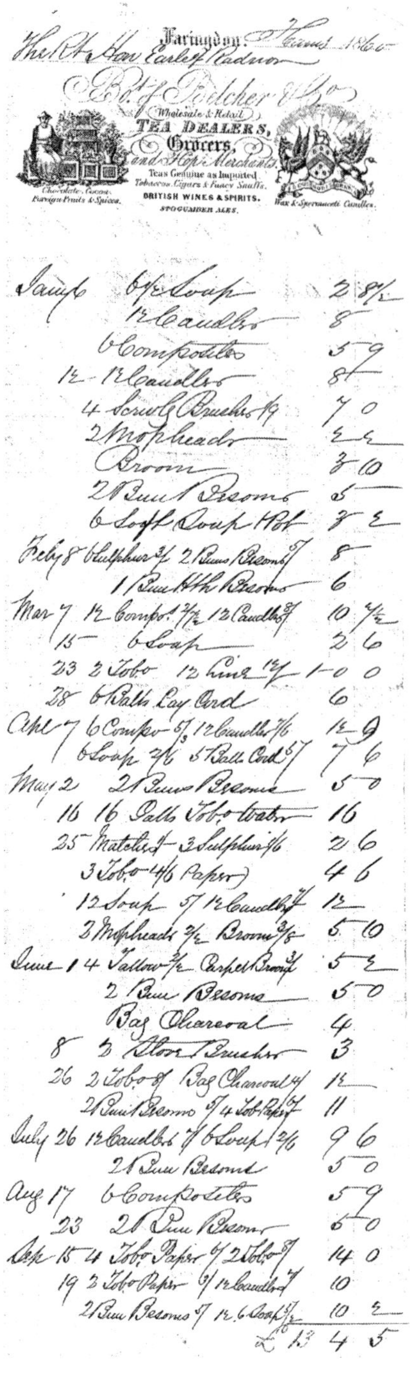
Faringdon: Xmas 1860
The Rt Hon Earl of Radnor
Bot of Belcher & Co
Wholesale & Retail
TEA DEALERS,
Grocers,
and Hop Merchants.
Teas Genuine as Imported
Tobaccos, Cigars & Fancy Snuffs.
BRITISH WINES & SPIRITS.
STOGUMBER ALES.
Chocolate, Cocoa.
Foreign Fruits & Spices.
Wax & Spermaceti Candles.

Jany 6	6lb Soap	2	8½
	12 Candles	8	
	6 Composites	5	9
12	12 Candles	8	
	4 Scrub Brushes 1/9	7	0
	2 Mopheads	2	2
	Broom	3	10
	2 Bun Besoms	5	
	6 Soft Soap & Pot	3	2
Feby 8	6 Sulphur 3/ 2 Buns Besoms 5/	8	
	1 Bun Hth Besoms	6	
Mar 7	12 Compos 7½ 12 Candles 8/	10	½
15	6 Soap	2	6
23	2 Tobo 12 Line 12/	1-0	0
28	6 Balls Lay Cord	6	
Apl 7	6 Compo 5/3 12 Candles 7/6	12	9
	6 Soap 2/6 5 Balls Cord 5/	7	6
May 2	2 Buns Besoms	5	0
16	16 Galls Tobo Water	16	
25	Matches 1/- 3 Sulphur 1/6	2	6
	3 Tobo 4/6 Paper	4	6
	12 Soap 5/ 12 Candles 7/	12	
	2 Mopheads 2/2 Brooms 3/8	5	10
June 1	4 Tallow 2/2 Carpet Broom 3/	5	2
	2 Bun Besoms	5	0
	Bag Charcoal	4	
8	2 Stove Brushes	3	
26	2 Tobo 8/ Bag Charcoal 4/	12	
	2 Bun Besoms 5/ 4 Tob Paper 6/	11	
July 26	12 Candles 7/ 6 Soap 2/6	9	6
	2 Bun Besoms	5	0
Aug 17	6 Composites	5	9
23	2 Bun Besoms	5	0
Sep 15	4 Tobo Paper 6/ 2 Tobo 8/	14	0
19	2 Tobo Paper 3/ 12 Candles 7/	10	
	2 Bun Besoms 5/ 12.6 Soap 5/2	10	2
	£16 13	4	5

A Belcher & Co bill.

(£13-13-0) it cost, with a box of tools for 5/-.[26] (You provided your own pony). A prompt cheque from the Earl brought a 13/- discount. We'll see in a bit how they got on with it.

Cylinder mowers won't cut in the wet, so it was the pleasant summer of 1859 that probably spurred the purchase. In an autumn letter to the Earl, his niece wrote that she hoped he'd enjoyed 'the beautiful summer we have had'.[27] The Eckfords would have done, for it brought the safe arrival of their second child and another son, who they had baptised George on the sixth of August. George was the name of one of Henry's older brothers. It was a popular name though, and frequently heard. Other names, as surnames, that cropped up regularly around Faringdon, Coleshill and Great Coxwell were those of Gerring, Fidel, Belcher and Acott. The Gerrings were mostly tenant farmers. The Belchers were less fortunate, ranging from tradesmen to agricultural labourers.

A 'Belcher & Co.' were grocers at Faringdon, general providers from whom Henry bought various everyday items. The bill presented to Henry at Christmas, 1860 shows them attempting to move upmarket. Their bill-heading was now 'Chocolate, Cocoa, Foreign Fruits & Spices' beneath an attractive drawing of a Chinaman by a temple garden. 'Teas Genuine as Imported. Tobaccos, Cigars & Fancy Snuffs'. [28] Unfortunately, this end of rural Berkshire was proving something of a challenge, for the items sold – quoting at random – hadn't changed from before: January 12th, four scrubbing brushes, 1/9 each, two mop-heads, 2/2, broom, 3/10; May 25th, matches 1/-, twelve soaps 5/-, ... mop-heads (again); September 15th, tobacco and tobacco papers... Yes! No, this was insecticide for the greenhouses...

The summer of 1860, especially by contrast to 1859's was wet; exceptionally so. This was the summer of 'no sunshine but continuous rain' referred to earlier. The gardens' new mower wouldn't have been much use and they probably had to go back to using scythes. New parts were arriving regularly from Samuelson's, and in August it seems that someone came the thirty-odd miles down from Banbury to repair it. [29] Henry was getting the worst of both worlds. It was usually too wet to use it and when he could it needed repairing to a total cost of 19/6. But Henry and his fellow gardeners persevered.

The weather stayed foul. The lost summer was followed by a bitter winter, the worst for twenty-four years, and destroyed the spring of '61 in its icy breath. Right in the middle of this, Charlotte gave birth to their third child and first daughter, Isabella Sarah. Henry's mother and grandmother were both Isabel, and his younger sister and a niece were called Isabella. Charlotte's mother and probably her mother's mother too were called Sarah. Two of Charlotte's sisters had been called Sarah, and both had died. The church of All Saints, founded in the late twelfth-century, [30] lies in the centre of Coleshill. Isabella was baptised here on the 9th May. What passed for spring that year cast a cold spell over the infant girl. Engaged to be married in around 1892, her fiancé died. Aged thirty-one, she may have felt too old for another try at marriage. Who knows. She was to live on with her father and step-mother, helping to bring up two of her brother John's children there and helping her father and later, John, in running the family's nursery business. The 'desperate shortage of opportunity to survive socially outside the respectable home', writes Geoffrey Best, compelled 'its daughters either to accept whatever offer of marriage they could get or to put up with lifelong dependant subservience to father and mother at home'. [31] About the only positive comment one can pass is that at least Isabel was seemingly spared the rigours and dangers of childbirth. She was to live, as a spinster, to 1943.

The Eckfords now had three infants to care for. The census of 1861, taken on the evening of 7th April, shows that they also now had a maid-servant to help with the housework. (32) Or rather, to take over the housework. 'Few Victorians who could get a servant to do work for them' writes Best, 'would for preference work themselves...' ; almost every early or mid-Victorian 'hired servants the moment he thought he could afford to (and sometimes before he could really afford to do so!)...' (33) The Eckfords' maid-servant was Charlotte's sixteen years old sister Lydia, so there probably wasn't too much of a strain on the Eckford budget. Charlotte's mother Sarah Stainer was also here on the 7th April. She is described as a servant and a visitor. It was probably more than a visit. With her surviving daughters now grown-up, one imagines that she kept the company of two of them, and three grand-children, for as long as she could.

Nearby on Middle League Farm, the census reveals, were the Gerrings. Friends of Henry and Charlotte, and farming three hundred acres, they also had three children. There was Godfrey, Maria and Emily the eldest at seventeen. Charlotte had, it seems, given her age as twenty-seven although she could have been no younger than twenty-nine, which age she almost certainly was. The enumerator could have made a mistake but people didn't always know their exact age. Only baptisms were officially recorded by the Church of England. Although in practise baptisms seem to have been close to their birth-dates, they could, within convention, be two years afterwards. Unless births were noted in, say, a family bible, they were guessed at. Your age, and how you spelt your name, if you could write, were your affair. When the arbitrary date of your baptism was more important that the physical reality of your birth, it's easy to see how the standardised time of the new railway timetables could take precedence over time based on the movement of the heavens.

Charlotte Eckford, née Stainer.

Assuming that Charlotte did know her own age, it's likely that the first of only two known photos of her was taken this summer, as a special celebration of her thirtieth birthday. Her clothes and hairstyle place the photo in the early 1860s, with the plants in leaf indicating spring or summer. Charlotte's baptism had been on 28th August 1831, so her birth would have been a little earlier. Therefore a guess is that the photo was taken in their garden, near their house, in July or early August this year, 1861. If she had really thought of herself as only being twenty-seven in April, then the other guess for the photograph's date would be July or August of 1863.

On 23rd August 1861 a bill from London arrives to account for the sudden coal increase noted earlier. 'To making various alterations, improvements and additions to the Hot Water Apparatus (as per estimate) £67-3-6. To extra Materials and Labour' and the 'Carriage of Engineer from London' inflated the bill to £74-8-3. (34) A substantial sum, reflecting the heating infrastructure required by what was obviously a large increase in the number of heated greenhouses. The company involved, John Weeks & Co., of the Kings Road, Chelsea, were too big to be given the run-around. Their bill was paid a mere week from receipt, on the 30th August. Work continued the following winter, for another bill arrived in the spring. 'April 2nd, 1862. Received of Mr Henry Eckford, the sum of twenty three pounds seventeen shillings and nine pence for works in connection with Hot Water Apparatus at Coleshill Gardens'.(35)

Bernhard Samuelson was to gain fame, fortune and a knighthood by providing Britain with the American McCormick Reaper and other industrial muscle to the agricultural boom years of the eighteen-fifties, 'sixties and 'seventies. One of his more reliable bread-and-butter lines though must surely have been mowing machine spares and repairs! Henry's mower had had to go back for repairs and was now ready for the start of the new season. 'April 7th, 1862. Repairing one 30 inch Lawn Mowing Machine, £1-15-0. Carriage to Banbury 5/-'. (36) Sending the mower back to Shrivenham station from where the garden cart would have gone to collect it, would have brought the total cost to £2-5-0.

Henry was probably expecting too much from relatively new technology. It may also have been that Henry didn't realise the importance of routine maintenance and stone free, even surfaces for cylinder mowers. If there were a lot of lumps and bumps in the ground, then, one of the gardeners may have reasoned, a narrower width, 12" or 16" say, would make a better job. Fair enough, the job would take longer but... a small mower, pushed by hand, was more manoeuvrable. Better for narrow widths. Cheaper too. Samuelson's 12" was just £5, whereas their 30" models were now fourteen guineas, and that was last year's prices... (37) You can almost hear the arguments being tossed back and forth. There's an old Samuelson mower, incidentally, on display in the little museum over the canal at Banbury. The size, the curator kindly measuring it for me (being too scared, myself, of setting the alarm off!) was 10", two inches shorter than advertised. Perhaps Samuelsons' had respecified it.

Being accessible by train, Henry may have even visited Banbury to select his mowers and see what else there was of interest. Samuelson's works covered much of Banbury then and some of their buildings remain. Visit the Victorian parts of the industrial estate by Swan Close Road and you'll get a feeling of the times, especially in the darker evenings when the present-day fades from view and the close-by buildings and the nineteenth-century close in around you.

A bill may be missing from the Radnor archives for it's expecting a lot from the boilers they had at Coleshill to power all those extra heating pipes. The engineers would have known this but Lord Radnor, who was paying the bills, may have jibbed at the extra expense. Henry may have anticipated this, and had the expensive infrastructure put in place and paid for, before asking for the (less expensive) boiler. It's rather silly, he might have reasoned, to have spent all this money and then spoilt the ship for a ha'porth of tar. Buying another boiler first, and asking for larger sums afterwards, the Earl may well have dug his heels in. Employers regularly complained of the 'extravagances' of their head gardeners (38) and the Earl might have settled for having a 'back-up' boiler, and left the heating (and glasshouse) increases to some future, indeterminate date. Henry wanted to experiment with the new foreign plants sweeping into Britain, the most fashionable of which were also the most tender. Those greenhouses were, for Henry, a must.

He got his way. That autumn, when it was obvious that the old boilers couldn't cope, a 'N°3 Weeks's upright Tubular Boiler, £12.00', was installed. Together with its parts, labour costs, carriage and travel expenses, the bill, presented on 13th February, 1863, came to £31-0-11. (39) It was paid within the month. 'Our engineer's time in firing, 10 days @ 7/- [a day] £3-10-0' shows their engineer earning at least as much as Henry, a head gardener, was and probably more. Such differences remain to this day. The boiler, a bill from Shrivenham Station shows, came from Bristol on the Great Western Railway. Their engineer's railway expenses of 15/- would have been the cost of his fifty-mile return journey from there. That 15/- ticket was one and a half times the average weekly wages of an agricultural labourer from this area at this time. [40] The railways were to broaden the horizons for many, but many, too, were excluded.

We're now taking a short break from Coleshill and returning to Stenhouse, the village where Henry was born. Henry had been the youngest son of the family but he wasn't the youngest child. This was his sister Isabella, three years younger. She had stayed in Stenhouse and married James Brunton, a tailor, born in the village of Yarrow, near Selkirk. The Bruntons must have moved to Liberton soon after the birth, for James had been a friend of Henry's since boyhood. By the time of the 1861 census they were living in a 'public house' run by Peter Eckford. As mentioned earlier, this was probably the one called The 'Robins Nest' in Upper Stenhouse, the name later adopted by a twentieth-century roadhouse (still there) in nearby Greenend. [41]

Unusually the house had seven main rooms, i.e. rooms with one or more windows in them; a large nest. It would have had to be, for as well as serving as a pub it was home for the Bruntons, with their four children Isabella, James, Peter and David. Also there was the children's great aunt Barbara. Peter Eckford was a bachelor. A neighbour was the local G.P.,Thomas Peacock, with his doctorate in medicine from Edinburgh University. [42] Incidentally, James and Isabella's son James S.Brunton [43] was to move south to Lancashire where he founded the *Horticultural Trade Journal* for commercial gardeners, together with the Hortus Printing Company at Burnley. In 1923 he was the prime mover in the international florists' scheme that eventually became known as 'Interflora'. He also held various offices in the new twentieth-century National Sweet Pea Society, including those of President in 1924 and long-serving editor of their annual journal. [44] A lad o'pairts!

Another one of the writer William Howitt's new 'country excitements' had recently surfaced here in Stenhouse. The fifth annual 'Liberton and Newton Horticultural Society's Show' took place 'in the schoolroom at Stenhouse' on Saturday 13th September, 1862, ran the report in the October edition of *The Scottish Gardener*. [45] 'The flowers and vegetables exhibited by [professional] gardeners, amateurs, and cottagers, were arranged on tables and stands placed round and down the centre of the schoolroom, one-half of the space being devoted to gardeners, and the other half to amateurs and cottagers. The articles exhibited, notwithstanding the past very unfavourable season, were considered fully equal to those of any former years, and several of the specimens, both of flowers and vegetables, might have successfully competed at any provincial show. The wet and boisterous weather which prevailed from an early hour in the morning till between four and five in the afternoon, prevented several of the competitors at a distance from bringing forward their larger plants. The show was opened to the public at two o'clock', with a good attendance.

The fifty-odd prizes on offer to the professionals, gardeners in what we would now call 'private service', were dominated by just seven men. The similar number of prizes on offer to the 'amateurs and cottagers' were more widely shared. John Greig of Forkenford Cottage and Dr Peacock won three first prizes each, but everyone was well behind James Brunton of Stenhouse with seven firsts and one second. Henry's brother-in-law was one of those people who, as an exhibitor, you wonder with declining hope whether he might just give it a miss this year. Not a chance! Here he is: best two stalks celery – 1st. J.Brunton, best four onions (spring sown), best four leeks, best pint of pease, best three Asters, best six Asters, best two Fuchsias – all James Brunton. These shows were places where a tailor might compete with a doctor and beat him. It was only open to men but by the lights of its time was meritocratic and a hope for the future.

1863 was probably James's best year. On Saturday 12th September the Stenhouse schoolroom 'was very tastefully decorated for the occasion' with large displays of grapes and peaches. In good weather 'the showroom was crowded to excess during the afternoon; the proceedings were enlivened by the presence of a band of music. On the whole this was by far the most successful show the society had ever held'. [46] James won twelve first prizes and four seconds. Dr.Peacock was the only man near him with six firsts. The introduction of a 'prize for ladies' saw Mrs Greig, Forkinford, [47] John Greig's wife, take the prize for two bouquets. The following year James again had twelve firsts but John Greig was closer this time with nine. A welcome development saw the 'ladies prize' joined by another prize open to both sexes.

One of the judges in 1862 was William Thomson of Dalkeith. One of the best-known gardeners of the nineteenth century, Thomson was born in the borders at Bowden, just a few miles from James's home village of Yarrow. Thomson was soon to edit the national magazine *The Gardener*, bringing Henry to prominence with a mention for one of his new Verbenas, appropriately called 'Coleshill', in 1869. Merit will out of course, but knowing the right people (through your brother-in-law) helps. Henry was pleased with James's successes and encouraged him by naming some of his new plant varieties after his children.

We'll return to Coleshill now and just before we get there, about five miles before, we'll call in at the village of Little Coxwell. 'March 6. 1863. To the Right Honble the Earl of Radnor. My Lord, it is the general wish of the inhabitants of Little Coxwell that the aged poor people should have a plain substantial dinner together next Tuesday to commemorate the marriage of the Prince of Wales; and also that the children should have a little treat at the same time – I am therefore requested to ask if your Lordship would kindly assist the Committee with a donation towards the above object. The enclosed list contains the names of all the principal inhabitants of the parish (except those who have not yet been applied to) – The population of the village is about 300 people. I am My Lord, Your Lordships obedient Servant, C.Belcher (For the Committee).' (48)

Despite the flowery formality this is actually a quite abrupt letter. They were collecting donations, not requesting them. Lord Radnor made a note: '1863. Belcher. Sent £2.' (49) Queen Victoria's eldest, Albert Edward (Bertie; the future Edward VII) was about to marry the young princess Alexandra. 'There was never enough of anything, excepting food' (50) recalled Flora Thompson, born a few years later in nearby Oxfordshire. A nice new set of clothes would have been more welcome fare, one suspects.

~~~

April. Now is the time to get the mowers out and do all those routine maintenance jobs you meant to do in the autumn. Thomas Anns' bill for April 17th, 1863 reads: 'Bottle of Oil for Mowing Machine, 1/6d.' (51) In mitigation Henry had just bought a new mower or, rather, a 'noiseless lawn mowing, rolling and collecting machine' from Thomas Green's iron works at Leeds. The cost of this 16" wide mower was £6-10-0, with another 5/- for the packing case. (52) The following month Henry was purchasing, from Anns again, his local Faringdon suppliers, 'steel wire for spring for mowing machine.' (53) Despite the problems, Henry and his gardeners must have been pleased with their new mower that season, for they bought an identical one in the autumn. But they bought it locally this time, to avoid carriage costs. The pony-drawn 30" model was fine for the larger areas, but another smaller hand-pushed model like the 16" was needed for the fiddlier bits. A garden roller to squash in the stones and odd bumps in the lawn seemed a good idea too and was purchased at the same time for £2-12-6. (54) The garden was a quiet and pleasant place to work, when mechanisation merely meant ever quieter push-mowers.
~~~

There would have been plenty of noise in the Eckford household though! A fourth child, their third and probably their favourite son, John Stainer, (known as Jack) was born in the autumn of 1864. This would have been a particularly happy time for them, coming after still-born sons in September 1862 and August 1863. (55) Relief for Charlotte, surviving a sixth pregnancy, dangerous then. Relief and happiness for Henry, seeing Charlotte attractive and productive still; four lively children... and in 1866 his first mention, a passing one in the prestigious *Gardeners' Chronicle*. Verbenas had become a very fashionable flower and in September, under the heading 'Florists' Flowers', the magazine was criticising the many new varieties for their similarity. Their correspondent, William Heale of Upper Clapton, hadn't 'seen the varieties raised by Mr.Eckford, and sent out by Mr.Keynes, [so was] unable to say anything respecting them, but no doubt some of your readers... will tell us something...'. (56) The author was probably the 'W.Heale' who was known to have been a nurseryman in Wiltshire, (57) where he would have known John Keynes and, probably, William Dodds. His present address, Upper Clapton, was almost certainly that of Low's nursery where Henry had worked back in 1847. Enough said!

On the best estates, horticultural apprenticeships were so encompassing that Henry would have known of plant hybridisation from an early age. He would have had a hand too in crossing Dahlias and Camellias when he worked with William Dodds at Salisbury. Henry was also using the same nursery as Dodds, that of John Keynes of Salisbury to bulk up, market and sell his plants. So, Henry's first solo venture had begun and it was to be with Verbenas and, shortly afterwards, Dahlias. For a beginner wanting to make a name for himself, these were safe and sensible choices. The new half-hardy plants from central and southern America had become popular because of their long flowering season. Their crossing together had given rise to summer bedding schemes, big showy displays of hybrid Verbenas for the grand estate gardens of the well-to-do. These were the places that Henry and others like him knew and where they worked. The only drawback with hybridising such plants was that you would be fighting for room in a crowded market, for competition was fierce.

The situation that Henry now found himself in seems to have occurred in a wholly conscious and rational manner. But there were unconscious influences at work too. Remember Henry's apprenticeship garden near Inverness with its 'rich display of Dahlias, and other showy flowers'. In 1832, *The Gardener's Magazine's* William Sanders had reported on a summer visit to the main seat of the Earl of Radnor; Longford Castle, near

Salisbury. The only plant mentioned by name was *Verbena chamaedrifolia* (now. *V.peruviana*) one of the species that became the parents of *V.xhybrida*, the bedding *Verbena*. (The others seemed to have been *V.phlogiflora*, *V.platensis* and *V.incisa* [58]). It may have been that Verbenas had became a favourite of the Radnors. Lady Radnor was, their correspondent stated, 'a warm encourager of horticulture [who] sent her gardener to Paris, in the summer of 1829, to see the principal gardens &c., there, and to collect what he possibly could that was new and rare'. [59]

Verbena incisa, with its common white 'eye', was, said the garden historian Richard Gorer, 'the species used by Henry Eckford... to produce his 'auricula-eyed' strain, which is what we know nowadays'. [60] That's a little generous to Henry, who, though prominent, was one of many hybridists, all using the 'eye' feature. 'Auricula-eyed', too, wasn't an appellation used at the time by such as *The Gardeners' Chronicle*.

Like Lady Radnor, George Fleming, who Henry had worked under at Trentham, was an enthusiast for Verbenas (as much else). Shortly after Henry had left there, Fleming was writing to *The Gardeners' Chronicle*: 'Having succeeded in keeping the different sorts of Verbenas in small pots through the winter, when my neighbours have failed, I beg to state the method I adopt...'. [61] Henry was to hybridise a number of different genera over the years and it's interesting to note how many of them were zygomorphs, plants whose flowers were irregular, having only one plane of symmetry. Many had the uneven (if common) number of five petals too. Scent also seemed important to Henry. Such plants included *Pisum sativum*, the garden pea; *Lathyrus odoratus*, the sweet pea; Verbenas, Primulas, Pelargoniums and pansies (*Viola*).

The long Age of Classicism that had held sway from the 1760s had evolved into the Romantic Era by the time of Henry's birth in 1823. Classicism's watchwords of simplicity and elegance had found horticultural expression in symmetry, where 'the central line of the mansion had been carried on into the formal geometric gardens and echoed in the parklands by an avenue of symmetrical trees' [62] to quote the garden historian Jennifer Davies. For the largest and most fashionable estates, flowers were swept away to join the fruit and vegetables in the kitchen garden. In horticulture as elsewhere, a reaction had already set in. Emotion would break down the importance of form. Colour returned with a vengeance to arrest the eye with the *en masse* planting of the new bedding schemes.

Another factor worth considering is, that although the head gardener had great authority, employers generally brought in an outsider when major design changes were contemplated. This must have annoyed head gardeners. It would have reminded them of those horticulturally bereft landscape 'gardens' whose authors might not even have seen the grounds they designed for. (Not a uniquely eighteenth-century practice!). J.C.Loudon had attacked Humphrey Repton for designing Valleyfield Garden in Fife from far-away Essex. (63) When Loudon in his turn produced designs for the grounds at Coleshill, Henry ignored them, producing glasshouses instead. For Henry and many like him, the focus was on the plants themselves, now seen as part of no greater whole beyond that of the flower bed.

At Coleshill, Henry's expertise with Dahlias and Verbenas seems to have begun in the early 1860s. On 30th May 1861 he purchased from 'The Royal Nurseries, Slough' a 'selection of Choice Dahlias and Verbenas' for an extravagant £5, in a 2/- hamper. (64) The following April he bought one rather expensive Chinese *Primula* for 10/6 and a 'Collection of Perry's Verbenas' at £3 from William Bull's nursery at Chelsea. (65) The new tender Primulas were plants that Henry was also to find success with. Charles James Perry was the leading *Verbena* grower of the day. William Dodd's prominence with Dahlias may have caused Henry to focus more on Verbenas. But Henry was never to quite get the measure of Mr. Perry, who generally stayed a step or two ahead. In the summer of 1863 Henry was buying Verbenas from Bull again. It included some French varieties this time, *V.* 'L'av de Ballent' and some specimen bedding geraniums. (66)

Two years later, a plant not mentioned in the '*Chronicle's* columns before surfaced in a letter from a seedsman. 'Sweet Peas. Who does not love this sweetest of flowers?... it is perhaps not impossible to produce something grander than we are already familiar with…' (67) If Henry saw this he would reasonably have thought well, fine, yes; but I've got plenty on my plate right now. Something for the future, perhaps. And he stored the thought away…

One might now reasonably ask, what was Henry doing, indulging in plant experiments and neglecting his duties as head gardener? As we saw earlier, Peter Loney, head gardener at Fingask, was no looker-on but led by example through the varying practical tasks of the year. He contrasts with the view you sometimes hear that 'as soon as a hierarchy of servants was established in a big house, the upper ones made as easy a life of it as they could – until, in the smartest and wealthiest establishments, the butler and the housekeeper, the head groom and head gardener would be putting their feet up and relaxing…' (68) At Coleshill, it's likely that the answer lay somewhere between these two examples. Henry was expected to supervise and administrate, duties that would have taken him from practical work whether he'd wanted them to or not. No practical person likes these sides of a job, and Henry was a practical man. He may have felt that what practical work he did manage to do would be something he really enjoyed, if only as compensation for those aspects of the job he didn't. Also, as long as the cook had her fruit and vegetables, the house had its cut flowers and the grounds looked neat and decent, I don't suppose anyone worried too much about exactly what Henry got up to. And, too, there was a certain reflected glory for the Earl in his head gardener making a name for himself.

The next mention of Henry in *The Gardeners' Chronicle* came in 1867. 'Florists' Flowers... The meeting of the Floral Committee at South Kensington on September 3rd brought together a goodly number of seedling Dahlias from... well-known raisers...'. There followed a long list of exhibition varieties, with a mention at the column end for 'The Hon. Mr.Bouverie (Eckford), buff, slightly streaked and pencilled with rosy violet...'. An Edward Playdell-Bouverie was, as we've seen, the retired vicar of Coleshill, related to the Radnors. The article ended: 'Mr Eckford, of Coleshill, produced a stand of his new Verbenas, though somewhat behind the fine quality of Mr Perry's new varieties of the present year. The most striking flowers of the batch were Lord Derby, brilliant deep orange scarlet, with a conspicuous lemon eye, but the pips were cupped so much as to give it an unsightly appearance; and Bravo, cerise, changing to pale rose with age, with a small lemon eye, the pip of good form and stout, though a little rough, and the ultimate loss of the bright cerise hue gives it a somewhat washy appearance. R.D.' [69] Damning with faint praise? 'R.D.' was Richard Dean, who overcame his initial reservations to become a lifelong friend and champion of Henry's. The two may have already met. Dean had been at the Royal Nursery at Slough in the mid 'fifties when Henry was one of their regular customers. Lord Derby, incidentally, was the Conservative Prime Minister. Henry, in common with other hybridists, favoured the Vicar of Bray approach to politics, naming his plants from the leading lights of the day, whatever their party. One 'leading light' however was noticeable by his absence. Benjamin Disraeli was the century's foremost Conservative leader but whereas Henry was to name a sweet pea after Catherine Gladstone there were no Verbenas or Dahlias for the Disraelis now. It's just possible that a sweet pea Henry would call 'Lady Beaconsfield' in 1892 would be for the, by then, late Mary Anne Disraeli, ennobled as Viscountess Beaconsfield in 1868. After 1847 the Liberal Party were extraordinarily dominant in Scotland, gaining a majority of seats there in every general election until 1900. (70) William Gladstone's sobriety and his connection to Midlothian would have won Henry's support. Especially so in comparison with the raffish and reckless Disraeli, a man of a different stripe to Derby, his fellow Conservative.

The Floral Committee of the Royal Horticultural Society had more Dahlias to look at a fortnight later. 'Mr Eckford, The Gardener, Coleshill, had Eliza and the Hon. Mrs Bouverie… and Mr Eckford exhibited the following new kinds:- Achievement, Lord Derby, Julius, Isa Key, and Novelty'. (71) Henry and Charlotte both had older sisters called Elizabeth. Lord Derby was now the name of both a *Dahlia* and a *Verbena*. Such uninspiring practice was quite common and, one imagines, equally confusing. Perhaps Lord Derby appeared attractive because he was a rather apolitical aristocrat, a more eighteenth than nineteenth-century figure who preferred billiards and whist to affairs of state. (72)

The next *Gardeners' Chronicle* report came in March, 1868. A Woolwich nurseryman had field-trialled the previous season's new British and continental *Verbena* varieties. This gave a better indication of their overall worth, compared to judging then in an exhibition hall. As a result of this, and probably also because Henry was improving as a plant raiser he received a glowing report. 'From Mr Eckford we have Fanny Martin, a beautiful clear rose, with remarkable fine trusses, good substance, and altogether a first-class variety. Mr Ellice is a fine improvement on King Charming, sent out the season before, very attractive in colour, being a peach pink or a bright salmon, with small green eye: it is really a very beautiful colour, and rather a moderate grower. Earl of Radnor is a beautiful clear rosy scarlet, with a lemon eye; it has smooth pips and large trusses, and is the best of this shade of colour'. (73)

This was a nice pat on the back, a double celebration for it came with the safe arrival of the Eckfords' second and last daughter, Charlotte Elizabeth. Charlotte was to be their only child's name distinctly from the mother's side of the family. Elizabeth, as we've seen, was from both parents' sisters and, in all probability, Charlotte's grandmother and Henry's great-grandmother too. And Henry's aunt! There was something of a pattern to it. The first daughter was often named after the mother's mother; the first son after the father's father. With her mother also called Charlotte, it seems the child became known as Eliza.

The summer of that year, 1868, was to be the hottest summer until 1911. It was talked about throughout the press, even getting into books. In *The Amateur's Flower Garden* of 1871, Shirley Hibberd wrote that the summers of 1868 and 1870 had 'not a drop of rain for months together, but tropical heat instead'. (74) It would have been difficult to keep those huge estate bedding schemes watered and probably hastened the eclipse of the *Verbena* in favour of the more drought tolerant Pelargoniums, the bedding geraniums. It certainly made up for that 1860 summer of 'no sunshine but continuous rain'. Despite originating from Mexico, Dahlias suffered too. In September, Richard Dean noted that the new seedling varieties inspected by the floral committee of the Royal Horticultural Society showed deficiencies 'fairly chargeable to the hostile character of the season'. Nevertheless, Henry picked up two first-class certificates for 'Memorial... pale rose, a very pleasing shade of colour... Mr Brunton... pure white ground, with heavy lacing of deep purple... very fine…'. His 'Firebrand' picked up a second-class certificate.

'Memorial' might have been named after the old Earl Radnor, who died this year. 'Mr Brunton' was of course either James Brunton the father, or the son who'd married Henry's sister Isabella. At the same meeting 'Messrs. C.J.Perry and Eckford each staged a fine lot of seedling Verbenas, the latter being especially strong with new varieties. First-class certificates were awarded to Mrs Eckford... white, with pale rose centre...; Master Jacob... crimson maroon...; Conspicua... bright crimson, with pure white eye...; Ace of Trumps... bright salmon rose...; Lady Folkestone... rose suffused with violet'. (75) Another six Verbenas appear to have missed certificates.

The next R.H.S. floral committee report in the '*Chronicle* ran: 'With new Verbenas, Mr. C.J. Perry may be said to have regained at this meeting the temporary loss of his ascendancy, wrested from him on the last occasion by his worthy rival, Mr. Eckford'. After listing Mr. Perry's new varieties, 'Mr Eckford', it remarked, 'received a First-class certificate' for Anna Keynes and 'a second-class certificate for Lotty Eckford... very bright looking…'. Descriptions of 'other good varieties' of Henry's were also mentioned. (76) Anna Keynes would have been the wife or daughter of John Keynes, Henry's nurseryman. Lotty is the diminutive of Charlotte, the Eckfords' new baby girl.

A fortnight later the 'Chronicle was describing another of Henry's Dahlias, 'Miss Ruth'. 'A large and somewhat coarse-looking yellow flower, tipped with white, which invariably comes hard-eyed, but effective for border decoration…'.(77) (Hard-eyed?). A Chester nursery was more sympathetic, praising its garden value; still 'in fine condition about the middle of September', and standing up to being exhibited as far afield as Liverpool and Dublin. (78) In the early spring of 1869 the '*Chronicle's* list

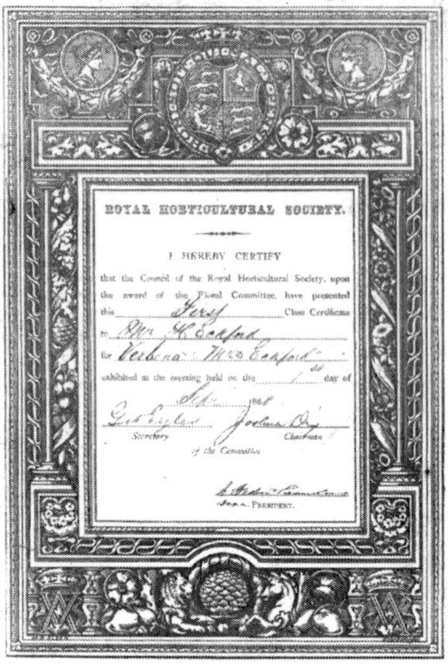

ROYAL HORTICULTURAL SOCIETY.

I HEREBY CERTIFY

that the Council of the Royal Horticultural Society, upon the award of the Floral Committee, have presented this ... Class Certificate

to ...

for ...

exhibited at the meeting held on the ... day of ... 1868

Secretary ... Chairman

of the Committee

... PRESIDENT.

RHS Verbena certificate, 1868,
"Mrs Eckford."

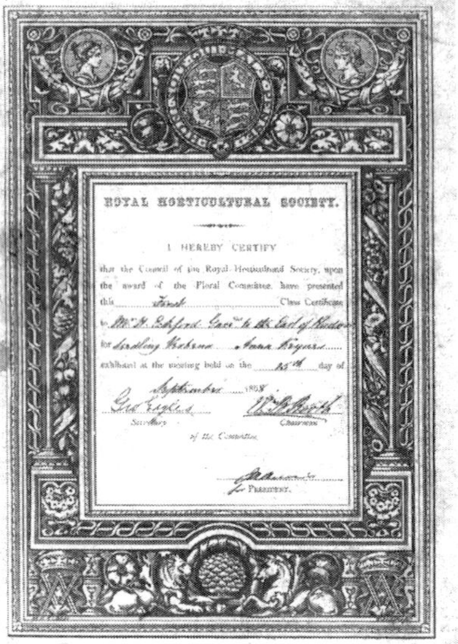

ROYAL HORTICULTURAL SOCIETY.

I HEREBY CERTIFY

that the Council of the Royal Horticultural Society, upon the award of the Floral Committee, have presented this First Class Certificate

to ...

for Seedling Verbena Anna Keynes

exhibited at the meeting held on the ... day of September 1868

Secretary ... Chairman

of the Committee

... PRESIDENT.

RHS Verbena certificate, 1868,
"Anna Keynes."

of most promising *Dahlia* 'selfs' (flowers of mostly one colour) included two of Henry's Show Dahlias – 'Memorial' and 'Mr Brunton'. [79] In 1870, Richard Dean highlighted these two varieties again but with this proviso: the *Dahlia* 'appears to have sunk in public estimation', as the *Verbena* did too about now. For these two half-hardy plants, their moment in the sun was nearly over and they would have to find their new niche in the garden. The year is instructive; 1870 saw the publication of William Robinson's seminal book *The Wild Garden* and British horticulture was off on a new tack.

Popular plants of the future would be less exotic and ostentatious. An article in the '*Chronicle* attempted to catch the changing mood. New forms of the sweet pea, it said, 'are within reach... No flower garden or drawing room need be without the sweet adornment of these lovely flowers from May to November.' [80] There was now a revival of interest in the *Dahlia*, in its less showy single and brand-new cactus types. As mentioned earlier in passing, Robinson explained what had been going on on the ground in his book *The English Flower Garden* in 1883. 'The reason why [Show and Fancy groups] are less valuable than many other kinds of Dahlia in the garden is because of the weight of the flowers. There is little graceful beauty about them, the stems being bent with the burden of a too heavy blossom, hence the greater popularity of the many lovely cactus varieties'. [81]

'The value of Single Dahlias as beautiful garden flowers was not considered until a reaction set in against the show blooms, and then the elegant single kinds became popular'. [82] It was a glance back to the Romantic Era, when the natural beauty of the recently discovered *Dahlia* was a fit subject for the French artist Eugène Delacroix's 'A Vase of Flowers' in 1833. Incidentally, Robinson's list of recommended Show Dahlias include one called 'Richard Dean' and in Fancy Dahlias, 'H.Eckford'. They come after 'Queen of the Belgians' and 'General Gordon' (another Eckford variety) respectively. All equal in the world of horticulture!

One thing that may have struck you, reading through all these plant descriptions, is their potential for comedy. Lotty Eckford probably was herself 'very bright looking'. Lord Derby however, an unsightly appearance occasioned by a conspicuous lemon eye and brilliant deep orange-scarlet features... And pity poor Miss Ruth, whoever she was, having her namesake described as 'large, hard-eyed and somewhat coarse'. Let's hope the recipients were content to have a plant named after them and didn't look too closely at the trial and exhibition reports. Such incongruities would have been lost on our hybridists but we in our islands have always laughed at anyone taking themselves (too) seriously. Ruth Duthie in *Florists' Flowers and Societies* [83] speaks of a writer on *The Tatler* way back in 1710 supposedly overhearing a conversation about Kings and Generals. On discovering that they are actually talking about tulips he goes off with the company to see their tulip beds where he inadvertently admires all the wrong ones! The more things change... Plant names have changed a bit since then. Raisers still name them after family and friends, and royalty is still popular. But the nobility and the Generals have gone, and politicians have taken a back seat – to celebrities.

The second thing that strikes you about these lists is that they describe the flowers, not the whole plant. Floriculture was originally the production of any plant grown for its floral attraction. Over time, the epithet 'florists' flowers' became confined to a small number of around eight hardy plants. Each one had its own Society of mostly amateur florists who championed them and changed their appearance. In the nineteenth century the list expanded to include, among others, the new tender bedding plants like Petunias, Dahlias, Verbenas and Pelargoniums. A florists' flower, it was said, must be capable of sexual and asexual propagation (to create and maintain new varieties) and, quoting Duthie 'the flower had to have a circular outline', its petals 'to have a good texture and be smooth-edged, not fringed or jagged. If the flower were double the floral parts had to lie smoothly. Whether double or single, some variegation in the colour was desired; these colours had to be related to one another in a particular way...'[(84)] *The Gardeners' Magazine* of 1842 saw the wood from the trees. It was probably J.C. Loudon, the editor, who wrote: 'Florists' flowers differ from wild flowers and border flowers in being so entirely changed by culture as no longer to resemble their original type'. [(85)]

The other point to grasp is that the emphasis had become unhealthily focused on the flower itself. Such qualities as hardiness and disease resistance, or specific garden qualities like sturdy flower stalks, a long flowering period, scent, some acknowledgement of the plant's original natural beauty – such qualities were neglected. Competition arose to produce the showiest flower for the exhibition hall. Henry was one of a new breed of florists. A trend had been noted by the journal *The Scottish Gardener* in 1853, a little after Henry had left Scotland for London. Head gardeners, 'men at the head of their profession' had been 'ignorant of florists' flowers and their culture'. However, we now 'recognise the spread of Floriculture among [this] higher class'. [(86)] These men were attracted by the changes occurring among roses, Fuchsias and Camellias and to the new tender introductions from abroad. It was only to be expected that well-trained professionals would be looking at this plant-raising business in a new light. Changes would take place that would put the floristry of old into disrepute.

In the 1870s the word 'floristry' evolved into its present sense. It came to refer to women (it was usually women) who arranged and sold flowers in the new flower or florists shops that had begun springing up. [(87)] These shops were supplying a new market, that of the newly expanding middle-classes. This was the market that Henry would eventually arrive at too. But not yet. A few first-class certificates don't make a summer. The societal changes taking place weren't yet at that flood that can lead on to fame and fortune. Henry had a few more years in the service of their noble lordships yet. The years with the third Earl, however, had come to a close. Lord Radnor, Henry's employer for the last fifteen years, died at midday on Monday March 11th, 1869. 'The poor', said *The Gardeners' Chronicle*, have lost 'a kind and generous friend, and horticulture a liberal patron. He was a fellow of the Royal Horticultural and Royal Botanic Societies of London, and President of the Wilts. Horticultural Society... [and] a very good judge of all kinds of fruit and flowers'. [(88)]

Jacob, the new, fourth Earl, was unlikely to be as committed to horticulture as his father. In fact, he was to spend most of his time at the Radnors' other seat of Longford Castle, forty miles away at Salisbury. Like his father though, he 'swung between the two' places, 'the distance considered to be completely rideable'. [(89)] Four months after the death of the old Earl, the Eckfords' baptised their sixth and last child, Peter William, on the nineteenth of July. (Henry's two great-uncles had been Peter and William). Luckily for Charlotte, she had carried him in the summer that lay between two of the century's hottest. Charlotte's side of the family may have come up from Queen Camel now, seventy miles away. As we saw earlier, Charlotte's mother was visiting and her sister Lydia living-in as a 'servant' at the time of the 1861 census. One imagines her mother there now, fussing and happy with her six grandchildren. It must have been a contented time for them all, a contentment that was to last for a few short years more.

References to Chapter Five

(1) *Burkes Peerage and Baronetage*, (1897), p.1009.

(2) Emma Fox, ed., *Coleshill 2000*, (2000), p.4.

(3) Herbert A.St.John Mildmay, *A Brief Memoir of the Mildmay Family*, (1913), pp. 117-118.

(4) Census return, St. Martin's ward, Salisbury, Wiltshire, 1851.

(5) RHP/1983/1-13, The Clerk of Penicuik Papers, National Records of Scotland. A 'plan for new gardens', post 1874 includes 'Amy Hogg Geraniums'.

(6) 'Florists' Flowers', *The Gardeners' Chronicle*, (15-4-1871), p.486.

(7) Geoffrey Best, *Mid-Victorian Britain, 1851-75*, (1971), p.203.

(8) George Frederick Pardon, *The London Conductor*, 1851, (1984), p.iii.

(9) E.R. Kelley, ed., *The London Post-Office Directory*, 1857.

(10) Eric A. Willats, *Streets with a story: the book of Islington*, (1988), pp.264, 185.

(11) Highbury & Islington Ordnance Survey, 1871, (1990).

(12) Charles Knight, *Pictorial Half Hours of London Topography*, 1851, (1984), p.116.

(13) K. Bryant, R. Giles (eds.), *Queen Camel, Now and Then*, (1990), p.17.

(14) 'Club Day', *The Agricultural Economist*, (7-1904), p.215.

(15) William Howitt, *Rural Life of England*, (1845), p.497.

(16) D/EPb, A28, (1857), The Pleydell-Bouverie archives, Berkshire Record Office.

(17) D/EPb, A28/25, (1857), The P.-B. archives, B.R.O.

(18) D/EPb, A28, (1857), The P.-B. archives, B.R.O.

(19) 'The Agricultural Gazette' section, *The Gardeners' Chronicle*, (3-7-1858), p.529.

(20) Flora Thompson, *Lark Rise to Candleford*, 1945, (1948), p.411.

(21) D/EPb, A28, Bdl. 13, (1858), The P.-B. archives, B.R.O.

(22) D/EPb, A28, (1859), The P.-B. archives, B.R.O.

(23) Shirley Hibberd, *The Amateur's Flower Garden*, (1871), p.268.

(24) D/EPb, A28/25, (9-4-1853), The P.-B. archives, B.R.O.

(25) Christopher Thacker, *The Genius of Gardening*, (1994), p.251.

(26) D/EPb, A28/25, (1859), The P.-B. archives, B.R.O.

(27) D/EPb, A28, (17-10-1859), The P.-B. archives, B.R.O.

(28) D/EPb, A28, Bdl. 14, (1860), The P.-B. archives, B.R.O.

(29) D/EPb, A28, (1860), The P.-B. archives, B.R.O.

(30) *A History of Berkshire*, Vol. IV, (1924) of The Victoria History of the

Counties of England, p.521.
(31) G. Best, op.cit., p.281.
(32) Census return, Coleshill parish, Berkshire, 1861.
(33) G. Best, op.cit., p.223.
(34) D/EPb, A28, (1861), The P.-B. archives, B.R.O.
(35) D/EPb, A28, (1862), The P.-B. archives, B.R.O.
(36) D/EPb, A28, (1863), The P.-B. archives, B.R.O.
(37) 'Samuelson's... lawn mowing... prices for 1861', The Samuelson Papers (no ref. nos.), Banbury museum.
(38) *Diary of a Victorian Gardener: William Cresswell and Audley End*, (2006), p.128.
(39) D/EPb, A28, (1863), The P.-B. archives, B.R.O.
(40) Pamela Horn, *Labouring Life in the Victorian Countryside*, (1976), p.259.
(41) Rev. Campbell Ferenbach, *Annals of Liberton*, (1975), p.39.
(42) Census returns, Stenhouse, Liberton, 1851, 1861.
(43) 'National Sweet Pea', *The Gardeners' Chronicle*, (4-11-1933), p.352.
(44) 'Obituary', *The Gardeners' Chronicle*, (27-4-1935), p.352; Jennifer Davies, *Say it with Flowers*, (2000), p.72.
(45) 'Liberton and Newton Horticultural Society's Show', *The Scottish Gardener*, (10-1862), pp.309-310.
(46) 'Liberton and Newton Horticultural Society', *The Scottish Gardener*, (10-1863), p.308.
(47) Ibid, p.309.
(48) D/EPb, A28, Bdl. 14, (1863), The P.-B. archives, B.R.O.
(49) Ibid.
(50) F. Thompson, op.cit., p.96.
(51) D/EPb, A28, (1863), The P.-B. archives, B.R.O.
(52) D/EPb, A28, Bdl. 12, (1863), The P.-B. archives, B.R.O.
(53) Ibid.
(54) D/EPb, A28, (1863), The P.-B. archives, B.R.O.
(55) Letter from Harry Eckford to Bernard Jones, c1980. Still births on 15-9-1862 & 22-8-1863.
(56) 'Florists' Flowers', *The Gardeners' Chronicle*, (15-9-1866), p.879.
(57) R. Desmond, *Dictionary of British and Irish Botanists and Horticulturists*, (1994), p.330.
(58) Richard Gorer, *The Growth of Gardens*, (1978), p.145.
(59) 'Observations on several gardens in England', *The Gardener's Magazine*, (10-1832), pp. 548-551. The visits appear to have been in

July 1830.

(60) R. Gorer, op.cit., p.145.

(61) 'Floriculture', *The Gardeners' Chronicle*, (2-9-1854), p.567.

(62) Jennifer Davies, *The Victorian Flower Garden*, (1991), p.11.

(63) A.A. Tait, 'Scottish Gardens' in J. Harris, ed., *The Garden*, (1979), pp.177-178.

(64) D/EPb, A28, (1861), The P.-B. archives, B.R.O.

(65) D/EPb, A28, Bdl. 4, (1862), The P.-B. archives, B.R.O.

(66) D/EPb, A28, Bdl. 4, (1864), The P.-B. archives, B.R.O.

(67) 'Home Correspondence. Sweet Peas', *The Gardeners' Chronicle*, (22-7-1865), p.676.

(68) G. Best, op.cit., p.223.

(69) 'Florists' Flowers', *The Gardeners' Chronicle*, (7-9-1867), p.931.

(70) E.W. McFarland in C. Williams, ed., *A Companion to Nineteenth-Century Britain*, (2004), p.514.

(71) 'Societies', *The Gardeners' Chronicle*, (21-9-1867), p.978.

(72) Anthony Wood, *Nineteenth Century Britain, 1815-1914*, (1984), p.230.

(73) 'Florists' Flowers', *The Gardeners' Chronicle*, (7-3-1868), p.241.

(74) S. Hibberd, op.cit., p.62.

(75) 'Florists' Flowers', *The Gardeners' Chronicle*, (5-9-1868), p.946.

(76) 'Societies', *The Gardeners' Chronicle*, (19-9-1868), p.994.

(77) 'Florists' Flowers', *The Gardeners' Chronicle*, (3-10-1868), p.1043.

(78) 'Florists' Flowers', *The Gardeners' Chronicle*, (10-10-1868), p.1069.

(79) 'Notices to Correspondents', *The Gardeners' Chronicle*, (10-4-1869), p.394.

(80) 'A chapter on Sweet Peas', *The Gardeners' Chronicle*, (3-4-1869), pp.360-361.

(81) William Robinson, *The English Flower Garden*,1883, (1905), p.529.

(82) Ibid, p.530.

(83) Ruth Duthie, *Florists' Flowers and Societies*, (1988), p.91.

(84) Ibid, p.33.

(85) 'What Constitutes a Florist's Flower?',*The Gardener's Magazine*, (9-1842), p.454.

(86) 'Floriculture and Plant Growing', *The Scottish Gardener*, (3-1853), pp.72-74.

(87) J. Davies, *Saying it with Flowers*, (2000), p.17; passim.

(88) 'Obituary', *The Gardeners' Chronicle*, (16-3-1869), p.343.

(89) Letter to the author from Lord Radnor, (9-1-2006).

Chapter Six

The activation of minds by a good education was still largely the preserve of the well-to-do. Occasionally however, a leg-up in early life allied with a drive and determination to succeed could overcome poverty's otherwise insurmountable barriers. Horticulture, in its many manifestations, could only benefit from minds made active by the broadest of educations. One such mind belonged to Johann, later Gregor, Mendel, from Moravia in Central Europe. It is an odd coincidence that Henry's work was first noted in 1866. It was in 1865 that Mendel read his famous paper on hybridisation to the Brünn Society for the Study of Natural Sciences. (1) It contained his discovery of what later became known as genetics. Published the following year by the Society, Mendel sent copies to various addresses abroad, including four in the British Isles. It remained equally famously ignored and where it wasn't its meaning lay undiscovered until 1900. Had it been acknowledged earlier, Henry and other hybridists would have had a scientific foundation for their work.

While researching in the Cambridge University Library I discovered a then uncatalogued letter of Mendel's, dated 27th October 1866(2), that had accompanied the copy sent to the Royal Observatory at Greenwich. Mendel had previously lectured on astronomy to the Brünn Society and continued to send copies of their journal to Greenwich. The Royal Observatory wasn't perhaps the best address to arouse interest in garden pea experiments! He also sent copies to the Dublin Natural History Society, the Royal Society and the Linnean Society. I couldn't find it noted in the Accessions Register of the Linnean Society, though they received it. Dozens of journals arrived monthly at such places, and this one was no doubt accidentally omitted from the register. But one can speculate that this meant it also missed the promotion that new journals get in a library.

Various reasons have been advanced for the paper's thirty-five year neglect. That it was written in German is one, but many scientific papers were and continue to be. It is also said, somewhat parochially, that he came from a mid-European backwater. From central Europe, from the gardens of the Mendelianum, say, one can equally see Britain as a collection of mostly uninhabited islands lying off the north-west mainland. In 1865, Brünn, capital of Moravia, had a population of some 65,000. The seventy miles to Vienna, capital of the Austro-Hungarian Empire, had been linked by train since 1839. The Brünn (now Brno) – Prague railway line had been completed in 1849. (3) Moravia had long before that felt the cross-currents of European thought.

One explanation among others is this. The great natural-scientists of the age, rich or poor, acknowledged or ignored were united by their reception: incomprehension. That was the case when Mendel gave his lecture to the Brünn Society in 1865. (4) Arguably, it was also the case when Charles Darwin and Alfred Russel Wallace's new theory of evolution by natural selection had been presented to an audience at the Linnean Society in 1858. 'If the thirty or so people in the [London] audience had any idea that they were witnessing the scientific highlight of the century', writes Bill Bryson, 'they showed no sign of it. No discussion followed. Nor did the event attract much notice elsewhere'. (5) Where there was comprehension, there was also fear of the consequences. To ordinary people with an interest in natural-history, the likes of Darwin and Mendel were speaking in unknown tongues. Initially you pick up the odd word, understand a phrase or two, and hope one day to be fluent. But you wonder if you ever will be.

Henry Eckford and Johann Mendel were born within months of each other. Though from different cultures, they came from similar backgrounds. Mendel was from a village in a province on the northern border of the Austrian Empire. Neither the Eckford nor the Mendel families were rich, but as farmers in fertile countryside they were kept from poverty. Indirect evidence implies that Henry received a good village education until he was fourteen. Johann Mendel we know had a good education in his village too, one especially strong in the natural sciences. Christian doctrines would have been emphasised in both. Henry went on to receive a horticultural apprenticeship that would have been both thorough and wide-ranging. Johann's studies, which he couldn't really afford, were, critically to his later unacknowledged success, wider than Henry's, though patchy. Mendel's love of learning took him to a monastery in Brünn where he could continue to study and teach. To his scientific studies, including his hybridising work with the garden pea, he was able to bring both a practical and theoretical outlook. For by then he had studied aspects of agriculture, horticulture, natural history, biology, physics, mathematics, astronomy, philosophy, theology and Greek.

To us, perhaps, philosophy (and Greek and theology) stands oddly apart. Theories born of contemplation do not sit well with the British love of empiricism. That at any rate was the view of Emerson, the American writer travelling around Europe. In his *English Traits* of 1856 he considered the English to be 'jealous of minds that have much facility of association, from an instinctive fear that the seeing many relations to their thought might impair' it. 'They are impatient of genius, or of minds addicted to contemplation, and cannot conceal their contempt for sallies of thought, however lawful, whose steps they cannot count...'. (6) The English respect only performance, he observed, they 'value ideas only for an economic result... I suspect that there is in an Englishman's brain a valve that can be closed at pleasure, as an engineer shuts off steam'. (7) Again: 'The Englishman has accurate perceptions: takes hold of things by the right end... He loves the axe, the spade, the oar...'. He is materialist and mercantile. He wants 'muffins and not the promise of muffins'. (8)

'The eminent benefit of astronomy' for example, 'is the better navigation it creates to enable the fruit-ships to bring home their lemons and wine to the London grocer'.

'Beauty, except as luxurious commodity, does not exist'. (9) Well, beauty had certainly been strangled out of florists' flowers. Increasing numbers of Lowland Scots like Henry, settling in England (internal Highland migration was more towards the cities of Glasgow and Edinburgh in the Scottish central belt) would only reinforce this so-called 'English' identity of Emerson's.

Henry the gardener and Gregor the scientist (Gregor was Mendel's new monastic name) were both hybridising plants. Although their origins were similar the paths they took ensured that their two approaches would be dissimilar. Henry's was the traditional one, enjoying the challenge of turning wild plants into garden flowers. Based on the work of the florists' societies, and by trial and error, he found out what worked in practice. As a head gardener he had the time but not the inclination, apparently, to look any further. Bedding plants were the fashion. Everyone wanted them, and the wider the choice the happier everyone was. In the days before pensions the study of science held out no practical means of saving for security in old age. For families it fed no mouths. It brought no kudos, only suspicion.

The age of science made slow progress while it remained the interest of those wealthy enough to pursue it. To those like Mendel, or England's Alfred Wallace, another exact contemporary of Henry's, it was a struggle.

Should Mendel have considered the possible misuse of his experimental findings? Mendel's life, with his love of learning and teaching, though it took him out and about was essentially a cloistered one. The daily exposure to human nature in all its variety was limited. He wouldn't have considered such a question because he probably couldn't. Such considerations, too, usually proceed *from* actions rather than precede them. The work of Henry and his fellow gardeners, in contrast, involved no great leaps into the dark. Their work increased crop yields and led to greater choice among improved garden flowers. It seemed to satisfy them.[10]

~~~

One 'sweltering summer's day', one of many in that hot dry summer of 1870, Richard Dean decided to leave London for the countryside. Arriving at Shrivenham station, he was soon at the entrance to Coleshill, notebook in hand. *The Gardeners' Chronicle* waited until December and copy was slack before printing his report. The winter would have been a nice time too, to have reminded their readers of summer sunshine but the hot summer of 1870, like 1868's, were ones they would rather forget. Richard Dean gave us our best picture of Coleshill under Henry's stewardship and is worth quoting more or less in full.

> '... The park of Coleshill Place, as it is termed, is some 400 acres in extent, set down in a pure country district, with but little of the busy hum of town life. The mansion was designed by Inigo Jones, and is a solid substantial structure, facing due south, and commanding extensive prospects. Beech and Elms, fine pyramid Yews, with Limes dotted among them, form a cordon round the mansion, with outer cordons and irregular lines beyond, till the open country, and arable land, with attendant breaks of woodland, are reached. In the pleasure ground may be seen fine specimens of Abies Pinsapo, and Taxodium sempervirens; [*Sequoia sempervirens* now, the Californian Redwood] these last flourish as if in a very congenial home.
~~~

On the right of the mansion, and close to its northwest corner, is a short avenue of Elms, which was at one time the old coach highway; and the London coaches, before the counties were strapped together by the iron roads, used to pass under these Elms, but some years ago the road was diverted, to the manifest improvement of the property.

On the lawn, on the south front of the house, lies what there is of flower garden, and it is a single strip formed of two sets of choice beds. A new hybrid Ivy leaf Pelargonium, after the type of Wills' new kinds, but raised by Mr. Henry Eckford, the gardener at Coleshill, and bearing trusses of crimson flowers tinged with purple, made a charming bed, and was edged with Sempervivum californicum. All the freedom of bloom of the ordinary Ivy-leaved Pelargonium seems to have been transmitted to this new hybrid. Coleshill, a rich scarlet zonal, also raised by Mr. Eckford, together with Merrimac, with fine cerise-crimson coloured flowers, made fine beds also. A good and pleasing bed was formed by Iresine Herbstii and Mrs. Pollock Pelargonium, mixed together; and yellow Calceolarias, edged with Pyrethrum Golden Feather, furnished a much better contrast than many might be supposed to imagine.

The kitchen garden lies near the mansion, and is of some 6 acres in extent. There are four vineries and other houses besides, but as the Earl of Radnor rarely occupies the mansion, but little is done in the way of plant growing. There is a small but good collection of Orchids. Of these Saccolabium Blumei majus [now *Rhynchostylis retusa*], Cypripedium stonei [now *Paphiopedilum stonei* and then a recent discovery by the son of Henry's former employer, Hugh Low, in Borneo], and others, were in flower; and a fine plant of Peristeria elata [the Holy Ghost or Double Orchid] was throwing up several blooming spikes. Of Saccolabiums [a now defunct genus] there was a good collection, all in fine condition.

Mr Eckford does not appear to be so ready to condemn Mrs Pince's Grape as some other cultivators. He finds that it materially aids the Grape to give the Vines heat during the time of setting the berries, and as soon as it is in bloom he takes the top air off [i.e. he closed the ridge vents] and keeps all draughts out of the house as much as possible. In one of his houses Mr. Eckford has a White Grape, which he states he raised from seed of the Black Hamburgh without any fertilising process. It seems to possess the merit of earliness, as it was over in an early vinery before the Black Hamburgh was well ripe, both vines being started at the same time.

A few Pines [i.e. pineapples] are grown, principally Queens, Black Jamaica, and Smooth Cayenne. Barr's Monarch, a new green-fleshed Melon, was commended by Mr. Eckford as being a reliable and certain variety for a main crop. Coleshill Cucumber, one of the Sion House type, is a capital forcing variety, yielding abundance of good table fruits. There is a fine collection of pyramid Pears at Coleshill, all in rare health, and, in season, laden with fruits. Duchesse d' Angoulême, Jargonelle, Marie Louise, Van Mons Léon le Clerc, Napoléon, Ne Plus Meuris, Beurré de Capiaumont, and Bon Chrétien were all very fine. The bush cherry trees on either side of a walk in the kitchen garden are a fine feature, and at Coleshill they do ten times better so grown than they do on walls, as they are apt to die, owing to the heat. Mr. Eckford does not summer prune these bushes. Some old Mayduke Cherries on walls went very bad, and nearly died. These were transplanted and grown as bush trees, and are now doing well. There were very fine crops of all kinds of fruit at the time of my visit, and Figs on walls were very fine. A low wall of Morello Cherries had enormous crops of fine fruit. All the shoots are not cut back or nailed in, only the leading ones, but thinned out, and the old wood throws plenty of young shoots that give abundance of fruit the following year; Mr Eckford finds this mode keeps the wall cool, and the branches give some protection to the fruits.

It is more particularly as a florist that Mr. Eckford is known. As a raiser of Verbenas he can hold his own against Mr. Perry, even though he cultivates under totally different conditions. Mr. Perry grows his plants under glass; all of Mr. Eckford's are in the open ground, and have to battle with sun, wind and rain alike. In several parts of the flower and kitchen garden could be seen patches of seedling Verbenas, and self-sown plants had sprung up on all

> hands... The various types of Pelargoniums are hybridised, and with Ivy-leaved forms Mr. Eckford is fully abreast of Mr. Wills.
>
> The visitor to Coleshill should not fail to look at the village of Coleshill, with its pretty, commodious, well-arranged cottages... R.D.' (11)

Richard Dean's report makes a number of interesting observations. Firstly, what there is of flower gardens are little more than trial beds for Henry's latest hybrids. For the first time we see he's also experimenting with Pelargoniums, including the ivy-leaved types, which puts him in competition with the grower John Wills (1832 – 1895). Wills it was, incidentally who suggested that the new Albert Memorial in Hyde Park should be enclosed in glass to protect it from 'the vitiated atmosphere of London'. (12) Prescient, that. He'd helped lay out the grounds around the Crystal Palace, which must have given him the idea. *Paphiopedilum majus* had been discovered growing in Borneo by the plant-hunter Hugh Low, sending it to his father to introduce in October 1862. (13) Hugh, the son, was probably the only one of the Low clan Henry hadn't met. Stuart Low had introduced the Holy Ghost orchid in 1864, (14) even more recently, so Henry and the Lows obviously kept in touch.

We can see how innovative Henry is. In the kitchen garden he's experimenting with ventilation, finding an early grape variety and an easy to force cucumber, as well as trying out the latest melons and new ideas of cherry production. They seem to work too. With Verbenas Henry also comes across as modern in his approach, seeing hardiness and other qualities of a good constitution as primary requirements of a good garden plant. Dean doesn't mention that, earlier in the century, village cottages once stood where the kitchen gardens and an orchard were. South of the mansion, much of the estate is on an escarpment and the then Earl wanted more flat and water-retentive land, such as that of the cottages on the south side of the village road. So he knocked all their houses down and took it. (15) It wasn't really so long ago that lords of the manor could, and did, do exactly as they liked.

~~~

'Verbena. – There is not in all the catalogue of bedding plants one that more perfectly answers to the requirements of the garden colourist than this. Its trailing habit, forming a close carpet of vegetation, its well-sustained umbels
~~~

of brilliantly-coloured flowers glittering above the suitable groundwork of dark green leaves... the continuousness of its intensity of colour…'. So wrote Shirley Hibberd in his *The Amateur's Flower Garden* of 1871. 'And yet', he cautioned, 'The Verbena has been steadily declining in popularity during many years past, in consequence of frequent failures...' He ended the entry thus: 'In a collection of over a hundred kinds grown in our experimental garden in the burning summer of 1870, the following were the best', and he listed twenty-two varieties. Five of them were Henry's: 'Isa Eckford', 'Mrs Eckford', 'Lady Folkestone', 'Polly Perkins' and 'King Charming'. (16)

Problems with Verbenas had been noted earlier. John Douglas, gardener at Kilkea Castle, County Kildare, wrote to the '*Chronicle* in August 1866: '... I have the same thing to complain of as many other gardeners – the failure of my bedding Verbenas. I do not think that frost has anything to do directly with the malady, as some that were planted out late, and had no frost, failed, as well as those planted out early... Some sorts are more delicate than others. Some of the old scarlet varieties seem to be particularly hardy...'. (17) The newer sorts, it seems, were not as hardy as the old ones. If disease was a problem it fastens first on the weaker plants. If the disease was mildew then it might well appear in the wetter Irish climate first.

An editorial in the '*Chronicle* had noted a 'mysterious disease' affecting not only Verbenas but Petunias, Fuchsias, Deutzias and Lobelias as well. 'It is perhaps worth enquiring whether in some cases mischief is not done by some of the remedies against thrips and mildew... There is so much empiricism in the application of some of the vaunted remedies…'. (18) Bad cultural conditions were blamed too; cuttings struck in too warm a house. Vegetative propagation was the only way then to retain a particular variety. Plants could be raised from spring-sown seed but they wouldn't come true to type. By 1893, *Thomson's Handy Book of the Flower Garden* noted that Verbenas 'are very subject to green-fly and mildew in winter, and these pests must be kept under by tobacco-smoke and flowers of sulphur'. A bigger threat was to come from competition. Thomson's *Verbena* entry concluded: 'The Pelargonium, with its every shade of colour, has very much displaced the Verbenas for beds, but some of the following varieties are very effective…'. Five of Henry's varieties featured among their select list of bedding Verbenas: 'Master Jacob', 'Mauve Queen', 'Conspicua', 'George Peabody' and 'Eclipse'. (19)

Back in the early 'seventies however, Verbenas were still very much the thing. In the spring of 1871 Richard Dean was again singing their praises in the columns of *The Gardeners' Chronicle.*

'That during 1870 the Floral Committee should have given Mr. C. J. Perry four, and his scarcely less successful and worthy floricultural brother, Mr. H. Eckford, three, First-class Certificates for Verbenas, is a pretty good proof of the excellence of the new flowers of the past year. Of Mr. Perry's flowers, Mr. Turner will shortly distribute a batch of 13 varieties; and of Mr. Eckford's, Mr. Keynes will send out a batch of nine varieties'.

The three flowers that Henry received the first-class certificates for, from an entry of nine were 'George Peabody', 'Mrs Dodds' and 'Peter William'. A 'Miss Charlotte Mildmay' received a second-class certificate. 'Mrs Dodds' would have been named after Jane Dodds, William Dodds' wife. 'Charlotte Mildmay' would probably have been named after a baby daughter of the Mildmays of Queen Camel and therefore a niece of Lord Radnor's second, late wife, Judith Anne, who had been a Mildmay.

With 'Mr. Eckford's flowers', Dean continued, 'glass is repudiated; the seedlings are all planted in the open ground, and there they take their chance. In walking round the kitchen garden at Coleshill, one stumbles upon a bed of seedling Verbenas here and there, fully exposed to all the vicissitudes of weather, just as if they were plantations of strawberries. A good habit is a *sine quâ non* with Mr. Eckford, and he has every opportunity of studying the habits of growth, as well as the floral qualities, of his new flowers. It was curious to notice how in different parts of the grounds, kitchen and flower garden alike, seedling Verbenas – self sown – would thrust themselves up in various places, in some instances on spots where they might least be looked for, not excepting gravel paths.

As a matter of course, any flower that shows signs of a fine development is protected and shaded, in case it might be requisite to send it to the Floral Committee, but that is all. About all the plants nearby there appears to be a general robustness of habit, which was, perhaps, not to be wondered at, seeing that a strong habit is of such prime importance in Mr. Eckford's estimation. It was a rare thing to see a plant with a weak, spare habit. The

> seed is sown in March, in pans or boxes, and covered with soil from one half to three quarters of an inch in depth. These are placed in a vinery, the seed soon germinates, and when the plants can be handled, they are pricked off into 24-sized pots, about 50 plants in a pot, and as soon as they begin to draw roots, the pots are placed in a cold frame. The next stage is to dib them out in beds prepared for their reception, advantage being taken of a shower, immediately after which planting out is done. It is no exaggeration to say that they grow like weeds, and after any one has seen how well the plants succeed at Coleshill treated in this way, a kind of uncomfortable feeling springs up, that as a general rule the Verbena gets far too much "coddling".
>
> The difficulty which besets many plant growers, that of keeping Verbenas through the winter, does not appear to trouble Mr. Eckford. He asserts that as a general rule Verbenas are kept far too dry during the winter, and in consequence, many of them die. He gives water sufficiently to keep the plants growing, but avoids a luxurious growth, and they do well. Any named sorts he is anxious to preserve he propagates by taking cuttings at the beginning of September, which are placed in store pots [broad, shallow pots]. About the middle of April cuttings can be taken from these in plenty, in five days they root in heat, and are gradually hardened off in the vineries.
>
> In front of one of the vineries there was a bed of named Verbenas, growing in a hot, dry, shelterless spot, fitted to test their endurance to the last degree'.

'That well-known and successful raiser', [20] Dean called him. Henry was successful by his own efforts but becoming well-known was thanks to Richard Dean. A week later he was writing again. 'Mr. Eckford reminds me that his fine orange-red Verbena Grand Monarch received a First-class Certificate at the Royal Horticultural Society's Great Provincial Show at Oxford last year. I hasten to supply the omission, especially as Mr Eckford thinks it to be the very finest of the set of his raising Mr. Keynes is now sending out…'.[21] Two columns to the right and Mr. Dean was off on a new tack...

> 'The Wild Garden. I thought of Mr. Robinson's book on this subject, and the many excellent suggestions and illustrations it contains, when visiting Lord Radnor's seat at Coleshill, Wiltshire, a few days ago. There is a considerable extent of shrubbery walks on one side of the park, and beneath the trees, where the underwood was scarce, were to be seen very large patches of the common white wood Anemone, A.nemorosa, growing luxuriantly, and flowering profusely. In the immediate neighbourhood of these... were patches of plants, apparently much younger in development. Were these the produce of seed, carried there it may be by the passing breeze?' he wondered.
>
> 'Along the summit of a low wall overgrown with grass', he continued, warming to his theme, 'which forms the boundary of the park, with a ditch on the park side, could be seen a minute growing Myosotis, scarcely rising above the level of the dwarf grass about it, and so profuse of small deep cerulean blue flowers as to form a radiant carpet. What species is this? Mr. Eckford said he had not seen it elsewhere than at Coleshill. Though so minute, its early blooming quality seemed to make it valuable, and I bought away a few plants…'.

'We can't pretend to name any plant without seeing it, least of all a Myosotis' (22) was the '*Chronicle's* editorial put-down. A little thought all round would surely have fixed on the early forget-me-not *Myosotis ramosissima*. But such thought was lacking. Gardeners had taken plants, changed and used them in the garden without due regard to the bigger picture, of how nature herself grows and displays her flowers. The time was right for change, not so much from formal to informal as towards more naturalness in the artifice. And what better way to start than by observing the flowers on the doorstep, or in this case among the grasses on an old stone wall. 'Mr. Robinson' was William Robinson (1838 – 1935) whose book *The Wild Garden* had recently appeared. Robinson, an Irishman, together with Gertrude Jekyll (1843 – 1932) an Englishwoman, now picked up the reins from J. C. Loudon the Scotsman to be the main and appropriately diverse driving forces in nineteenth and early twentieth-century British horticulture.

Robinson now started up *The Garden*, spotting an opening for gardening magazines aimed at amateur gardeners and their smaller urban and suburban patches of ground. One of *The Garden's* first contributors was Richard Dean, riding two horses as it were. Here he is in December of 1871, more didactic than in his *'Chronicle* submissions: 'There is nothing like the open ground on which to test Verbenas. There is nothing more likely to interfere with their well being than coddling them in pots or houses. One of the foremost Verbena raisers of the day, Mr. H. Eckford, of Coleshill, grows all his seedling Verbenas in the open ground; and there is this great advantage about it, that it is a most favourable, and, at the same time, fitting mode of testing the habits of growth of the seedlings. A Verbena without a good free stocky habit of growth is, after all, of but little value'(23). Substance, as well as style, marked *The Garden's* difference to the other magazines. For example, one can't imagine the following from the rosarian the Rev. Samuel Reynolds Hole appearing without distancing comment in the columns of the *'Chronicle*: 'The Cry of the Labourer against Landlord, Farmer and Priest'.

> 'It is affirmed to be the duty of landlords, as owners or builders of houses, to provide dwelling-places in which men and women may live in decency and comfort. It is stated that, in many instances, the stables of the rich are far more carefully, commodiously, and expensively constructed than the habitations of the poor, and that the healthful conditions of the beasts which perish is accounted of more importance than human life…'.(24)

A week after his article in *The Garden*, Dean was back writing in the *'Chronicle*. 'When a raiser of the Verbena exhibits some nine or ten seedlings in the course of a season, and succeeds in obtaining First-class Certificates for five of them, it must be assumed that the improvement of the Verbena goes on steadily and surely from year to year'.

> 'It is Mr. Henry Eckford, of Coleshill, who has this season won five First-class Certificates for seedling Verbenas, and in a batch of seven varieties to be distributed by Mr. Keynes next spring four of these will be included; one named Lady Braybrooke being held over, probably for want of stock'. The other four were 'Lady Edith', 'Lady Gertrude', 'Mauve Queen' and 'Pluto'. Those failing certificates were 'Grand Duke', 'Isa Brunton' and 'Royalty, dark crimson-scarlet, considered by Mr. Eckford to be one of the

most promising bedding varieties he has yet distributed, having a dwarf, compact, short-jointed habit, and being very profuse of bloom: the colour also is likely to stand exposure'.

This description is a good example of Henry's emphasis on the garden qualities of a plant.

'Mr. H. Eckford recommends the early planting out of Verbenas, if they have been sufficiently hardened off in cold frames. The Verbena has a somewhat wiry growth, and if this is made tough by due and gradual hardening in cold frames, it is made capable of withstanding considerable exposure. Even if but little growth is made, owing to the weather being inclement, the degree of harm done to the plants will be small…'.

'The Verbena cannot yet be spared as a bedding plant' [25] meaning that it soon will be. 'Isa Brunton' incidently was named after Henry's younger sister Isabella, married to James Brunton in Stenhouse. Or, more likely, after a daughter they had. For amateur growers, it was now considered that the disadvantages of Verbenas outweighed their advantages. In April of 1872 The *Garden* was cautioning: '...a few years ago the Verbena was a universal favourite, easily propagated... But if we ask any gardener of twenty years experience, he will tell us that the Verbena does not grow nowadays as it used to, and therefore he dare not trust it'[26]. The professionals weren't so easily put off, it seems. In the summer Dean visited Charles Perry's grounds at The Cedars, Castle Bromwich and wrote in the *'Chronicle*: 'Mr. Eckford, Mr. Banks, Mr. G.Smith, and others, have sent out many fine varieties, but Mr. Perry has stood pre-eminent for a long time, and

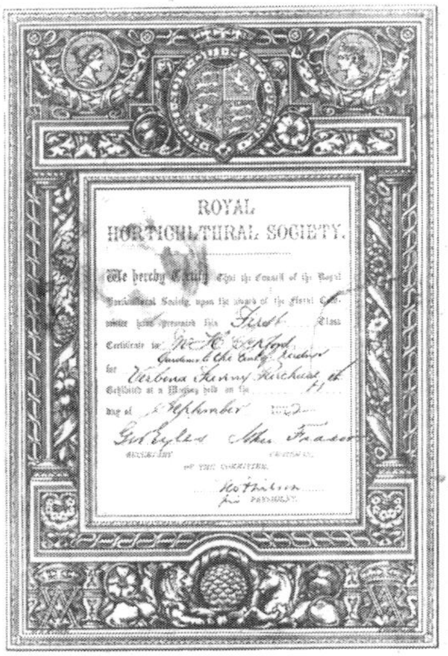

RHS Verbena certificate, 1872, "Fanny Purchase."

> after again seeing his collection this year, and the superb seedlings to be sent out next spring, there can be no question that Castle Bromwich is the home of the Verbena'. [27]

Richard Dean comes across as a good-hearted and open-minded enthusiast, quoting poetry and seeing the best in everyone. He admired Henry's rational horticultural approach while standing in awe of those exhibition blooms of Perry's. The following January, 1873, he was reviewing the new flowers of the previous year; Primulas, Phloxs, roses 'the Queen of Flowers', Penstemons, Antirrhinums, stocks, marigolds... and Verbenas. 'First-class Verbenas, as represented by Certificates, were sparingly produced during 1872, in comparison with previous years. Crystal Palace, Lady Bradford, and Mrs S.R. Hole, of Mr. Perry's raising; and Fanny Purchase, Mrs. Levington, and Star, of Mr. Eckford's raising, were among the fortunate flowers'.

> 'It was a pretty and novel fancy of one of the Persian poets that the gaudy hues seen in the flowers of the Tulip resulted from their having drunk the life-blood of the great, while the modest purple Violets of the field represent the beauties slain by the hand of time. From whence come the varying colours of the flowers he raises is a question the painstaking florist rarely puts to himself. He does not speculate as to their causation, but his heart yearns for new types. And the new flowers of a given year represent the products of Nature, controlled to some extent by the intelligence of men, but mainly the outcome of her own large-heartedness…'[28]

Dean ended by quoting some lines from Emerson's 'Song of Nature'. He could have quoted 'Beauty is its own excuse for being' [29] a line from another nature poem of Emerson's, but with its implied 'leave well alone' he didn't. Knowledge of the pigments concerned in flower colouration was to come, in part, from investigating the genes controlling their development[30]. Whether the painstaking florists wondered or not, the possible answers in Mendel's paper continued to lie unread in the great libraries of science.

Three months later, on Good Friday, Charles Perry died. He was fifty-one. [31] With his main competitor gone, Henry seemed to lose interest in hybridisation. It would be easy, from our own age, to see an intense rivalry among the florists when there was perhaps little. But competition had begun its ascendancy in the middle years of the nineteenth century. Partly it grew out of the industrial age but again it was a feature of the rising middle-classes, that phenomenon that Henry aspired to join and which shaped his life from the 'eighties onwards.

In its redefinition of manhood, families would have fewer children, while regularised leisure-time and rising incomes saw energies directed towards hobbies and sports unmotivated by cruelty and killing. Rule-based amateur sports like athletics, football, rugby and tennis, channelled through club competition, would become some of its principal social manifestations. It contrasted with the non-competitive pursuits of hunting, fishing and shooting which, together with coursing, horse-racing and their associated gambling, combined, if unequally, the aristocracy with sections of the remaining rural proletariat. Sports clubs too, usually male, countered, though aped, elitist gatherings like the London men's clubs of the aristocracy. This isn't to say that people with similar interests haven't always clubbed together. Henry's 'club-mates' were his fellow florists. Their flowers, the elite ones that is, competed for attention at the prestigious Royal Horticultural Society's premises in London's South Kensington.

Henry Eckford at Fifty

~~~

There are just two surviving photographs taken of Henry before he was old. The first, an early portrait in the history of photography, is of him as a young man. The second was taken when he was fifty. The later photo might have been taken for his fiftieth birthday, that milestone in our lives. He was already wearing the full beard that he had in later life. He had probably started it in his thirties, for it was a mid-century fashion, a little dated now, in May 1873 and completely unfashionable when he wore it among the later clean-shaven or moustachioed Edwardian males. It would have had the effect then that my grandparents, born in the 1870s, had on us children when they arrived out of the dark winter evenings into our bright twentieth-century house in the nineteen-fifties and sixties. He with
~~~

his pocket-watch, chain and waistcoat and she in her fox-fur, complete with fox's head and tail. An arresting sight. The fashions of your youth, however, can keep you feeling young and Henry, in his portrait, looks out with confidence, quite Jack the Lad and up for everything life can throw at him. Which, in 1873, was just as well.

It was also in May that year that Charlotte's father Job Stainer died. He was buried in his home village of Queen Camel on the 29th May. Charlotte was three months pregnant at the time but doubtless travelled the sixty-odd miles to Somerset for the funeral. It was just six months later, on Sunday 30th November that Charlotte, Henry's wife for the last sixteen years, died also. An Ellen Didcock was recorded as being present at the death. Cardiac disease was given as the official cause. Henry's entry in their family bible, 'My beloved wife died on the 30 Nov 1873 H.Eckford' [(32)] also shows that she died seven days after giving birth to a still-born son. Childbirth was fraught with danger and seems to have been the most likely cause of Charlotte's untimely death. In the absence of reliable contraception and with both partners fertile, sexual intercourse inevitably led to a steady run of pregnancies which only death or the menopause would stop. Charlotte's age at her death was officially recorded as thirty-nine. Her baptism however, on 28th August 1831 meant that she could not have been younger than forty-two, which age she almost certainly was.

Premature death, its focus on mothers and children, its capricious strikes across the barriers of age and class, made death impossible to ignore. You might think that people became hardened to it but the evidence is to the contrary. Following the death of her husband in 1861, Queen Victoria took solace in the poet Alfred Tennyson's *In Memoriam* ("'Tis better to have loved and lost/Than never to have loved at all"[(33)] is one of its best known lines). Though both popular and beautiful, by the 1870s comfort for the less cultured majority was provided by the new cheap press in the form of little sentimental condolences in prose and verse. These could be neatly copied out into keepsakes and journals, such as that of a young Fraserburgh woman of the time:

Friend after friend departs,
Who hath not lost a friend?
There is no union here of hearts
That finds not here an end.

Were this vain world our only rest,
Living or dying none were blest.
Beyond this flight of time,
Beyond this vale of death,
There surely is some blessed clime
...

Some religious doubt? Another verse went:

Weep not for her – in her spring-time she flew
To that land where the wings of the soul are unfurled
And now like a star beyond evening's cold dew
Looks radiantly down on the tears of this world.[34]

This was a bit like the recently deceased Charles Dickens' *A Child's Dream of a Star*. Henry and Charlotte's families would find many such sentiments to compliment the traditional comforts of friends, chapel or church and the bible. None of this was perhaps enough for Charlotte's mother Sarah who herself died three years later and almost to the day. She was buried in the parish churchyard of St. Barnabas, Queen Camel, on the 28th November, the same day that she and Job had been married there, back in 1822.

The Radnors' would most probably have provided a horse-drawn hearse to convey Charlotte's coffin to the church-yard of All Saints. A wheeled hand-bier would have been too difficult to navigate on Coleshill's steep main road. You can still see Charlotte's gravestone. It's the second on the left past the lych-gate. The elements have all but defaced the inscription:

IN MEMORY
OF
CHARLOTTE ECKFORD
THE BELOVED WIFE OF
HENRY ECKFORD
WHO DIED NOV.30.1873
AGED 40 YEARS

One's actual age obviously wasn't too important.

Born in 1876, Flora Thompson wrote movingly of the simple village funerals of her youth in neighbouring Oxfordshire. The Eckfords weren't poor but neither were they rich, not yet 'middle class', so there would have been many similarities. In 'walking funerals' the mourners followed the coffin on foot. 'Sometimes there would be but three or four mourners, perhaps a widow supported by her half-grown children'. In longer processions, the women would be 'in decent if shabby and unfashionable mourning, often borrowed in parts from neighbours, and the men with black crepe bands round their hats and sleeves. The village carpenter, who had made the coffin, acted as undertaker, and the cost of the funeral, but £3 or £4, was covered by life insurance. Flowers were often placed inside the coffin, but there were seldom wreaths; the fashion for those came later'.

'Very little food would be eaten in a tiny cottage while the dead remained there; evidences of human mortality would be too near and too pervasive. Married children and other relatives coming from a distance might have eaten nothing since breakfast. So a ham, or part of a ham, was provided... a ready-prepared dish which was both easily obtained and appetizing'. As the mourners grew calmer at the funeral meal, 'the gentle persuasion of those less afflicted than the widow or widower or the bereaved parents, for the sake of the living still left to them, should take some nourishment. Then their gradual revival as they ate and drank. Tears would be wiped away furtively, but a few sad smiles would break through, until, at the table, a sober cheerfulness would prevail. They had, as they told themselves and others told them, to go on living, and what greater restoratives have we poor mortals than a good meal taken in the company of loving friends?' At Charlotte's funeral, the Eckford household may have run to sherry and biscuits, 'provided in more prosperous households'. [35]

Understandably enough, Henry now seems to have lost interest in his hybridising, content instead to lose himself in the everyday duties of a head gardener of a large country estate. Fourteen months to the day of Charlotte's death, Henry married again; to one of those 'loving friends', Henry and Charlotte's best friend Emily Gerring. Ten or eleven, Emily was, when the newly-marrieds had arrived in Coleshill. They had watched her grow up. Emily was a thorough local, born in Coleshill in the early hours of 19th April, 1843. Her mother Mary had come from just over the border, from Purton, near Swindon. Mary Plummer was thirty when she'd married the already widowed farmer Godfrey Gerring, a year younger than herself, on New Year's Day, 1842. These were Henry's new in-laws. There were new sister and brother-in-laws too, Maria and Godfrey, both younger than Emily and both at the wedding.

The wedding of Henry and Emily took place in the parish church of All Saints on Saturday 30th January 1875. Henry would have passed by Charlotte's grave on his way into church and passed it again with Emily as they left. The closeness of the living to the dead was easier to acknowledge, perhaps, when people lived their lives closer to the land and the natural rhythms of nature; of life.

~~~

Winter, followed by spring. Henry was lucky. Here he was, fifty-one years old, marrying a woman some twenty years younger. Young, attractive, someone to keep him young and all being well someone to care for him in his old age. Childbirth would be a threat here... Henry may also have been blaming himself for Charlotte's death. The 'great fertility decline' of the 1870s, a change likely to have been felt by Henry, was more than a concern for women in childbirth however. It was part of what the historian Martin Daunton calls 'a shift in male identity' where 'the man with a large number of children was defined as irresponsible and feckless'(36). The practical means must have been a decline in sexual intercourse within marriage, and there were to be no children for Emily. From Henry's point of view he had already had his family and in a man's world the final decision on whether or not to extend it would have been his. Emily may, of course, have been infertile. Whatever maternal desires Emily may have had were destined to be unfulfilled. She would, however, live to a great age, not succumbing until just before her ninety-ninth birthday on 25th March, 1942.

Manhood may now, in the 1870s, be viewed as responsible and respectable but lay open to the familiar charge of Victorian hypocrisy. In London in 1857, 80,000 women, one in every sixteen women living in the capital, had been considered by the British medical journal *The Lancet* to be prostitutes. The numbers were commented on by foreign visitors though perhaps not those from Paris, where, with half London's population, prostitution was thought to be twice London's number. (37) On a wholly conscious level, Henry was reasoning astutely. Godfrey Gerring pops up occasionally as an 'overseer' in the papers of the Radnors'archives in the Berkshire Record Office, and as one of the biggest farmers, features in the commercial directories of the area. A loving relationship Henry's and Emily's may well have been but from Henry's point of view an advantageous one too.

On 27th March, 1875, the ' *Chronicle* thought it opportune to nudge Henry back towards the *Verbena*. 'This once popular plant', it said, 'sharing the fate of many other florists' flowers, has become much more neglected
~~~

than it should be...'.[38] In the August 7th edition there was a long article on 'The Verbena Past and Present'. Of the parents of the new bedding Verbenas, the author, the botanist Leo Grindon (1818 – 1904) wrote: 'The grand appearance of the scarlet Verbenas upon the plains of their native countries is said to be unimaginable. There they are "bedded-out" by nature, or after the same artless and unsophisticated way in which in old England she displays the Bluebell and the wood Anemone'. It was however 'a very different thing to have flowers disposed after patterns that Nature never dreamed of, and would abolish in a twelvemonth were she emancipated'.

Bedding-out had always had its critics, but such prominent criticism here seems to be heralding a change of taste. A perfumer's scent called 'Verbena' was popular, said by Grindon to be 'on a thousand toilette-tables, stored in little scent bottles'. Verbenas had nothing to do with it, however, being apparently 'prepared from one of the fragrant East Indian Lemon Grasses'[39] (*Andropogon* sp.), the odour of which it resembles. Another more odorous plant with *Verbena* - like scent is *Lippia* (then *Aloysia*) *citriodora*, the lemon or sweet-scented *Verbena*. As a shrub it is quite different to the *Verbena*, though of the same family. Interestingly, such scent as the bedding hybrids have is derived from one of its parents, the white night-scented *V.teucrioides*. It is considered to have a soft sweet perfume, rather like Heliotrope. [40] So, the nearer the hybrids were to white, the stronger their scent was. This must have been known about but seems to have gone unreported, perhaps because gardens are rarely visited at night and Show plants are exhibited and judged during the day! Back in 1849, the *'Chronicle* had noted that to gain the scent more strongly you crushed the plant slightly, which released its essential oils.[41]

Another scented-leaved plant that Henry had some later success with was the already popular *Pelargonium*. As we've seen, Henry had raised at least two hybrids from the actually non-scented ivy-leaved *P.peltatum* by 1870. The *' Chronicle* had drawn attention to the zonal types, the 'bedding Geraniums' four years before that. 'The Zonal Pelargonium is generally speaking of free and easy growth', they wrote, 'possessing colour in flowers of almost every hue; it is as well suited for the cottager's window of six panes of glass, as for the Italian terrace of some noble Lord's mansion. What would our gardens be without it?' [42] Henry may have been working on Pelargoniums now but not commercially, for none were mentioned. The main influence towards Pelargoniums was to come later, when Henry had left Coleshill to work for a Dr. William Sankey. By February of this year,

1875, Dr. Sankey had joined the newly formed Pelargonium Society. (43) A keen amateur hybridist, Sankey followed the crowd; Henry, in contrast, tried to anticipate its movements.

Richard Dean's words back in March appear to have had an effect, for in August 1875 it was probably Dean again who was able to report, in *The Garden*, on two first-class certificates for 'Verbena The King (H.Eckford) – This is a bright rosy variety... a fine grower and flowers well'. The other was 'Verbena George Brunning... a rich velvety-purple variety... one of the finest in its class'. Henry must have worked with George Brunning (1830 – 1893), a Suffolk man, after he came down to London. Brunning had emigrated to Australia as a nurseryman in 1853 (44) so it's likely that he and Henry had met at Hugh Low's nursery in 1847. The report concluded: 'A very showy stand of Verbenas also came from Mr. Eckford, gardener to Lord Radnor, Coleshill, Berks. The individual flowers of some of these varieties were nearly an inch in diameter, the colours being scarlet, purple, white, rose and lilac'. (45) According to Henry's own C.V., he thought he'd left Coleshill in 1874. He obviously hadn't!

In September, Dean was able to report on another first-class certificate for Henry, 'Lady Ann Spiers... a large flowered variety... pure white'. Also present at the show was one of Henry's Dahlias, a crimson maroon variety, the colour inviting the name 'Willy Eckford', after Henry's son Peter William. (46) More family connections were on display that month. Edinburgh was hosting an 'international fruit and flower show' at the Royal Winter Gardens, where a prize-winner in the *Dahlia* 'Show' section was 'Mrs Eckford', an 'excellent' new pure white. Another of Henry's winners here was 'Cremorne– ground yellow, heavily suffused lurid crimson'. Among twelve prize-winning 'New Fancy Dahlias' was 'Maggie Gerring – charming in colour, delicate blush, with flakes of lilac deepening into purple; size large, form excellent'.

In person, Maggie Gerring was possibly a new baby daughter for Emily's brother Godfrey. 'Considering the state of perfection that many of the [*Dahlia*] varieties have reached', it was 'really difficult to supersede them, so what we have in the main to look for is new colours in seedlings, and improvements on the existing forms'. Sensibly describing the new Dahlias 'not from the stands in the show-room, but from the open ground where grown' was one known to be Henry's – 'Royal Queen... colour cream, tipped irregularly crimson-purple... a large perfectly formed show-flower'(47).

We now come to the first of Henry's few surviving articles. 'The Verbena as a Hardy Annual' appeared in the February 19th, 1876 edition of William Robinson's *The Garden* magazine. The title is deliberately contentious. The bedding *Verbena – V.x hybrida* – is, after all, a half-hardy perennial. Seed could throw up anything in the way of flower. It wasn't possible to maintain varieties that way then, the varieties that Henry, Charles Perry and others had so painstakingly produced. As now, varieties were treated as half-hardy annuals. Then, the varieties were maintained by overwintered cuttings. Now, they're usually just seed-grown. It was a lot easier for amateurs to grow plants from seed then too, rather than to take and overwinter cuttings in expensive heated glasshouses. It was an attitude that Henry could easily favour, perhaps empathise with. At any rate, he always valued hardiness in a plant. Henry had chosen an amateur gardeners' magazine to speak to the new middle-classes, amateur gardeners with their small urban and suburban gardens.

THE VERBENA AS A HARDY ANNUAL.

In the cultivation of the Verbena as a hardy annual, glass structures and fire-heat are quite unnecessary, and the amateur without any glass whatever may have beds of Verbena quite equal to those in gardens in which ample accommodation is provided. The seed must be sown early, and in the beds in which the plants are to remain throughout the season. It should be sown at the beginning of March. Some time beforehand dress the beds with suitable manure and dig them up to a fair depth, for a deep and moderately rich soil is most essential to the production of a fine display of flowers throughout the summer. The surface soil will require to be broken rather fine, in the same manner as for Mignonette or other annuals sown in spring; after the sowing has been made the surface should be raked over lightly to cover the seed with soil.

The young plants will make their appearance all over the beds by the end of April, and from the first will grow with great vigour, commence to bloom very early, and continue in flower until quite late in the autumn. If any thinning is necessary it should be done before the plants suffer overcrowding, but there should

> be no hurry to remove the plants when they are of a small size. The seedlings will vary more or less in colour; the mixture of colours will produce a most pleasing effect, and the perfume will be most delightful, for nearly every light flower will be as sweetly scented as the blossom of the Honeysuckle. To ensure a thoroughly good display, seed saved from flowers of first-class quality must be sown, and therefore, in purchasing, care must be taken to obtain seed from a strain of Verbenas known to be good. When the beds are required to be filled with any one particular colour the stock must be raised from cuttings, but unless they form part of a geometrical scheme mixed beds have certainly the most attractive appearance.

William Robinson – it was certainly he – wrote a footnote.

> 'The above excellent suggestion is well worth carrying out. Verbenas when growing well and free from disease are so beautiful that one cannot but regret the way in which they have disappeared from flower gardens during the past few years. Beds of good mixed Verbenas are among the loveliest in the choicest garden, and Mr. Eckford's plan not only points out how to raise them more simply and inexpensively than of old, but also how to secure a much more free and beautiful growth and one more free from insect-pests'. (48)

It was sound, practical advice from Henry and uncommercial. Ignore the hybrids and grow from seed! But Henry wasn't a commercial grower. His interest in hybridisation, one feels, was more in the process. As one might, with practice, cook and present a perfect meal. Something to be enjoyed for the moment, not for all time. It's doubtful that anyone took much notice of Henry's advice and it was probably just as well. It was one thing for Henry, gardening on a large scale and picking out the hardiest plants. A few March-sown seeds sown in an average garden, on the other hand, could all die in a cold snap and money would be wasted. Not everyone could then afford to buy bedding plants from a nurseryman. And why should they? What was needed, Henry was beginning to realise, was a genuine hardy annual, one where the hybrids could be maintained through the seed. Incidentally, Henry was the only person to write that the sweet *Verbena* scent came from the lighter-coloured flowers, i.e. those hybrids closest to *V.teucrioides*.

Four years later, when Verbenas were said to be in decline, John Keynes told the *'Chronicle* that bedding Verbenas were still much in demand; he grew 50-60,000 annually. (49) But many of these would have been Henry's varieties, hardier and more disease resistant than most. And Henry was *very* persuasive! No other growers wrote in to back Keynes. It's just possible that Verbenas, though unfashionable, were still popular. But their time under the spotlight had passed. In the 'eighties, the *'Chronicle* was reminiscing: 'Mr. Henry Eckford, in the days when he was a raiser of new Verbenas... Mr. Eckford's practice was to plant his seedling Verbenas out in the open ground on a well-prepared border…'. (50)

'The names of Sankey,... Perry, Eckford, Keynes, and others have been associated with the flower in times past…'. (51) A further mention, in 'eighty-five was followed by a last, in 'eighty-nine: 'Cut Verbenas at Flower Shows. At the larger flower shows held in the West of England, prizes are still offered for stands of these; but the varieties shown are generally of poor quality, and unworthy of the money prizes offered. The fact is, the Verbena, as an exhibition subject, and for other purposes, is a declining flower, and ere long it will be banished from the flower garden. Now, if compilers of schedules would offer prizes instead for bunches of Phlox Drummondii, of which there are now many fine varieties, and also for bunches of different varieties of Sweet Peas, the number having been considerably augmented of late, the gain would be very great indeed. Phlox Drummondii and Sweet Peas are charming subjects for show purposes and they have high claims to recognition in schedules of prizes'. (52)

David Thomson, Henry's fellow Scot, was more accommodating in the 1887 edition of his *Handy Book of the Flower Garden*. 'The older sorts are still the best for bedding' was his opinion. His recommended varieties 'suitable for mixed borders' included five of Henry's: 'Conspicua', 'Eclipse', 'George Peabody', 'Master Jacob' and 'Mauve Queen'. 'Conspicua' and 'Eclipse' also made it as two of the five best crimson and 'Master Jacob' as one of the two best maroon bedding Verbenas. That 'the older sorts' were better indicates that the cause of the Verbena's decline lay in bad horticultural practice. The popular 'Mrs Pince's' Muscat grape had also suffered thus. David Thomson again, writing in his magazine *The Gardener* in September 1882 spoke of the variety 'regaining its original constitution, this being impaired by over-production'. (53) Meaning that, people were so keen to propagate it they chose any stalk, however weak and inferior, as cutting material. It led to weak and inferior plants.

With Verbenas it's likely that the choice of seedlings was over-determined by the flower, and the plant's vigour neglected. Attacked by thrips and mildew, they succumbed. Henry, professionally trained, knew the danger. Amateur growers like Charles Perry did not. Today the Verbena has happily regained its rightful place, being a particularly common and cheerful sight in window-boxes and hanging-baskets.

Returning to the 1870s, a short piece in *The Garden* of 24th May 1876 may have caught Henry's eye. It was on a relatively neglected hardy annual, the sweet pea. 'Sweet Peas in Pots. – These are now seen in abundance in the London flower market, and are well worthy of some attention in private gardens. If well grown they are very useful and welcome, especially earlier in the year, before Sweet Peas bloom out-of-doors'.(54) It was reminiscent of the puff given to another plant, twenty-two years before: 'Pot culture of the Verbena. The Verbena is unrivalled and ought to be more generally grown in pots... especially now that the numerous varieties are so much improved...'(55) It was about now however that Henry left Coleshill, and he may have missed the sweet pea article, having other things on his mind.

It was now twenty-two years since Henry had arrived at Coleshill. A lot had happened since. He'd married and had six children; in an age when head gardeners were important and respected he'd gone one better, winning a name for himself with his hybridising work... And Charlotte died... and he'd married again, young Emily Gerring, while his first wife lay, a daily reproach, in the churchyard in the middle of the village... Perhaps too, Henry felt the winds of change moving around the old aristocracy. They were belt-tightening now (though don't feel too sorry for them – '80% of the land in the United Kingdom in 1873 was owned by fewer than 7,000 people'). (56) For Henry it would have been a gardener gone and not replaced, a piece of equipment needed and not bought. Leave the past behind, yes. Coleshill had been good to him but... a new wife, a new start. That was the idea. But where?

References to Chapter Six

(1) Gregor Johann Mendel, Versuche über Pflanzen-Hybriden, Verhandlungen Band 4, *Der Naturforschende Verein in Brünn*, (1865), pp. 3 – 47.

(2) Part 1, RGO 252, letter from Gregor Mendel, (27-10-1866), The Greenwich Observatory Archives, Cambridge University Library.

(3) Dr. Vladislav David, ed., *Brno*, (guidebook), (c1972), p.6.

(4) Bill Bryson, *A Short History of Nearly Everything*, (2003), pp.347-348.

(5) Ibid, p.344.

(6) Ralph Waldo Emerson, *English Traits*, 1856, (1902), p.46.

(7) Ibid, p.129.

(8) Ibid, p.135.

(9) Ibid, p.144.

(10) The following select list of works were consulted on Mendel: *Accession Registers of the Linnean Society*, 1865, 1866, 1867, (3 vols.); The Royal Society, compilers, *Catalogue of Scientific Papers*, Vol. 8, 1864 - 1873, (1879), p.378; *The Mendel Journal*, Nos. 1 - 3, (10 - 1909 to 9 - 1912); Jacob Bronowski, *The Ascent of Man*, (1974), pp. 379 - 388; *Dictionary of Scientific Biography*, (1974), p.283; Vitěslav Orel, *Mendel*, (1984); Robert Olby, *Origins of Mendelism*, 2nd. edn. (1985); R.N.Jones, A.Karp, *Introducing Genetics,* (1986), pp. 58- 60; Anna Matalová, *Mendelianum,* (guidebook), (1990); Robin M.Henig, *A monk and two peas, (2000).*

(11) 'Garden Memoranda', *The Gardeners' Chronicle*, (31-12-1870), pp. 1734 - 1735.

(12) Ray Desmond, *Dictionary of British and Irish Botanists and Horticulturists*, 2nd edn., (1994), p.745.

(13) *Curtis's Botanical Magazine,* (1-12 -1862), no paging.

(14) *Curtis's Botanical Magazine,* (1-3-1864), no paging.

(15) J.A.B. Heslop, unpublished mss. on the history of the park and gardens of the former Coleshill House, (c .1990), Berkshire National Trust, p.11.

(16) Shirley Hibberd, *The Amateur's Flower Garden,* (1871), pp. 90, 92.

(17) 'Home Correspondence', *The Gardeners' Chronicle*, (11-8-1866), p.756.

(18) Editorial, *The Gardeners' Chronicle,* (11-3 -1865), p.220.

(19) David Thomson, *Thomson's Handy-Book of the Flower Garden*, (1893), pp.36, 38.

(20) 'Florists' Flowers', *The Gardeners' Chronicle*, (15-4-1871), p.486.

(21) 'New Verbenas', *The Gardeners' Chronicle*, (22-4-1871), p.517.

(22) 'The Wild Garden', *The Gardeners' Chronicle*, (22-4-1871), pp.517-518.

(23) 'How to Grow Verbenas from Seed', *The Garden*, (2-12-1871), p.27.

(24) 'The Cry of the Labourer', *The Garden*, (6-12-1873), pp.468-469.

(25) 'Florists' Flowers', *The Gardeners' Chronicle*, (9-12-1871), p.1588.

(26) 'Loam as a cure for the Verbena "disease"', *The Garden*, (6- 4-1872), p.441.

(27) 'Florists' Flowers', *The Gardeners' Chronicle*, (31-8-1872), p.1169.

(28) 'The Florists' Column', *The Gardeners' Chronicle*, (25-1-1873), p.122.

(29) R.W. Emerson, 'The Rhodora', line 12, in G.Moore, ed., *The Penguin Book of American Verse*, (1979), p.73.

(30) L.J.F.Brimble, *Intermediate Botany*, (1962), p.410.

(31) 'Obituary', *The Gardeners' Chronicle*, (19 -4-1873), pp.547 - 548.

(32) Letter from Harry Eckford to Bernard Jones, c.1980.

(33) Alfred, Lord Tennyson, *In Memoriam*, verse XXVII, lines 15-16, 1850,(2004), p.24.

(34) Robina of Fraserburgh's Journal, 1869 - 1870. I am grateful to Francis Anderson and Catriona Nurse for allowing me to quote from their unpublished family heirloom.

(35) Flora Thompson, *Lark Rise to Candleford*, 1945, (1948), pp.465 - 466.

(36) Martin Daunton in C.Matthew, ed., *The Nineteenth Century*, (2005), p.69.

(37) Reader's Digest, *Life in the Victorian Age*, (1994), p.25.

(38) 'The Culture of the Verbena', *The Gardeners' Chronicle*, (27-3-1875), p.408.

(39) 'The Verbena: Past and Present', *The Gardeners' Chronicle*, (7-8 -1875), pp.162 - 163.

(40) Roy Genders, *Scented Flora of the World*, (1978), p.473.

(41) 'On the odours of plants, and the modes of obtaining them', *The Gardeners' Chronicle*, (10-11-1849), p.709.

(42) 'Florists' Flowers', *The Gardeners' Chronicle*, (8-12-1866), p.1168.

(43) 'The Pelargonium Society', *The Gardeners' Chronicle*, (20-2-1875), p.244.

(44) Ray Desmond, op.cit., p.112.

(45) 'Societies and Exhibitions', *The Garden*, (21-8-1875), p.166.

(46) 'Societies and Exhibitions', *The Garden*, (4-9-1875), p.208.

(47) 'Gossip About Plants', *The Gardener*, (1 - 1876), pp.41-43.

(48) 'The Verbena as a Hardy Annual,' *The Garden*, (19-2-1876), p.174.

(49) 'The Verbena as a Garden Plant', *The Gardeners' Chronicle*, (10-1-1880), p.50.

(50) 'Verbenas from Seed', *The Gardeners' Chronicle*, (12-6-1880), p.747.

(51) 'Florists' Flowers', *The Gardeners' Chronicle*, (21-6-1884), p.799.

(52) 'Cut Verbenas at Flower Shows', *The Gardeners' Chronicle*, (7-9-1889), p.279.

(53) 'Notes of a Horticultural Tour', *The Gardener*, (9 - 1882), p.406.

(54) 'Sweet Peas in Pots', *The Garden*, (24-6-1876), p.588.

(55) 'Pot culture of the Verbena', *The Gardeners' Chronicle*, (12-8-1854), p.519.

(56) Colin Matthew, *The Nineteenth Century*, (2005), p.16.

Chapter Seven

'After a short interval', wrote Henry in his *Curriculum Vitae*, I 'accepted an appointment to Dr Sankey at Sandywell, Glos...' (1) Sandywell Park was a lunatic asylum, so Henry's reticence about his recent whereabouts could mean that he'd become a patient of Dr. Sankey's, before becoming his head gardener. Charlotte had died just two years before and he may now have had a breakdown of some kind. Henry's name, however, doesn't appear among Sankey's surviving patient lists and other possibilities offer themselves.

'Did Harry tell you that grandfather Henry was Head Gardener to the Duke of Buccleuch in Scotland?' wrote Dorothy, one of Henry's granddaughters, in a letter to Bernard Jones in 1979. (2) The Buccleuchs owned and still own land and a number of properties in the Scottish borders. Bowhill House is the family seat, six or seven miles, as the crow flies, from Traquair. In 1876, Henry's mother was almost certainly the 'Isabella' Eckford who was dying from 'cardiac dropsy' at Traquair Mill, at that time part of the Traquair estate. Despite having been left a widow in her early thirties, Isabel had never remarried. Once her children, or those that survived, had all grown up she left Stenhouse. It seems she went south to the borders. Despite irregularities in the census evidence, she seems to be the Isabella Eckford who in 1861 was farming twenty-four acres at Traquair Mill, employing two men, one woman and a boy. She later became a housekeeper. If this had been at Traquair House, she would have had a large, light and airy room of her own at the rear of this famous old house.

The surnames Eckford, Tait and Brunton crop up regularly in the census returns and parish registers for Traquair at this time. Henry's grandmother on his father's side was born Jean Tait (Henry's older sister was Jean Tait Eckford) and you will remember that his younger sister Isabella had married James Brunton. It would appear that these particular Tait, Eckford and

Brunton clans all came from this area. You might expect the Eckfords to, with Eckford itself being a village in the borders. Our Isabel seems to have come to be with the relatives of her in-laws. Returning to her own family roots, the Pirries and Willoxes in Aberdeenshire would have put her too far from her children in the Lothians. Following a six-month illness she died, aged eighty-one or eighty-two, at 10am on Saturday 12th August, 1876.

Despite a lack of concrete evidence, a return to Scotland seems the most likely answer to Henry's whereabouts during his 'short interval' between his Coleshill and Sandywell appointments. Why he should choose to be so mysterious about it is a mystery in itself. The most likely explanation is that his mother had gone south partly to be with the progeny of her late husband James, the results of affairs he had had here before marrying Isabel and making an honest man of himself. Such dirty linen, Henry must have thought, was best washed away from the public gaze.

One further possibility remains, via a casual footnote in a local Kent newspaper of 1906. 'The late Mr Henry Eckford, who, perhaps, has done more than any other man in this country towards the improvement of the sweet pea, was at one time head gardener at Bromley Palace to the late Mr. Coles Child'.[3] Unfortunately, no supporting evidence has surfaced. Mr. Coles Child had died in 1873. It's possible that Henry went to Kent, working for Child's descendants, though it seems a little odd to leave the Cotswolds, only to return to them. Also, Henry never arrived anywhere 'out of the blue'. His English employers were mostly fellow Scots or obvious contacts made through the trade. Or, as we shall see, a florist acquaintance. Henry usually named plants after his employers and no 'Coles Child' is known to exist. What is more likely is that Henry's sweet peas were being grown there, trialled perhaps, and that Henry came to inspect them as he did elsewhere. This may have given rise to the thought that he'd worked there which, technically, he would have done.

Bromley Palace still exists. It is now owned by the London Borough of Bromley, former Parks employers of mine, and part also of Bromley Civic Centre, in whose registry office my brother was married. It is a pity, therefore, that I can't make a stronger case for it!

~~~

Sandywell Park lies twenty miles north of Coleshill, near the village of Dowdeswell, a 'grey village on a green slope of the Cotswolds' [4] about five miles east of Cheltenham. Sandywell Park is so named after its well or spring and because the park, c.1588, was there before the mansion, or at
~~~

Sandywell Park, near Cheltenham, Gloucestershire

least of the present one, built around 1704. The house has an interesting history, being, it is said, owned by the grandson of Sir Walter Raleigh, before being given to that man of letters Horace Walpole. 'A square box of a house, very dirtily situated'(5) was Walpole's opinion. Although wings were added later he had other properties and lived elsewhere. In 1848 it became a lunatic asylum for the monied upper or professional classes. So it was when Dr. Sankey took it over in 1865.

Dr William Henry Octavius Sankey Esq., M.D. was, wrote Henry, 'an ardent breeder and lover of Florists Flowers and who has himself raised some very good Verbenas and was at that time a member of the Floral Committee of the R.H.S. and a great admirer of the fine varieties of Verbenas it had been my good fortune to originate while at Coleshill…'. (6) An ideal employer in fact. A fellow enthusiast, if an amateur one, a man likely to indulge Henry in what had become the passion of his life.

'Sandywell Park is highly ornamental', informed the local directory of 1879, 'and about 150 acres in extent, with avenues of beeches and limes. The soil is clay and gravel; subsoil, clay and gravel'.(7) The estate itself appears to have contained no agricultural land, being mostly a deer-park or 'pleasure ground' to the front and sides of the house. In front of the house, a broad avenue of trees indicates that the original entrance to the house was here, westwards, and a narrower sweeping serpentine path from the north-west indicates a mid-eighteenth-century change. By Henry's time, the main entrance was a utilitarian one from the north-east, via the house's new gas-works. Behind the mansion, to the east and south, was the three-acre-odd flower garden, beyond which lay a slightly smaller, one acre plus, walled kitchen garden.

The gardener's home, Sandywell Park.

In 1923, the estate was up for sale. It is likely that it was little different from Henry's day, for there wasn't much 'old money' about now for major changes. The acreage, at 118 acres, shows that part of the park must have been sold for needed income. The sale particulars included two entrance lodges, a gardener's cottage which 'is stone-built and stone-tiled and contains living room, kitchen, three bedrooms, &c. Adjoining is a laundry with water laid on'. The kitchen garden, walled on all sides, had in its centre a small pool supplied by a spring 'never known to fail'. Many 'choice fruit trees' were noted, along with forcing pits, a fruit house, tool-shed, glasshouse, forcing house and an orchid house. (8) Four glasshouses – and there were four glasshouse ranges in Henry's time.

A major upheaval during Henry's tenure was the arrival of the Gloucester to Banbury railway, part of a longer line to run from Swansea to Newcastle-upon-Tyne. Between 1873 and 1881[9] a tunnel and cuttings were dug through the north of the estate, affecting both entrances to the house. A station appeared just south of the estate. Unfortunately for Henry, the line wouldn't be completed until 1887, well after he had left. Even then it was a mostly goods service, taking iron ore from Banbury to Swansea and returning with coal. The only passenger service would be a once daily Ports Express. So something like the old Gloucester-London stage-coach, which had stopped at the Frogmore Inn at nearby Shipton, would have been welcome, whether any such still ran or not.

Locally, most people, being poor, went by foot or got lifts in carts. The better off had their own horse or pony-driven private carriages; the less well-off, a goat chaise. Those rich enough or those who could claim travel expenses through work took one of the taxis of the day. The 'Commissioners in Lunacy' inspectors visiting Sandywell in 1881 did, spending 18/- on a round trip from the Queen's Hotel, Cheltenham; 14/- for the carriage and pair of horses and 4/- for the driver. [10] This bill, being found among the Sandywell estate papers, makes it look as if Dr. Sankey had to pay for it. Let's hope he didn't have to foot their stay at that grand hotel too.

Speaking of feet, as it were, Henry had to find his in his new environment, and may have preferred dealing with the hundred and one tasks familiar to head gardeners than setting aside time for his hybridising work, passion though it was. He had had something of a break from it, and it takes time to re-engage. Besides, he had unique problems to consider here. One of the unfortunate patients 'threw herself down in a flower bed in the garden', an August, 1880 Commissioners' inspection report noted, adding, gravely, 'which had been planted out'.[11] 'Planted out' sounds like bedding. One or two plants to replace, a bit of tidying up, no great problems really. If it was a common occurrence the gardeners would have had to forget formal planting and think more along the lines of informal designs and resilient plants.

By 1881, Dr. Sankey's brother W.Arthur C.O. Sankey, destined to die, just thirty-four, in 1886, [12] was joint superintendent at Sandywell, along with his father. Arthur appears there in the 1881 census, along with his wife Annie, their baby son Philip and a brother of Arthur's. There were twenty-five 'inmates' or patients (nine men and sixteen women) and twenty-three 'officers' (seven men and sixteen women). An almost one-to-one staff/patient ratio, which makes it look as if some of the 'officers' were in fact domestic servants, particularly as none are mentioned in the census. Working in the grounds, the gardeners probably had more company than they needed, yet doubtless had their favourite patients to talk to. In the circumstances such exchanges would have been on an even footing, and more welcome that that with the so-called officers, whom, in an age of servant hierarchies, one suspects they found distant.

Henry and Emily are there, in that April 11th, 1881 census, though with none of their children, mostly grown up now. 'Lottie' (Charlotte Elizabeth) and 'Willie' (Peter William), thirteen and eleven respectively, may have been away with aunts and uncles in Somerset, or with Emily's relations in Berkshire. In later years, Henry had his grandchildren staying or living with him. Now, the Eckfords had a 'visitor' staying or living with them. This was Emily's younger sister Maria, thirty-six and unmarried in that mid-Victorian 'dependant daughter' tradition. Her occupation? 'Lady Help'.

Whatever Henry was up to at Sandywell, it attracted no comment in the horticultural press. What there was was mostly reflection on the past. Henry's Show Dahlias 'Cremorne' and 'Royal Queen' were still well-spoken of. His old Verbenas too were occasionally alluded too but without enthusiasm now. ('Mr Henry Eckford, in the days when he was a raiser of new Verbenas...').(13) There seemed to be a lull, a waiting for the next 'new thing' to arrive. In 1878 a '*Chronicle* article on the old traditional sweet pea mentioned, along with the old 'white' and 'Painted Lady' varieties, some new ones; a Suttons' 'Butterfly' and a Carters' 'Violet Queen'. There was mention of a 'Scarlet Invincible' too.(14) Perhaps the seed companies were onto something...

One new thing that had arrived to make a mark in horticulture, and more than a mark elsewhere it has to be said, was spoken of with awe in the pages of *The Gardener* for November, 1879: 'The Telephone in Horticulture. Messrs Dicksons & Co., Nurserymen and Seedsmen, Edinburgh, have established telephonic communication between their seed warehouse in Waterloo Place and their Pilrig Park Nursery, Pilrig Street – a distance of fully a mile. The instrument they have adopted is Crossley's Patent Transmitter, which is one of the best for using in towns... It must prove highly advantageous to Messrs Dickson & Co., bringing as it does the two branches of their business into instant communication. So far as we know, this old established firm is the first in the trade that has taken advantage of this recent invention'.(15) Other than plants, anything new like this had to be tried and tested before Henry used it. Years later, Henry ran a successful nursery business without a telephone, the firm only acquiring one after his death, in 1909.

Another arrival in 1879 was the electric light bulb. Add these two profound technological advances with other previous and current mid-Victorian events: Charles Darwin's *Origin of Species* (1859), challenging religious acceptance; the flight from the countryside, urban expansion and the rise of suburbia; mass emigration; universal education (the 1870 Elementary Education Act) to name a few... It's hardly surprising that reactions occurred, desires for fixed points in uncertain seas. One now detects a 'back to nature' movement which, horticulturally speaking, was a nostalgia for cottage gardens, old-fashioned flowers and home-grown fruit and vegetables. The 'cottage garden', in fact, was destined to become more common and popular than it had been when Britain's population was primarily rural. Witness their depiction in the over-romanticised scenes of artists like Miles Birkett Foster, Helen Allingham and their school, popular from the 1860s through into the Edwardian era.

Among more serious artists there was a vogue for vegetable garden subjects in countryside browns and greens. The style – of narrow focus, limited palette and high horizon found widespread popularity among European and European-influenced American artists from the 1860s to the 1880s. Consider for example the girl among cabbages in the Scot James Guthrie's 'A Hind's Daughter' (16) or, cabbages again in 'A Flemish Kitchen Garden' by Henri de Braekebeer. (17) Influential gardening books like William Robinson's *A Wild Garden, Hardy Flowers* and *The English Flower Garden* all appeared between 1870 and 1883.

Reactionary romanticism and a preference for native or long-established plants was a mood whose outlook could be narrow and insular, a mere banging of the national drums. It has modern resonances with politicians, then as now, quick to harness it. Benjamin Disraeli had written in his novel *Lothair* of 'all those productions of nature which are now banished from our once delighted senses: huge bushes of honeysuckle, and bowers of sweetbriar and jessamine clustering over the walls, and gillyflowers scenting with their sweet breath the ancient bricks from which they seemed to spring... banks of violets which the southern breeze always stirred, and mignonette…'. (18) Good old British names, with just a hint of exotic climes. When *Lothair* was published in 1870, Disraeli became Prime Minister of a new Conservative administration.

Not to be outdone, the Liberal Party leader W.E.Gladstone spoke in Wales at the Hawarden Horticultural Society's annual show in the cold, wet summer of 1879. He thought that we could grow more of the fruit and vegetables, onions say, that we currently import. Anticipating increasing mechanisation, he extolled the virtues of 'garden spade cultivation'. [19] The 'Chronicle responded with a leading article. On climate alone, it argued, foreign food has the advantages at particular periods of every year, more so in bad years (like the present one). There was a consumer preference for Spanish and Portuguese onions in the autumn, both for size and flavour. [20]

Superior, rational arguments, the editors must have felt, but the 'Chronicle was swimming against the tide. Out of office at the time, Gladstone was returned to power the following year. As we've seen, this is an issue that has reappeared. Gladstone is in tune with David Thomson (see chapter four) and they both chime with current concerns to reduce energy demanding imports by becoming more self-sufficient and by being content to enjoy fruit, flowers and vegetables in their season.

In a bad year, it was more than the plants that suffered. Premature human deaths have long since receded from their mid-Victorian heights, as too have higher deaths in winter. They were both common occurrences until well into the twentieth-century, however, and all were subject to them. George, Henry's second eldest son had become a clerk. Whether this was here at Sandywell, Gloucester or further away, by October 1879 he was at Sandywell, for he died here on Friday the seventh of November. *Phithisis pulmonalis* was the official verdict; lung disease, pneumonia probably. And probably long-drawn out too, which the cold wet summer did nothing to help. George was buried in the churchyard of St Michael's church, Lower Dowdeswell.

Arriving there now, from the busy A40, an avenue of trees surrounds you on a quiet road and time shifts back a gear. You take the entrance to the churchyard by the south-western gate and go down the steps. By the path near the entrance is the usual Eckford burial spot and here the gravestone of George Eckford is clearly visible, the second on the right. The village of Whittington was actually slightly nearer Sandywell but with a slightly less imposing church it appears to have been avoided. Also, the church there is called after Saint Bartholomew, to whom Henry seems to have had an aversion!

George Eckford's headstone

~~~

The Sankeys were tenants, not owners, of Sandywell and in 1882 they were obliged by the owners to leave. There was another place available near Shrewsbury. It would mean leaving the Cotswolds, Emily's birthplace and where Henry had lived for most of the last twenty-eight years. It was one thing to move about when you were young, quite another at the age of fifty-eight, nearly fifty-nine. Emily might have agreed to move, and she might not, but wives had no power beyond their tongues to object. What probably clinched it for Henry was that so long as he could continue his plant breeding, then, well, seedlings grew much the same in one county as another. His focus was plants, and different people and different landscapes could filter through to his consciousness as they might.

On 31st March, 1882, the Lunacy Commissioners were reporting that 'some of the furniture of the establishment has been removed to Boreatton Park, the new licensed premises which have been taken by Dr. Sankey'. (21) This may have been effected initially by a three-horse furniture van into Gloucester. The hundred miles north-west to Baschurch would have been via Shrewsbury on the Great Western Railway, Baschurch already having its own country railway and telegraph station, on the Chester line. It took a long time to move everything and everyone but eventually it was done. On 13th October 1882, William Sankey could finally sit down at his desk in his new study at Boreatton Park and send off a letter to Gloucester's Shirehall on his new headed-notepaper. 'All the patients remaining under my care have been transferred from Sandywell Park to this House…'.(22)

During this time of upheaval, Charles Darwin died in April at his Kent home at Downe. Henry never named a flower for him, not even a 'Mrs. Darwin'. For another British icon, General Charles Gordon, who was to be killed in the Sudan three years later, Henry had long since named one of his 'fancy' type Dahlias. Following his death, he named a (short-lived) sweet pea for him too. The two men's history make interesting comparisons. Darwin had often had letters in *The Gardeners' Chronicle*, happy to cross between botany and horticulture. The deference to him in the '*Chronicle* is palpable and his stature since his death is only the greater. The standing of icons of Empire like Gordon was also high until the Empire that sustained it dissolved. A loss of faith in scientific enquiry would also, one supposes, see Darwin's name become one half-remembered, half-forgotten. Darwin has a few flowering plants named for him. You will remember the Australian genera *Genetyllis* that was renamed *Darwinia*. There are the popular
~~~

Darwin tulips introduced in 1886, and the later Darwin hybrids in 1943. There is the Brazilian *Abutilon darwinii* and the lovely Chilean evergreen shrub *Berberis darwinii* which Darwin himself discovered; especially lovely when it's allowed to grow unsheared. Among the modern 'Old English' roses there is now a yellow 'Charles Darwin' variety. Not an over-generous list, really.

~~~

The mansion of Boreatton Park had been built, self-admiringly, on top of a hill between 1854 and 1857. Like Sandywell, it was named after the park that was there before the house. It stands rurally, beyond Baschurch and just west of the village of Stanwardine in the Fields. Views west to the Welsh mountains and south to the Shropshire hills would have put Henry in mind of the Lothian landscape of old. Northwards and less familiarly is Baggy Moor, a flat boggy area whose drainage ditches run into the nearby River Perry. In front of the north side of the house and adjoining it were the stables, a cool spot for the animals. What must have been new kitchen gardens and glasshouses were some distance north again of the stables. Always large, their size here was four times that of the mansion's ground area. They don't show up on the Ordnance Survey map surveyed for between 1875-1880, but they do on the 1902 one. Either Dr. Sankey had requested the owners to have them in place before they all moved in, or they were put in under Henry's directions after 1882. You feel certain that Henry would want to be there, supervising things, and construction during 1882 would have involved him in a lot of to-ing and fro-ing between Sandywell and Boreatton. So the years 1883 or 1884 seem better bets.

The house faced east, with a courtyard in front and a driveway running out through parkland. Behind the house, a terraced lawn looked out over the pleasure-gardens and the deer-park beyond. At the south end of the house was a parterre which here were geometric flower beds, designed, as a 1989 guide to Boreatton reminds us, 'to be looked down on from upstairs [bedroom] windows, [a Renaissance practise revived by the Victorians], as well as providing colour and paths for a short walk in the fresh air without going too far afield'. (23) Out in the deer-park were 'fish ponds', lakes really, old marl-pits. One had a boat-house, the ones without, swans. So, fish for the house and for recreation boating, swimming and, in a cold winter, ice-hockey and skating.

One likes to imagine the facilities being open to all, the place being rented and no living-in owners hogging it all. But the servant pecking
~~~

order would have ensured that someone missed out. Boreatton Park was run as a stately home with contemporary furniture and decoration, potted ferns, Aspidistras and stuffed wild-animal heads, the last being the property of the owner. As well as a matron and nursing staff there was said to be a full complement of servants, with 'an elegant butler' the guidebook tells us, who 'must have looked very impressive in his long tail coat and starched shirt when he rang the gong for dinner. This was a very formal affair, for... the patients were from the upper classes, and even under these strange circumstances, everyone wore evening dress for dinner'. (24) (The more surprising thing about evening-dress, if *The Diary of A Nobody* is any guide, was its occasional appearance, as 'full-dress' at the domestic dinner-parties of the lower-middle classes. (25) 'Everyone [in England] is on his good behaviour, and must be dressed for dinner at six' was Emerson's view too). (26)

Henry should now have had nothing to worry about. The grounds – formal gardens, shrubbery, deer-park, kitchen gardens to come – all run-of-the-mill. The place – similar to Sandywell. His employer – the same fellow enthusiast he'd worked for at Sandywell. As for the district, well, Shropshire was new but Staffordshire wasn't and Trentham was only thirty miles away. Henry may have made a mental note to look up old friends there from thirty years before. George Fleming had recently died but there would be one or two old faces he'd remember. When you're nearing sixty, thirty years isn't such a long time. Henry would have been quite at ease if it wasn't that his youngest daughter, Charlotte Eliza, had been taken ill since just after her fourteenth birthday. She died, at home, on Saturday 4th November of that year, 1882. It was the same lung disease, akin to pneumonia, that had killed her brother three years before and almost to the day.

Charlotte Eliza Eckford's grave

Charlotte, George and now little Charlotte Eliza had all died in the month of November. How Henry must have dreaded the arrival of autumn. Charlotte Eliza was buried in the churchyard of All Saints, Baschurch. An English girl with a Scots name, lying among mostly Welsh names in this English border churchyard. Like George's grave in Lower Dowdeswell, Eliza's plot is just on the right as you pass through the, here, south-eastern gate. Her gravestone stands, or rather lies now, following nearby tree removals, next to an old yew. This tree, incidentally, is said to be older than the church, which indicates that the site was already one of worship when the Normans came and built on it.

In *The Beauties*, a contemporary short story of Anton Chekhov's, the action partly revolves around a girl at a wayside railway station. One thinks of Charlotte Eliza and the railway station at Baschurch.

> 'Standing at the [carriage] window talking, the girl, shrugging at the evening damp, continually looking round at us, at one moment put her arms akimbo, at the next raised her hands to her head to straighten her hair, talked, laughed, while her face at one moment wore an expression of wonder, the next of horror, and I don't remember a moment when her face and body were at rest. The whole secret and magic of her beauty lay just in these tiny, infinitely elegant movements, in her smile, in the play of her face, in her rapid glances at us, in the combination of the subtle grace of her movements with her youth, her freshness, the purity of her soul that sounded in her laugh and voice, and with the weakness we love so much in children, in birds, in fawns, and in young trees.
>
> It was that butterfly's beauty so in keeping with waltzing, darting about the garden, laughter and gaiety, and incongruous with serious thought, grief, and repose; and it seemed as though a gust of wind, blowing over the platform, or a fall of rain, would be enough to wither the fragile body and scatter the capricious beauty like the pollen of a flower'. (27)

~~~

On the plant-breeding front, Henry had been quiet at Sandywell. But he had been busy. It was difficult to anticipate new fashions in flowering plants. Fruit and vegetables were a safer bet. What was going to be popular was
~~~

what would grow quickly and easily in the new suburban gardens. That line of thought favoured vegetables. The most popular of these were the legumes - peas and beans. Especially peas, the 'prince of vegetables'. One or the other, anyway. Then, he could gamble with flowers. He could pick on one or two or he could pick a number. The mood of the times was against more exotic novelties. Sweet peas had been popular in Scotland. He'd always grown them there but apart from the commercial cut-flower trade you didn't hear much about them. They were difficult to hybridise, which kept them from the limelight. Hardy though, easy to grow; no greenhouses, no pricking out, no growing on. No real pest and disease problems either. Just right for an amateur and just right for a small garden. It was still a gamble. Not that that mattered. It wasn't a risk as he was employed. If it all took off though, he could go off on his own, run his own place... He'd never thrown caution to the wind and he wouldn't, couldn't start now. He'd experiment with a few others too. Something new out of something old.

The pansy was an old Scottish favourite; Dr. Sankey's too. Here in the south the hotter and drier climate was said to be against it. But the common pansy or heartsease, *Viola tricolor* grew everywhere, especially in the cornfields where it was protected from the wind and the midday sun. These amateur gardeners would just have to replicate that. It shouldn't be too difficult for them. Pansies were naturally variable, they lent themselves to hybridisation. Not that he was the first to think along those lines, and the last thing he wanted to do now was to pick on plants that others were hybridising. It was time to lead now, not follow. Henry really needed a crystal ball. By 1882 the picture was clearer. Henry had begun work on the garden pea together with his calculated gamble with sweet peas soon after his arrival at Sandywell, in 1879.[28] He'd kept very quiet about them and was now reaping the benefits. Well, age and experience has to count for something!

'Pedigree Sweet Peas. Under this heading Mr. Henry Eckford, gr. to Dr. Sankey, Boreatton Park, Baschurch, has forwarded a collection of new varieties from seed after carefully crossing certain varieties. They are all very handsome, and, on the whole, distinct; but another season's culture against the best of the varieties already grown is needed to demonstrate that they are distinct enough to be denominated new varieties. Sweet Peas are decidedly sportive in character, and time is required to ascertain and fix the characteristics of a new form'. So began the article, the first to connect Henry to the sweet pea, in the columns of *The Gardeners' Chronicle* for Saturday 2nd September, 1882. The author was none other than Richard Dean. It was just like old times.

> 'The Floral Committee of the Royal Horticultural Society', the article went on, 'have already acknowledged in the most practical manner their approval of Mr. Eckford's work by awarding him a First-class Certificate of Merit for Bronze Prince. This is a very fine form of the black Sweet Pea, with shining bronzy-maroon standards of large size, and rich purple-blue wings; the flowers are very large and striking in appearance. Blue King is in the way of the purple Sweet Pea, the standards large, stout and bold, as in the case of the preceding variety, and of a showy bronzy-crimson hue dashed with purple, bright pale blue wings, very fine and attractive, and particularly pleasing from its fine shade of blue. Grandeur has fine crimson-rose standards, the wings pale mauve, very fine and showy, the colour of the standards deeper in hue altogether than in the case of the scarlet Sweet Pea. This variety requires to be grown by the side of a fine type of the scarlet Invincible, but we think it will prove distinct from it.
>
> Lottie Eckford is like Butterfly, and, we think, not sufficiently distinct from it, as Butterfly, though opening very pale and delicately tinted when young, becomes deeper as well as more varied in colour with age. Lottie Eckford is a variety charmingly tinted with blue. Princess has pale standards, slightly suffused with magenta and pale purple, the margins slightly beaded with purple, the wings of young flowers white, with a fine wire beading of azure-blue on the wings; but later flowers have the wings and the standards in some parts, but not so heavily, striped and flaked with blue. In any case, it is a very pretty variety. Duchess of Albany has pale standards dashed with delicate magenta and blue, and slightly bearded with purple; the wings white, margined and flaked with blue. As sent from Mr. Eckford this variety comes very near to Princess, but when growing side by side there may be sufficient differences to warrant the two being regarded as distinct. The examples received were a little old, and they had come a long journey through the post.

One thing is quite certain, that Mr. Eckford has obtained a very interesting and valuable break. Further crosses cannot fail to give something of a valuable character. Years ago Mr. Eckford made his mark in raising Dahlias, zonal and nosegay Pelargoniums, Verbenas, &c., so that he is by no means new in the work; and it is as true of floriculture as of any other department of human work, that what men have done and are doing is but an earnest of what they shall accomplish in the future; there can be no limit to the possibilities of production, for the universe will always be wider than the largest imaginings of the human mind. R.D.'(29)

It's good to see that Richard Dean, master of the flowery finish, has lost none of his panache. So, these were Henry's first sweet pea offerings. There was some later mutterings in the press to the effect that Bronze Prince was unfixed; and so it was. Although Henry eventually got 'Bronze Prince' to breed true it initially gave rise to so many 'sports' that people thought they'd got mixed seed. The error preyed on his mind and it wasn't until shortly before his death in 1905 that he could openly acknowledge it. 'Like other beginners', he told *The Gardening World's* reporter, 'I had to learn and unlearn some of my work... At first, the matter was difficult of explanation, but the true cause of its failure was discovered afterwards'.(30)

The sweet pea, *Lathyrus odoratus*, comes from a country as renowned for the arts of man as of those of nature. It originates in southern Italy; in Lucania, Calabria and Sicilia (Sicily)(31). It has been said that the sweet pea 'was a cultivated garden plant around the Mediterranean and had a very long pre-Christian history of being used for garlands and wreaths...'(32).The first written description of it in the modern era seems to be by a Sicilian monk, Francisco Cupani, in the late seventeenth-century. (33) The first description is sometimes attributed to John Bauhin in his earlier *Historia Plantarum*. There are, however, no references to scent (odor) in any of the *Lathyrus* descriptions in Volume 2, (1651), the volume where it would be, and there appear to be no later editions. Scent was not, however, always referred to in pre-binomial days. It wasn't noted by the botanists John Ray and Heinrich Rupp, both referring instead to its Sicilian provenance. Soon after Cupani's description, it was illustrated by Casper Commelin in his Amsterdam Flora, (34) which indicates that Cupani's probably was the first written description. Its first appearance in Great Britain also appears to be thanks to Cupani, sending seed to Robert Uvedale, an Enfield schoolmaster and botanist in whose garden it flowered in 1700.

The sweet pea's five-petalled flowers, like those of all the pea or Leguminosae family, are unusual and distinctive. A relatively large central petal, the 'standard' stands behind and between two smaller 'wing' petals which themselves partly hide the two smaller clasped, basal 'keel' petals that enclose the reproductive organs. This joined arrangement ensures that the flower is self-fertile. It also makes artificial cross-pollination difficult, though all things come with practice. In its original habitat it climbs and scrambles about in the manner of our own related peas and vetches.

Horticulturally, it had nothing more than they have to recommend it, apart from its scent, with which among *Lathyrus* species it is almost wholly unique. It has, therefore, acquired the specific name *odoratus* meaning odorous, scented. Its generic name *Lathyrus* is more problematic. It is Latinised Greek, the 'thyrus' part meaning 'powerful' or 'vigorous' and appears to indicate the purgative properties of the seeds of *Euphorbia lathyris*, another plant entirely. (35) (We are looking at centuries of confusion here). Despite his primness, or perhaps because of it, J.C. Loudon, writing in 1829, thought that it had 'been applied to this plant in consequence of certain aphrodisiacal qualities ascribed to it'.(36) Alas, no John but this belief might account for its early popularity. Incidentally, as the ingestion of the seed from at least one *Lathyrus* species is known to cause muscle paralysis it would be as well not to experiment!

The first English Flora to note it, in 1700, spoke (in Latin, like all the early Floras) of the 'greater Lathyrus from Sicily, with very wide perfumed flowers, the standard being red and the blue petals encircling the keel'. (37) The Rev. John Ray put it into the 1704 edition of his *Historia Plantarum*: 'a very sweet-scented Sicilian flower with red standard; the lip-like petals surrounding the keel are pale blue. Its seed pod is hairy'. (38) An eighteenth-century Flora from Jena noted that 'sometimes it varies with a white flower'. (39) Thomas Fairchild, Britain's first known hybridist, chose to emphasise the scented quality of the sweet pea in recommending the plant for London gardeners in 1722. 'The sweet-scented Pea makes a beautiful Plant, having Spikes of Flowers of a red and blue Colour. The scent is somewhat like Honey and a little tending to the Orange-flower Smell'(40). In the same year the flower had become well enough known about in London to be offered for sale there by one 'Benjamin Townsend, at the sign of the Three Crowns and Naked Boy over against the new Church in the Strand.'(41)

Philip Miller was the long-standing head gardener of the still-thriving Chelsea Physic Garden and author of *The Gardeners' Dictionary*. The standard reference work of the time, it ran through eight editions. In his first, 1731 edition, Miller was the first to provide practical advice. 'These may be sown in March in the Places where they are to remain for good, being plants that seldom will grow if transplanted, except it be done while they are very young: These should be either sown near a Pale, Wall, or Espalier, to which they may be train'd; or if sown in the open Borders, should have Stakes plac'd by them, to which they may be fasten'd, otherwise they will trail upon the Ground and appear very unsightly, which is the only Culture these plants require, except the cleaning them from Weeds: They produce their Flowers in July, and their Seeds are perfected in August and September.

But the best method to have them very strong, is, to sow their seeds in August, under a warm Wall or Hedge, where they will come up in Autumn, and abide the winter very well; and these will begin to flower in May, and continue to produce fresh Flowers until July or later, according to the Heat of the Season: and one of these autumnal plants will be as large as four or five of those sown in the Spring, and produce ten times the Number of Flowers; and upon these Plants you'll always have good Seeds, when sometimes the other will miscarry: However, 'tis very proper to sow their Seeds at two or three different seasons, in order to continue their Flowers the longer, for the late planted ones will continue blowing [flowering] until the Frost prevents them'.(42)

By the time of his eighth and last edition in 1768 Miller had noted 'two other varieties of this sort, one of which has a Pink-coloured standard with a white keel, and the wings of a pale blush colour; this is commonly called Painted Lady Pea. The flowers of the other are all white…'.(43) The name Painted Lady is conceivably a connection of the sweet pea's butterfly-like flower with the 'Painted Lady' butterfly. (There was an 1855 Painted Lady runner-bean variety too; another legume and so with the same flower structure). James Justice, from Midlothian, had in 1754 noted that the Painted Lady sweet pea was less sweetly scented than the other two. (44)

From being noted primarily in London, by 1796 William Curtis could write in his *Botanical Magazine* that 'there is scarcely a plant more generally cultivated than the *Sweet Pea,* and no wonder, since with the most delicate blossoms it unites an agreeable fragrance.

'Several varieties of this plant are enumerated by authors, but general cultivation extends to two only, the one with blossoms

> perfectly white, the other white and rose-coloured, commonly called the Painted Lady Pea...'

With autumn sowing in pots, 'they can the more readily be secured from any severe weather, by placing them in a hot-bed frame, a common practice with gardeners who raise them for the London markets, in which they are in great request…'. (45) A return to hot-beds, ground heated by fermenting manure, might be an idea in our energy-conscious times, though finding the manure might be a problem. Curtis's coloured illustration of the sweet pea, incidentally, has two flowers on a stem showing claret/crimson-coloured standards with light-purple wings. The keel is coloured as the wings, with white undersides. It doesn't quite fit some of the earlier, variant descriptions, though it's actually quite like the recently reintroduced original flower. William Curtis covered himself by saying that his illustrations were 'coloured as near to nature as the imperfection of colouring will admit'.(46) In 1900, Henry's son John thought that the artist had either 'figured a well selected variety or else decidedly over-coloured his drawing'.(47) It's possible however that by 1900 neither John nor anyone else really knew what the original flower looked like.

John Abercrombie, in the second edition of his *The Universal Gardener and Botanist* in 1797 adds a scarlet to make four kinds to date. (48) Only three years later, one John Mason, trading at the sign of the Orange Tree, 152 Fleet Street, London, found room among his more fashionable dutch bulbs to advertise 'black, purple, scarlet, white and painted lady' (49) sweet peas, the black doubtless being a dark purple or red. John Kennedy, in his *Page's Prodromus*, (50) a list of plants cultivated in Southampton Botanic Gardens in 1818, adds a striped variety, to make six. J.C.Loudon in his 1822 *Encyclopaedia of Gardening* claims an unlikely 'nine sorts'. A more circumspect William Cobbett, in his 1829 *The English Gardener* stated that the sweet pea 'blows [flowers] a rose coloured flower of various hues…'.(51) Loudon again, in his 1829 *An Encyclopaedia of Plants* wrote that 'Lathyrus odoratus is one of our most esteemed border annuals, and is extensively grown in pots for decorating chambers and windows…'.(52)

Opposite a beautifully coloured illustration of sweet peas in Vol. 4 of Benjamin Maud's *The Botanic Garden*, 1831-2, Maud writes of 'the ever admired Sweet Pea. This is one amongst other bedding beauties, which never tires by its presence. It is not only always welcome, but always sought for…'. As we saw in an earlier chapter, this was a sentiment echoed by one John Fyffe of Milton Bryan, Bedfordshire, writing in J.C.Loudon's

Curtis's Wild Sweet Pea, 1796

Gardener's Magazine, the first, essentially, of the now familiar gardening magazines, in 1837. 'The Sweet Pea is esteemed by most lovers of the flower-garden for its rich profusion of flowers, and the delicate perfume which they put forth after a refreshing shower. We consequently meet with it in most of our flower-gardens, either in rows or patches, supported in the common way by brushwood stakes; though this method is very unsightly in the eye of a lover of neatness and order, making the flower-garden resemble the kitchen department…'.[53] This shows that sweet peas had become a popular flower among the populace, if a long way from being fashionable. We continue to grow them up brushwood stakes too, exhortations to the contrary notwithstanding...

Henry was a Beaufort Castle apprentice when the London seedsman and florist James Carter was offering seven varieties of sweet pea seed for sale. As we saw earlier, his catalogue for 1839 was displayed in Loudon's *Gardener's Magazine* that January and listed, in those days before Trade Description Acts, a black (a dark purple or maroon), a Painted Lady, a purple, a scarlet, a yellow (a cream) and a white. [54] This slow increase in varieties would continue until Henry began work on them in 1879. The little attention they did have was mostly the work of the Carters' seed firm. At least, they sent out the few varieties that there were.

The foregoing is, with one or two additions, a selective trawl through the known references and translations, with some corrections. It leaves out, for example, those of Carl Linnaeus, James Petiver and Henry Phillips, of botanical illustrators like Georg Ehret, Peter Casteels and Pierre-Joseph Redouté and of artists like Jean-Siméon Chardin. But hopefully it conveys a sense of the progress of the flower in Britain; of the very gradual development of varieties and of the attitudes produced from an understanding awakened between its arrival here in 1699 and Henry's acquaintanceship with it in the 1830s. Incidentally, the period between the sweet pea's arrival and Henry's attention to it in the 1870s spans the golden age of coloured natural-history book illustration. Horticulture's golden age in the nineteenth-century may have fed, in part, from the beautiful plant illustrations of these much sought-after books.

~~~

As a garden favourite in a primarily rural society, sweet peas may have been overlooked in the way that the commonplace and familiar often is. For example, there remains to be unearthed a comprehensive depiction of authentic labourers' cottage gardens, once so common. (Much as
~~~

now, there were prescriptive depictions but little descriptive reporting of commonplace gardens). It relates to the observation by the contemporary poet and gardener John Clare (1793-1864) that 'the poor man's lot seems to have been so long remembered as to be entirely forgotten'. Considering the qualities of the sweet pea it's surprising that it hasn't drawn more attention from poets and versifiers. The earliest attempt seems to be that from a chap-book of 1794, entitled 'The Sweet Scented Pea':

> Masters, search every garden round,
> And Miss, pray tell to me,
> If any flower can be found
> To beat the scented pea.[55]

John Keats was born the following year. He was often drawn to the then simple garden flowers and an early reference to sweet peas is frequently quoted. Keats was taken with the word 'sweet' and used it another five times in the poem!

> Here are sweet peas, on tip-toe for a flight:
> With wings of gentle flush o'er delicate white,
> And taper fingers catching at all things,
> To bind them all about with tiny rings.[56]

The narrative poem 'Endymion' ('A thing of beauty is a joy for ever') is another early work, if one can refer to 'early' and 'late' with Keats. Showing his developing interest between nature, myth and art, sweet peas appear in its early scene-setting:

> Many and many a verse I hope to write,
> Before the daisies, vermeil rimm'd and white,
> Hide in deep herbage; and ere yet the bees
> Hum about globes of clover and sweet peas,
> I must be near the middle of my story.[57]

The Londoner Keats, sensual and romantic, was, in contrast to John Clare, the outsider looking in. For the 'peasant poet' as contemporaries knew him, Clare's poetry arose from nature's natural rhythms and the rural life he was born into and grew up in. In one of a series of 'Scraps of Tragedy' he wrote:

> I told the sweet pea that some angry bee
> Had drove a poor butterfly
> From its original blossom; who still

Felt enamoured of its sweets and was content
To settle on the stem and change into a flower
For very love[58]

The 'sweet-scented pea' was now the sweet pea, Clare and Keats both noting down the common usage. Keats' allusion to tendrils – 'taper fingers' and Clare's conjunction of the habit and appearance of butterfly and flower is captured (along with runner beans) by the late poet Geoffrey Hill in 'Pisgah':

.....
around you the cane loggias, tent-poles, trellises,
the flitter of sweet peas caught in their strings,
the scarlet runners, blossom that seems to burn
an incandescent aura towards evening.[59]

The idea of giving symbolic meanings to plants is believed to have begun in the harems of the Ottoman Empire. An attempt at a systematised 'Language of Flowers' was first attempted in France in the early nineteenth century and achieved enormous success in early and mid-Victorian Britain. It's all great fun, though the logic founders on the fact that different plants have different meanings to different people and you rarely find two writers in agreement. In *The Country Flowers of a Victorian Lady* Gill Saunders says that because the sweet pea faded quickly after being cut 'the various symbolic meanings that have been ascribed to it relate to its fragile, ephemeral character. In the Victorian Language of Flowers some writers suggested its meaning as Departure', while for another 'it symbolised Delicate Pleasures'. In a poem from her beautifully illustrated journal, Fanny Robinson (1802-1872, the 'Victorian Lady' of the book's title) drew a 'comparison between the delicate pinks and purples of sweet peas and the colours of sunset-tinted clouds. These are flowers that share the softness and transparency of watercolour'.

Delicate pink,
Purple, and snowy white are thy wings
Fair Butterfly of Flowers. Thy many tints
Are dyed as if the sunset evening clouds
Had fallen to the earth in sudden rain
And left their colours. [60]

Later, when postcards were popular, they too would depict flowers like the by then fashionable sweet pea in picture and verse. 'The Song of the Sweet Pea Fairies' was one of Cecily Barker's delightful *Flower Fairies of the Garden*, [61] one of her ostensibly children's Flower Fairies books published from the 1920s onwards.

What were the varieties that Henry had to work with? John Eckford later claimed that his father began with just five varieties, obtained from the Lees's nursery at Hammersmith – 'White, Scarlet, Black, Painted Lady and Butterfly'. [62] This may well be so, but Richard Dean claimed others, as we'll see, and with John seemingly getting his dates wrong, a *Gardeners' Chronicle* article by Dean in the summer of 1882 is more reliable. In it, Dean sets the scene and describes the varieties likely to have been available to Henry, as Henry introduces his first sweet pea variety 'Bronze Prince' to an unsuspecting public.

> 'Sweet Peas. Probably no other common flower is so useful in the garden during summer as the Sweet Pea, and it is as indispensable to it as Mignonette; and yet, while it is so useful and so commonly grown, it appears to escape, to a considerable extent, the attention of writers. Perhaps it is assumed that nothing that is fresh can be written about it...'.

Interrupting Dean for a moment, there is an argument that music in the domestic sphere in the nineteenth-century was associated with the feminine and thus undervalued.[63] Sweet peas have what were considered feminine qualities and being a cut-flower for the house may have remained similarly overlooked by male hybridists and commentators.

> 'New varieties', Dean continues, 'occasionally come into cultivation, and they are of undoubted novelty and quality; but as they are so seldom met with in gardens, they appear to gain a footing there but slowly. Formerly we had but few varieties of Sweet Peas, now they have grown into something like thirteen or fourteen varieties, every one of which well deserves a place in the garden...
>
> In all probability the original form has been considerably improved upon, and it has either sported into new forms, or yielded them by means of seed. In later years new varieties have been obtained in this way: - Among the plants raised from seed of any one variety,

a new departure has been discovered in the case of a plant or two. Those whose practice it is to grow from seeds largely are aware of the tendency in many annuals to break into different characters, and when one appears it is marked, the surrounding plants are pulled out to give the new type space in which to develop itself, and the seed is carefully gathered and sown for another season. Sports of this kind are often very difficult to fix in a permanent character; they will appear for a year or two or more, as if they would do so, and then they will revert to their original form, to the great disappointment of the cultivator. On the other hand, such sports can be permanently fixed after a few years' selection, and when the durability of the new character is assured the variety can be sold in the ordinary way...'.

Of those in cultivation, beginning with 'the purple we get an exceedingly bright and attractive variety with a crimson standard, with very pleasing blue-violet wings. If only three varieties were grown this should be one of them, for the blue-violet tint on the wings is most attractive. The purple-striped, as the striped form of this is termed, has the standard and wings much streaked and spotted with white; but while the striped forms afford variety, they are scarcely so pleasing to the eye as the self-coloured flowers. The scarlet Sweet Pea has a standard of a deep scarlet or red hue, the wings being paler and brighter in colour, approaching magenta, with a white keel; indeed, the white keel appears to be characteristic of all the varieties.

The scarlet Invincible is a larger and finer selection of this, being more intense in colour in all its parts. It is now much grown for cutting, indeed, more so than any other variety, because of its fine appearance. The scarlet Invincible has striped forms also, in which pencillings of white are thrown across all the parts of the flower. Whether the black Sweet Pea was derived from the purple, or *vice versâ*, the black so-called has a maroon crest and deep purple wings, but it is not so bright in appearance as the purple. But a tendency to become purple will be found among the black, and to become black among the purple; in fact they are apt to run into each other, and need rigid selection to keep them true to character. The Painted Lady is a very pretty and distinct

Sweet Pea, the standards scarlet and the wings white. This variety, too, should have a place in every collection.

We have seen in Messrs. Carter & Co.'s collection a very fine striped form of the black Sweet Pea, in which the dark standards and the deep purple-blue wings are striped with white, and this is very pretty indeed, and may not materially differ from the ordinary striped form of the black. The white is well known from being perfectly white in all its parts; it is an exceedingly attractive variety, and like the scarlet Invincible is highly grown for cutting from. What is known as Sutton's Butterfly, is a white flower tinted in the most pleasing manner with delicate lilac-blue on the margin of the standards and wings, changing with age to deep lilac. It originated as a sport among some white Sweet Peas at Messrs. Sutton & Sons' trial grounds at Reading, and it is also known in catalogues as the blue-edged. Fairy Queen appears to have been another sport from the white variety, the standards being white, flaked with rose. This is a very pleasing variety indeed, and quite distinct in character.

Crown Princess of Prussia, which we believe to be another of Messrs. Carter and Co's raising, [no Richard, German: Haag and Schmidt] is also a very pretty and distinct variety, the crest salmon-pink, the wings delicately tinted with pink, quite novel, and deserving to be generally grown. The Queen [Henry was to bring out another with the same name in 1893] has the standards scarlet feathered with white on the edges, something in the way of a Tulip that is so marked, and with pencillings of the same in the centre, the wings slightly flaked with bright rosy violet; this is also very attractive and novel, and with Violet Queen, now to be described, originated at the St. Osyth seed farms, and, we should think, in both cases came up as sports from Painted Lady. Violet Queen has lovely rose-pink standards, with bright pale violet wings, and is very fine in appearance and novel in character. There yet remains the yellow Sweet Pea, which is not much grown, and which, no doubt, represents a cream-coloured form of the white; and, if this be so, it can scarcely be depended on for fixity of character.

An enormous quantity of Sweet Peas is every year grown for the trade. A wholesale house like that of Messrs. Hurst & Son, of Houndsditch, grows every year from 25 to 30 acres of Sweet Peas, but the produce is very variable. Last year, owing to the weather, there were many total failures; but a good average season and crop should produce about 20 bushels per acre, but this is seldom realised. Messrs. Carter, Dunnett & Beale grow large breadths of the different varieties of Sweet Peas at their St. Osyth seed farms, some sorts more largely than others, according as they are in demand, and there must be very heavy sales of Sweet Peas. Messrs. Hurst & Son put their annual sales at about 300 bushels, and the seeds are grown chiefly in Kent and Essex. The greater quantity is grown as mixed colour, separate colours being required only in comparatively small quantities.

Of late years Sweet Peas have come to be much grown for supplying cut blooms for market, the scarlet Invincible and the white in particular being cultivated for this purpose, as well as in mixed colours. A hedge of Sweet Peas of mixed colours is a very pretty sight indeed in any garden, and diffuses a most agreeable fragrance. The scarlet Invincible in conjunction with Tropaeolum canariense [the canary creeper, now *T.peregrinum*] is a charming combination, as delightful as it is novel. A garden without Sweet Peas is a garden without one of the most useful of flowers that can find a place in it. R.D.'(64)

~~~

For an unusual reference to hybrid *Dianthus*, gardening books sometimes quote an exchange between Perdita and Polixenes in Shakespeare's *The Winter's Tale*.(65) The twenty-four line passage shows, in as far as imaginative drama can be relied upon for historical accuracy, that both sexual and asexual plant propagation was understood here by 1611.

I have heard it said that
There is an art which in their piedness shares
With great creating nature.(66)

We live and work within the confines of our age. The later religious excesses of the 17th Century may have so poisoned the atmosphere that a
~~~

century on, when Thomas Fairchild and others were seeking the whys and wherefores of hybridisation, they are said to have been worried that their work opposed God's will, as expressed in nature. (Fairchild began with *Dianthus*, replicating what he may well have seen around him).

By the time that Henry Eckford began hybridising, such societal pressures had lessened, and Henry and others could work without compunction. With the advent of genetic modification, such pressures, secular this time, have returned. For the likes of Henry, hybridisation was a serious business but at heart it was a bit of fun. It was merely turning a plant into an acceptable addition to the garden and the flower vase. If everything went wrong, or fashions changed, the wild plant was still there. You could begin all over again, or leave it alone. In our time, such options appear threatened.

References to Chapter Seven

(1) Henry Eckford, *Curriculum Vitae*, undated, author's possession.

(2) Letter from Dorothy Fairweather, née Eckford, to Bernard Jones, (3-11-1979).

(3) 'Bickley and District Horticultural Society', *Bromley and District Times*, (20-7-1906), p.3.

(4) Arthur Mee, *The King's England: Gloucestershire*, (1966), p.120.

(5) PA117/1, Major Bamford, *A Short History of Sandywell Park*, (c.1966), p.3, Gloucestershire Record Office.

(6) Henry Eckford, *C.V.*, op.cit.

(7) 'Dowdeswell', *Kelley's Directory of Gloucestershire*, (1879), p.633.

(8) D4858, 2/4, Sale Particulars, Sandywell Park, (1923), Glos. R.O.

(9) PA117/1, Major Bamford, op.cit., p.14.

(10) Q/AL, 40/63, (Bill), (11-10-1881), Glos. R.O.

(11) Q/AL, 40/60, 'Commissioners in Lunacy' report, Sandywell Park Lunatic Asylum, (18-6-1878), Glos. R.O.

(12) E.Scot Skirving, (compiler), *Cheltenham College Register*, 1841-1927, (1928), p.179.

(13) 'Verbenas From Seed', *The Gardeners' Chronicle*, (12-6-1880), p.747.

(14) 'Sweet Peas for Cutting From', *The Gardeners' Chronicle*, (20-7-1878), p.84.

(15) 'The Telephone in Horticulture', *The Gardener*, (11-1879), p.531.

(16) The Scottish Collection, National Gallery of Scotland, Edinburgh.

(17) The Ionides Collection, Victoria and Albert Museum, London.

(18) R.Ash, B.Higton, *Gardens In Art*, (1994), quote, no paging.

(19) 'Mr Gladstone on Garden Cultivation', *The Gardeners' Chronicle*, (6-9-1879), p.296.

(20) Editorial, *The Gardeners' Chronicle*, (11-10-1879), p.464.

(21) Q/AL, 40/64, 'Commissioners in Lunacy' report, (31-3-1882), Glos. R.O.

(22) Q/AL, 40/64, Letter, Dr. W.H.O.Sankey, (13-10-1882), Glos. R.O.

(23) Yoland Brown, *Boreatton Park*, (1989), p.11.

(24) Ibid, p.30.

(25) George and Weedon Grosssmith, *The Diary of a Nobody*, (1892), passim.

(26) R.W. Emerson, *English Traits*, 1856, (1902 edn.), p.168.

(27) Anton Chekhov, 'The Beauties', 1888, in *Anton Chekhov; Stories*, Vol.

1, (1968 edn., the Constance Garnett translation), pp. 373-374.

(28) 'Mr Henry Eckford', *The Garden*, (2-1-1897), frontispiece.

(29) 'Pedigree Sweet Peas', *The Gardeners' Chronicle*, (2-9-1882), pp.298-299. (Lottie misspelt Louie).

(30) 'Occasional Interviews', *The Gardening World*, (8-7-1905), p.544.

(31) Letter to the author from the Botanic Gardens of the Università Degli Studi Di Napoli Federico II, (8-9-1997); Sandro Pignatti, *Flora D'italia*, Vol.1, (1982), p.693.

(32) Maggie Campbell-Culver, *The Origin of Plants*, (2001), p.143.

(33) Francisco Cupani, *Hortus Catholicus*, (1696), p.107.

(34) Caspar Commelin, *Horti Medici Amstelodamensis*, Vol.2, (1701), Fig.80, p.159.

(35) Greg Kenicer, unpublished doctoral thesis on the genus Lathyrus, (2007), The Royal Botanic Garden, Edinburgh.

(36) J.C.Loudon, *An Encyclopaedia of Plants*, (1829), p.620.

(37) Leonard Plukenet, *Almagesti Botanici Mantissa*, 1700, p.114.

(38) John Ray, *Historia Plantarum*, 1686 (1704 edn. as Lathyrus major e Siciliae), p.447.

(39) Albrecht von Haller, ed., H.B.Rupp's *Flora Jenensis*, 1718, (1745 edn. as Lathyrus Siculus Ravini), p.260.

(40) Alice M. Coats, *Flowers and their Histories*, (1956), p.138, quote.

(41) S.B.Dicks, 'The Early History of the Sweet Pea', *The Sweet Pea Annual*, (1922), p.23.

(42) Philip Miller, *The Gardeners' Dictionary*, 1st edn. (1731), no paging.

(43) Philip Miller, *The Gardeners' Dictionary*, 8th edn. (1768), no paging.

(44) Alice M.Coats, op.cit., p.138.

(45) 'Lathyrus odoratus Sweet Pea, or Vetchling', *Curtis's Botanical Magazine*, (1796), plate 60.

(46) *Curtis's Botanical Magazine*, (1793), preface.

(47) C.H. Curtis, J.S. Eckford, in R. Dean (ed.) *The Sweet Pea Bicentenary Celebration Report*, (1900), p.24.

(48) J.Abercrombie, *T.Mawe's The Universal Gardener and Botanist*, 2nd edn. (1797), no paging.

(49) 'Sweet Peas', *The Gardeners' Chronicle*, (6-3-1897), p.160.

(50) John Kennedy, *Page's Prodromus*, (1818), p.239.

(51) William Cobbett, *The English Gardener*, 1829, (1996), p.307.

(52) J.C.Loudon, *An Encyclopaedia of Plants*, (1829), pp.120-121.

(53) 'A Mode of training the Sweet Pea in Flower-Gardens', *The Gardener's Magazine*, (10-1837), p.446.

(54) Advertisement, *The Gardener's Magazine*, (1-1839), pull-out supplement between p.xxii-p.1.

(55) 'The Bibliography of the Sweet Pea', *The Sweet Pea Annual,* (1907), p.10, quote.

(56) 'I stood tip-toe upon a little hill', lines 57-60, Poems by John Keats, (1817); 1924 edn. as *Keats's Poetical Works*, p.4.

(57) John Keats, Endymion: A Poetic Romance, lines 49-53, (1818); 1924 edn., ibid, p.58.

(58) John Clare, one of a series of 'Scraps of Tragedy', Poems of the Middle Period, 11, 89. I am indebted to Dr. P.M.S. Dawson of Manchester University for this information.

(59) Geoffrey Hill, Pisgah, lines 7-10, *Canaan*, (1996), p.52.

(60) Fanny Robinson, *The Country Flowers of a Victorian Lady* (c.1840s, pub. 1999), pp.56-57.

(61) Cicily Mary Barker, *Flower Fairies of the Garden*, 1944, (2002), no paging.

(62) C. H. Curtis, J. S. Eckford, in R. Dean (ed.), op. cit., p.27.

(63) Kate Flint in C.Matthew, ed., *The Nineteenth Century,* (2005 edn.), p.248.

(64) 'Sweet Peas', *The Gardeners' Chronicle*, (5-8-1882), pp.182-183.

(65) See H.Maxwell, Flowers: *A Garden Note Book*, (1923), pp.xiv-xv; M.Leapman, *The Ingenious Mr Fairchild*, (2000), pp.164-165.

(66) William Shakespeare, *The Winter's Tale*, (c.1611), Act IV, Scene III, lines 79-103.

Chapter Eight

It was said of Richard Dean that there was scarcely a horticultural movement of any kind in which he didn't take an active and generally prominent part. Having a finger in every pie naturally included the pie with the edible peas in. Thus it was, a year after his sweet pea article that he now had one out on 'New Garden Peas at Boreatton Park' for his *Gardeners' Chronicle* readership.

> 'Those who are acquainted with Mr. Henry Eckford and know something of his good work with the Verbenas, Dahlias, Pelargoniums, &c., in times past, will not be surprised to learn that he is still actively engaged in making judicious crosses, and seeking in every way in his power to improve some, at least, of the plants he cultivates so well. He has secured a very fine break of main crop Peas that appears likely to be of great service for the general garden, and especially for exhibition purposes. The fact that the Fruit and Vegetable Committee of the Royal Horticultural Society has recently awarded Mr. Eckford a First-class Certificate for Duke of Connaught, one of his new varieties, is ample proof that the strain is a valuable one. Duke of Connaught is a strong growing, free-bearing variety, with large, long, well-filled, pale green pods, square at the end, and very fine for the exhibition table. It grows about 5 feet in height. This resulted from a cross between Champion of England and G.F.Wilson'.

Remember 'Champion of England'? Henry might have tried his hand at a few crosses when he was growing it at Coleshill, twenty-nine years ago. Dean's article continues, like his sweet-pea article, with descriptions of Henry's new varieties, this time of the edible pea, *Pisum sativum*. Henry had called his new varieties 'Progress', 'Magnificent', 'Invincible', 'Victor' and 'Perpetual'. False modesty is not a fault one can generally lay at the door of a lowland Scotsman and Henry was no exception. He had been growing other people's varieties and found many of them wanting.

When it came to naming garden peas, it's interesting to note that the initials of most varieties rhymed with 'pea' (see Appendix). 'C' was highest with seven varieties, then came 'P' with five, followed by 'E' with four. With others like 'Jubilee' the name itself rhymed, while names like 'Magnificent' joined with 'pea' to make a name or a statement – Magnificent Pea! Henry had now grown his new peas for two or three years running and, declared Dean, 'it can be stated they are uniform and fixed in character. They deserve to win their way into the front rank. In prosecuting his crosses Mr. Eckford aims at attaining to certain definite results, and he has succeeded in a remarkable degree. It now remains for the public to endorse by an appreciative approval the work so carefully and successfully performed with such satisfactory results. They are dealt with in this paper only after a very careful personal inspection.

No doubt the fine development of the Peas grown by Mr. Eckford is due, to some extent, to his practice of thin sowing. All his new varieties are of a singular free branching character, and the seeds are sown well apart; but they produce marvellous plants, that fill out the line as densely as do varieties sown much more thickly, and these plants produce crops startling in their profuseness. R.D.'(1). Henry, incidentally, had sent 'Invincible' to the R.H.S. trials as 'Home Ruler' and had evidently been instructed or advised to change it. 'Home ruler' is one of the meanings given to the name 'Henry'. However, it could have been interpreted, and the interpretation may have been Henry's intention, as giving support to W.E.Gladstone, now prime minister again and representing Midlothian, Henry's birthplace, in his conversion to Home Rule for Ireland. It was a contentions political issue which split the ruling Liberals, reflecting divisions nationally.

Gladstone was to eventually steer a Home Rule bill through the (elected) House of Commons in 1893, but the bill was destroyed by the House of Lords. 'In his Irish policy' A.N.Wilson writes, Gladstone 'was more enlightened than any British leader before or since'.(2) Serendipitously, one of the few enlightened Lords was the bill's mover there, the Earl Spencer, from whose garden and name the sweet pea would shortly gain national prominence. Another example of the garden and sweet peas' overlooked role in progressive politics is that following Henry's naming of his pink sweet pea 'Mrs Gladstone' (for Gladstone's wife Catherine) in 1889 she became prime mover in persuading her husband to take office for a fourth time, from 1892 to 1894; [(3)] two horticultural connections unaccountably ignored by historians!

Although Henry would have grown the sweet pea at many of his former gardens he would have grown the garden pea at all of them and been thoroughly familiar with it. The garden pea was then and has remained since immensely popular. It is one of our island's, and Europe's, first plant introductions, a western Asian import that has been discovered in Swiss lake-side dwellings from the Bronze Age. It hasn't always been so commonly available. In Elizabethan days, as a rare and expensive import (from Holland – nothing much changed there) they were considered 'fit dainties for ladies, they came so far, and cost so dear'.[(4)] By the nineteenth century, things had changed. It was in Henry's youth that William Cobbett had brought out *The English Gardener,* an addition to the library of any well-run garden bothie.

'Pea. – This is one of those vegetables which all people like. From the greatest to the smallest of gardens, we always find peas, not to mention the thousands of acres which are grown in fields for the purpose of being eaten by the gardenless people of the towns...'.[(5)] As a garden writer, Cobbett wasn't in the same league as J.C.Loudon but the latter couldn't help but be irritated by him. In 1822 Loudon had had this to say of Cobbett's *The American Gardener*: 'Though the author shows great ignorance of botany and physiology, he has contrived by his style, by many shrewd remarks, and by curious and bold assertions at variance with facts, to make an interesting book...'.[(6)] But Cobbett had the thickest of hides and brought out his *English Gardener* seven years later.

When Henry lived in London he would have seen, among the 'hot eel and pea soup men', the green pea vendors, carrying large pans of hot peas cooked in their pods, customers removing the peas by drawing them from the pod through their teeth. Salt, pepper, butter and vinegar were all provided, on the street, as accompaniments. [(7)] Bunches of the old blue-purple sweet peas were also a common sight, offered for sale in many a market place.[(8)] Henry had only seen what others had seen, but differently.

As you might expect, Henry didn't quite have the garden pea field to himself. Probably the best-known hybridist of the garden pea then, and also working with beans, strawberries and funnily enough sweet peas was Thomas Laxton (c 1830-1893). Incidentally, it wasn't Thomas but 'Laxton Bros. (Bedford) Ltd.', a firm founded by Thomas but run by his sons, Edward principally, who were responsible for the famous Laxton plums, pears and especially apples like the three raised between 1904 and 1921, 'Lord

Lambourne', 'Laxton's Fortune' and 'Laxton's Superb'. There is a little story to be told about Thomas. Back in 1866 Charles Darwin had written to the '*Chronicle* to highlight his difficulties with crossing leguminaceous (then papilionaceous) plants. '...Some years ago I crossed the varieties of the Sweet Pea, and many more flowers dropped off unimpregnated than were fertilised. The difficulty arises from the anthers opening at so early an age that they must be removed before the flower expands...'.[9] As an answer he then described a complicated method of cross-fertilisation using thin paper tubes.

This brought a reply from Thomas Laxton who stated that, with the similar garden pea he simply emasculated 'the seed-bearing parent whilst the anthers were in an immature state, at once making a single application to the pistil of the keel-petal of the male parent containing mature pollen', giving him ninety pods and four hundred and thirteen peas out of one hundred and four crosses. [10] Bob's your uncle and what's all the fuss about? Nothing, except that that's what Darwin had thought he'd been doing. As Laxton seemed to be rather good at it the great man got Laxton to do some experiments on *Pisum* for him, publishing what must have been the results in 1872.[11] So, Henry wasn't exactly in competition with a beginner. But the market for peas was vast and there was always room for another, even a big'n like Henry.

Richard Dean appears to have visited Boreatton Park prior to his garden pea article. When it came out he was back again at Boreatton, this time to report on the sweet peas. His findings were duly published in the '*Chronicle* a fortnight later.

> 'Eckford's New Sweet Peas. The raiser of these was undoubtedly placed at a great disadvantage when he exhibited cut specimens in a somewhat gloomy tent on the occasion of a recent meeting of the Royal Horticultural Society. The flowers had travelled nearly 200 miles and had suffered in consequence. But though the Floral Committee may have failed to recognise... any distinctness and enlarged beauty in Mr. Eckford's productions there is no doubt that these qualities exist in them, and when personally inspected a fortnight ago at Boreatton Park, under good cultivation and in comparison with a group of varieties already in cultivation, we were struck with the distinctness of character and fine development of not a few of them.

It is but a few years since that Mr. Eckford took in hand the Sweet Pea, while gardener to that much respected florist, Dr. Sankey, at Sandywell Park, Cheltenham. On Dr. Sankey removing to Boreatton Park, Baschurch, Mr. Eckford accompanied him, and continued there the work so well begun at Cheltenham, and in which Dr. Sankey takes a lively interest.

It may be stated that Mr. Eckford first began operations with such well known varieties of Sweet Peas as Invincible Scarlet, Violet Queen, Captain Clarke's Princess of Russia, Butterfly, The Purple, and one or two others. The flowers of these were all carefully fertilised (a work of some delicacy, requiring great care), crossing the varieties in various ways'. Hybridists rarely revealed their practise in print! Dean's article continued with a long list of descriptions of the new flowers called 'Fascination', 'Victoria', 'Bronze Prince', 'Princess', 'Leviathan', 'Lottie Eckford' – named for Henry's daughter and now her memorial, its 'standard suffused with purple... having a wire edge of blue to the white wings: this is very pretty indeed;' 'Blue King', 'Indigo King', 'Salmon Queen', 'Duchess of Albany', 'Grandeur', 'Lavender Gem', 'Queen of Roses', 'Empress of India' and 'Blue Beauty' completed the list.

'Fertilisation is done when the flowers are quite young, and before they can be interfered with by insects. Some varieties appear to be more readily crossed than others, and, as in the case of garden Peas, so with these fragrant varieties, one pod will produce three or four varieties; and fertilised flowers do not produce so many seeds in a pod as those untouched by human agency, and it is not uncommon for the progeny to be altogether distinct in character from the parents.

And, as in the case of garden Peas, the seeds of these Sweet Peas are sown far apart; the result is, that a single plant will branch out in a remarkable manner, and form quite a bush. A single plant produced the astounding number of 2500 seeds. If gardeners were to sow their Sweet Peas a little more thinly and in good soil, they would be surprised at the results...' . [(12)] Richard Dean had boned up on some other new plants of Henry's too, during his last visit. A fortnight later he finished his Boreatton reports for 1883 with 'Fancy Pansies in the Mass. – Mr. Eckford has a marvellous

> display of these in the kitchen garden at Boreatton Park. A broad walk runs right across the walled-in kitchen garden, and on either side is a border 4 to 5 feet in width and of considerable length. These two borders are filled with selected fancy Pansies, a good portion with plants selected last year for their obvious fine qualities, divided in the autumn and planted out in early spring; the remainder seedlings of the present year.
>
> It is long since that we saw such a blaze of Pansy blooms, the plants doing so well, the flowers so large, and so richly coloured. Starting with a few of the best named varieties in cultivation Mr. Eckford has seeded and selected with judgment and determination, and the result is a progeny of the highest merit. Early in August these were literally sheeted with bloom. One could have gone through the seedlings of the present year and selected a dozen or twenty varieties, all well worthy of distinctive names. Dr. Sankey, whose old love for florists' flowers knows no diminution, takes a deep interest in these Pansies, and is justly proud of them. There is something in the soil and climate of Boreatton that appears to suit the fancy Pansy exactly. They grow luxuriantly, throw many and very large blossoms, and colour with a richness unknown to us in the South'.(13)

Pansies were Henry's secondary interest, a probable obligation to Dr. Sankey, but he took the trouble, as always, to achieve good results. These 'Fancy' or 'Belgian' pansies had been around for some thirty years, having been earlier developed on mainland Europe. Before them were the 'Show' or 'English' pansies, derived from crosses among our native Violas, frequently natural hybrids themselves; with, possibly, *V.altaica* from Siberia. These Show Pansies, reported William Robinson's *Gardeners' Magazine*, were now 'rapidly being pushed into the background to make room for their large brethren the Fancies'. (14) The Show or English pansy may have been bred with showing or exhibiting too much in mind, neglecting the importance of hardiness. A disease eventually 'carried off the plants by the hundred'(15) in the south of England. Paradoxically the English pansy did better in Scotland, where a wetter climate and a gentler sunlight suited it better. As it did the Fancy pansies incidentally, much of whose development took place in Scotland.(16)

Crossing the Show pansy with the new Fancy hybrids and with species like *V.cornuta* from the Pyrenees and *V.alpina* from central Europe, the late Victorian and Edwardian gardeners were soon to lay the path to the highly hybridised, continual flowering, garden pansy of today. All this was to climax a little later, in the 1890s and 1900s. (17) For now, to have the Boreatton borders 'sheeted with bloom' in early August was obviously unusual enough to be remarked on; a tribute to Henry and the doctor.

July 26th, 1884. 'Mr. Eckford's Sweet Peas. – A charming bouquet of these has just come to hand from Boreatton Park, Baschurch'. So began the first of two short articles on Henry in the following year's *Gardeners' Chronicle*. Henry at this time seemed to be working almost alone and unnoticed on the sweet pea. In November '83 there had been mention of Carter & Co's 'Adonis', a deep pink 'floral novelty for the new year;' Hurst & Co. had their 'Carmine Rose' out in August that year... 'Mr. Eckford and others appear bent on doing some acceptable work in this direction' said the '*Chronicle*, without enthusiasm. If there were others busily working away they were not getting written about. Henry had stolen a march.

'In the fine soil of that district', the July '84 article continued, 'the flowers attain to a large size, and are richly coloured. Whatever may be the different opinions as to the distinctness of these new varieties, Mr. Eckford must be commended for his persistent attempts to add to our lists'. The 'leading ones' named and described were 'Imperial Blue', 'Purple King', 'Cardinal', 'Queen of the Isles', 'Princess of Wales', 'Isa Eckford', 'fleshy-pink crest, the wings almost white' and named by Henry for his last surviving daughter, now twenty-three years old. 'Very delicate and pretty' they were 'and, in addition to the size and substance of the flowers, they were richly fragrant. A bouquet of such pretty flowers is a floral dish fit to set before a queen'. (18) (A dish Henry did send, eleven years later, to Queen Victoria).

The second article that year was as short as the amount of rain that fell that summer. 'Fancy Pansies from Boreatton. Mr. H. Eckford, who has raised a very fine strain of fancy Pansies, has forwarded a box of blooms, to show the value of the strain. The prevailing drought has been severely felt at Boreatton and the district around it; notwithstanding this the flowers are fine and brilliantly coloured. Some of the tints are novel, and very beautiful, and also extremely varied. A rich velvety-maroon self, suffused with blue, and having a narrow edge of warm rose-pink, is of fine proportions, and well deserving the name given to it. A posy of these fancy Pansies would prove an acceptable present to anyone interested in these charming and popular flowers'.(19) The term 'self', a flower of one colour, was in practise less defined. There's no attribution to the piece but Dean's flowery finish gives him away.

Eight months later and we are in the spring of 1885. The first of the year's two '*Chronicle* articles on Henry begin where they left off, with pansies.

> 'A box of blooms of beautiful fancy Pansies sent by Mr. Henry Eckford, from Boreatton Park, Baschurch, is sufficiently noticeable as illustrating what Mr. Eckford is doing in the way of producing flowers with distinct scarlet and crimson hues. He has been fertilising and selecting with a view of intensifying the hues of colour found in Thomas Grainger and other flowers of that class, and he has succeeded to a remarkable extent, for many of them are singularly bright and effective. And they are also remarkable for their size, stoutness, smoothness, and the size, density, and clearness of the blotches. Perhaps there is something in the soil and climate that helps both the lustre of the colours and size of the flowers. Whether it is the intention of Mr. Eckford to name any of these varieties is not made known. Fancy Pansies are now not only numerous, but very fine, and any additions to the named varieties should be made with caution and care. Mr. Eckford is too good a florist not to recognise this necessity; at the same time the development in the fancy Pansy, high as it is at the present time, is confined by no arbitrary limits, and there is no knowing what remains to be revealed in the way of novelty'.[(20)]

What novelties were there in sweet peas this year? Another dry summer; difficult for pansy growers, especially, and tricky for sweet-pea growers too, as Richard Dean explained in his review of 1st August:

> 'A charming lot of new Sweet Peas has been forwarded by Mr. Henry Eckford... the flowers of great size, brilliantly coloured, distinct, and very fine. The latter quality, no doubt, is partly owing to the increased size seen in the new varieties previously raised by Mr. Eckford, and partly to his method of culture. Instead of following the somewhat barbarous practice of sowing the seeds thickly in a line – too thickly, indeed, for the plants to have sufficient space in which to develop themselves, and there is such a drain upon the soil that when a time of severe drought comes like that through which we are now passing, the flowering time is very short, if the plants are allowed to set their seed-pods.

If the old method is to be followed it would be much better to sow in a well manured trench like that used for Celery, so that a mulching could be given in hot weather, with plenty of water when necessary, and the earliest decaying flower-stems be cut away. Mr. Eckford grows his Sweet Peas either singly or in clumps of two or three plants in well manured soil; the result is that the individuals grow freely, branching profusely, and become perfect pictures of floral beauty, producing very fine blossoms'. As before, Henry's new varieties – 'Charmer', 'Charming', 'Purple King', 'Imperial Blue', 'Duchess of Edinburgh', 'Rosalind', 'Isa Eckford' and 'Mrs Eckford' – were named and described. Henry would bring out a primrose self 'Mrs Eckford' in 1892; this one was described as 'the standard delicate pinkish-mauve, shaded to a distinct margin of mauve, the wings and keel creamy-white, with a distinct edge of mauve – very fine indeed, and extremely pleasing'.

Richard Dean may have forgotten that he'd described three of these last year.

'It is remarkable', he continued, 'how very few varieties of Sweet Peas put in appearance up to 1882, but now that Mr. Eckford is busy with crossing the best varieties, the results are numerous and striking. It would not now be difficult to enumerate twenty distinct varieties of Sweet Peas. The fragrant Peas are among the most valuable, as they are one of the commonest of hardy annuals. A garden, whether large or small, seems to be incomplete without some of these fragrant flowers. Mr Eckford's new varieties appear to be very rich in perfume. When the box containing the bunches of flowers was unpacked the scent was delightfully acute. The bunches of flowers sent from Boreatton have been in water for a few days; their fragrance is still rich, and the blossoms retain it until they decay. R.D.' (21)

The '*Chronicle* had one further item on Henry that year, in November.

'In addition to the new varieties of Sweet Peas of Mr. H.Eckford's raising which Mr. W.Bull will distribute, Mr. J.C.Schmidt, of Erfurt, announces one of his own raising, named Vesuvius...'.(22)

So, Henry was attracting company from Germany at least. Here, in Germany and across the industrial and industrialising world, a flower like the sweet pea had a particular appeal. For the newly urbanised proletariat and the developing lower-middle-classes it was cheap, easy to grow and grew in a small space. For those without gardens it could be bought to brighten and make homely an unfamiliar dwelling in a strange town. Let us follow in the footsteps of a contemporary observer as he saunters the streets of London one Saturday night in the summer of 1884.

> Here is the 'Cheap Jack', who 'standing upon his cart, which is laden with all kinds of plants in season, a couple of flaming lamps illuminating the varied colours of the flowers beneath... soon gathers an admiring throng. Fuchsias just now are favourite window plants, and taking [one] up... he invites his admirers... to offer him 1s 6d for it'. He then 'runs down the whole gamut of intervening figures till he seems to think 4d a fair one...' or '3d, at which figures he clears off his stock presently... Jack, late at night, goes home a little the richer... [unless] a wet Saturday night interposes... Besides Cheap Jack, vendors of plants almost by thousands may be found on certain evenings in the chief market thoroughfares of the metropolis... In this way myriads of poor homes get plants to help light and cheer them, amidst the gloom of the London desert'.(23)

The author Molly Hughes, recalling her London childhood then, wrote of the street hawkers of various goods she espied from the house window, 'each with an appropriate cry: "Flowers all a-blowing and a-growing"' was one, the old word 'blowing' for flowering kept alive in a street-call when it had died out elsewhere. 'The long, wailing cry was a signal for us to crowd on to the ottoman to watch. Seeing our faces, the hawker would stop, look up eagerly, and hold up his goods...'.(24)

Henry, not in business but with seed to sell, had to turn to others. His old nurseryman, John Keynes of Salisbury, had died back in '78 and Henry was now using the upmarket William Bull nursery, as far from street hawkers and market traders as you can get. You may remember Henry purchasing some quite expensive plants from William Bull's nursery in The King's Road, Chelsea, when he was at Coleshill back in the early 'sixties.

~~~

Even in the winter there's work to do in a garden, especially in large estate gardens. Even in this very cold winter of 1885-86, with William Robinson complaining that starving rabbits had eaten most of his young
~~~

carnations. [25] (Well, he did promote wild gardens!). Work that gets put off in busier times gets done now: greenhouse interiors need scrubbing down; there's the odd fence post to replace; thinking of which, are the boundary walls and fences generally in good rabbit-proof order? And the gates into the walled garden? There are the paths to keep clear still... the odd tree to come down and a bonfire to look forward to, a new rockery to start or renew – all jobs to keep the men active, warm (ish) and contented... Seed sowing starts early in the year under glass, a nice warm job out of the rain. The boilers are critical now. They must never go out. Are the coal-stocks adequate? Even so, gardening is seasonal and the winter is a slack time. It is a time of short days and long nights. A time to brood in.

It would be quiet now in Henry and Emily's house in the gardens of Baschurch Park. Their children, the four who had managed to avoid or survive dangerous childhood illness were all grown up now, or nearly so. They may all have left home though John and Isa were, in their different ways, to return. Harry, (Henry) the eldest, was twenty-seven and a hansom-cab driver in Cardiff. Twenty-one year old John, by contrast, was on a sober white-collar career path in 'the city',presumably London. Isa (Isabella) had just turned twenty-five. Sixteen-year-old Peter appears to have been the black-sheep of the family, as Harry may have been too, for in some ways Peter took after him. What conversations passed the lips of Emily and Henry on these long winter evenings? Were their sons mostly a disappointment? Perhaps they were, if they failed to fit their parents' expectations. Theirs seemed to be contrary out of cussedness... Take Harry, driving those two-wheeled, one horse cabs. Cabmen had a reputation. Always have had. Loose fellows, idle drifters, drunks most of them. Some were decent of course, the self-employed ones mostly but – it was young men looking for excitement in the city. [26] Cities were the problem. They'd been there, Emily and Henry had. Couldn't get out fast enough.

The agricultural labourer who provided the manual labour during the years that Britain's biggest industry, farming, had prospered, had been paid a pittance. Now, in what George Orwell was to later call Newspeak, an agricultural 'depression' saw their wages rise. The bright lights still beckoned, however, and it was to towns and cities that country people like the Eckford and Stainer children continued to flock for work and better wages. Even by the turn of the century, in 1901, more than half the children of English working families were living in destitution. That is forty per-cent of all the children then living in England.[27] Three of the late Charlotte's sisters had moved to the Roath and Canton areas of Cardiff, two of them married to bakers. [28] Having uncles and aunts there may have been what drew Harry across the border; friendly faces to lodge with. Whatever fears Emily and Henry had had for their sons were to be realised that summer, and in the most brutal fashion.

SWANSEA HERALD AND NEATH GAZETTE,
JULY 14TH, 1886.
SHOCKING STREET ACCIDENT IN SWANSEA.
A MAN RUN OVER AND KILLED.

A distressing accident occurred, on Thursday night, in De-la-Beche-street. The Mount Pleasant Sunday School children had been to Sketty Park for their annual outing, and the Hafod portion returned via De-la-Beche-street, the time being about half past ten. They were riding in a well filled tram, and while ascending De-la-Beche-street, the driver, who was by the side of the car, ran and jumped on to the foot-board, but missing his footing, fell under one of the wheels, which cut off one of his legs completely below the knee. The poor fellow was otherwise injured, and was altogether in a deplorable condition. Mr.Phelps, who lives near, went with great promptitude for Dr.Roberts, who came at once; someone else communicated with Dr.Griffiths, who gave his assistance readily, and Dr.Paddon also appeared hurriedly on the spot, followed by Mr.Sugrue and Inspector Rees of the Tramway Company. The unfortunate man was conveyed immediately to the hospital. His name was Henry Eckford, and he lived at 23, Gam Street. He had only been in Swansea three weeks, and was temporarily engaged by the Tramway Company. The unfortunate man died shortly after his arrival at the hospital. An inquest will be held.

THE INQUEST.

The distressing accident that occurred, on Thursday night, in De-la-Beche-street, Swansea, and which soon afterwards terminated fatally, on Friday afternoon, formed the subject of an enquiry held at the hospital before Mr.Edward Strick, and the following jury: Messrs. Wm. Manaton (foreman), Alfred Smith, Jeremiah Foresdike, Henry Hardwick, Jno. Evans, Thos. Evans, John Gibbs, John Griffiths, Hy. Buzocott, Thos. Bowen, Wm. Cross, Wm. Smith and Evan James. The circumstances of the accident have already been reported. The deceased, on this occasion, gave his services, but sometimes was temporarily employed by the Tramway Company, one of whose trams on the evening in question ran over him, and the injuries thus received very soon resulted in death. The deceased, Henry Eckford, it has been since ascertained, is a native of Baschurch, near Weston-super-Mare. Mr J.Viner Leeder (Simons and Leeder) watched the proceedings on behalf of the Tramway Company, whose manager (Mr. Sugrue) was also in attendance. The jury having been sworn and the body viewed, the following depositions were taken:

William Helps, a driver, in the employ of the Swansea Tramway Co., of 23, Gam-street, said the deceased had lodged with him for about a month. He had known him before at Cardiff. On his arrival at Swansea, the deceased obtained employment in the Swansea Tramway Company as a tramcar driver. He was employed off and on, and was not a constant hand. In Cardiff, he was employed as a cab driver. On Thursday night, about a quarter to eleven, witness was engaged on a special car from Rutland-street to the Hafod. There were three horses attached to the car. As they were passing the Albert Hall, deceased gave him the leading rein, and got off the car. After that he (witness) saw nothing more of him. After proceeding some twenty yards the car gave a jerk, and someone cried out that there was a man under the wheels. He immediately pulled up. By this time a large crowd had collected. In the car were a large number of Sunday School children, belonging to the Mount Pleasant Branch School, at Hafod, who were returning from their annual outing at Sketty Park. On pulling up he did not get off, fearing that the children might cause still further mischief with the horses, and he saw nothing further of [the] deceased until he came to the infirmary to enquire after him. Deceased, however, had died before he could see him. The witness did not see him attempt to get on to the car, nor did he see him on the car. He had no idea how he got under the wheels. The deceased was not intoxicated, and he was a steady, sober man. He was not, on the previous evening, in the company's employ. He gave his services voluntarily.

Dr. Humphreys, who was fetched directly after the accident, stated that he found deceased lying in the road, on his back. His right leg was smashed, but he was perfectly conscious. In reply to questions witness heard him say "No, no, it was nobody's fault. It could not be helped". He then asked witness to write down the address of his father, so that a telegram might at once be sent to him. Witness sent a cab to the High-street station for a stretcher, and deceased was thus conveyed to the hospital. The shock and bleeding were the cause of death. The lower part of the limb was ripped to pieces, and the body was lying in a pool of blood. There were other injuries also to the right arm and other parts.

George Clements, driver, who had charge of the Cwmbwrla car, and was returning home with it on the evening in question, gave evidence as to being ordered to stop, because the cars in front were delayed. The witness detached his horses to assist. The deceased, after the reins had been given to Helps, stood away from the car, in the road on the offside, and was talking to the witness as the passengers took their seats. The car had got about five yards when the deceased stepped on to the footboard, and was about to take hold of the rails, when he missed with his left hand, and swerved by his right, and turning round his foot slipped off the step, and he was dragged under the wheels. Directly the second wheel had passed over him the car stopped. Asked whose fault it was deceased said "It is no one's fault. I slipped." Witness did not hear him say anything else except ask for water.

Alex McConochie, Rodney-street, tramway inspector, corroborated as to the manner in which the accident occurred.

By Mr. Sugrue: It is a part of my duty to see that the drivers are in their proper posts, and, therefore, I know that the deceased was not employed upon this occasion.

Dr. Nelson Jones, house surgeon at the hospital, said the deceased was brought to the hospital about 11.30 on the previous evening. The muscle and bones of his right leg were smashed as far as the groin, and his right arm was broken, the bones smashed in several places, and the muscles torn. The deceased, was conscious, but was in a state of complete collapse. The loss of blood must have been great. Witness sent for assistance. They did what they could for the poor fellow, but he died about 1 o'clock. It was quite a hopeless case.

> Mr. Thomas Yorath, draper, High-street, superintendent of the Hafod Branch of the Mount Pleasant Sunday School, attended the inquest and asked to be allowed to make a statement, and was permitted to do so. He said he could not give any further information respecting the case, but of those that had given evidence he thought he was the only one not connected with the Tramway Company. He fully confirmed everything that had been said. He would like also to bear his testimony to the care and caution of the officials of the company, which could not possibly have been exceeded. Mr. Sugrue and Mr. Rees, the inspector, were present, and saw that the children were placed into each car, properly and comfortably, and Mr. Sugrue himself superintended the starting of the cars, and would not allow one to start until the others were ready. Since the accident had occurred, he thought it only due to the officials of the company that he, as an impartial person not connected with the company, should bear this testimony.
>
> The Coroner, in summing up, dwelt upon the absence of blame on the part of the tramway officials, and upon the risk of jumping off or on to the street cars while they were in motion.
>
> The jury endorsed the observations of Mr. Yorath, and concurring also with those of the coroner, they gave a verdict of "Accidental death," no blame being attributable to anyone.(29)

What do you say? Harry's (Henry's) death certificate says that his father was present at the end. But if Henry senior had been in Baschurch it would have been almost impossible. It was a tragedy for Harry, his family and friends, certainly. But a bigger tragedy was the continuing unconcern, indifference and all-round cavalier disregard of the dangers posed by the industrial era, a hundred years in the making by 1886. Charles Dickens had long before noted this peculiarity of the public: 'That they are not accustomed to calculate hazards and dangers with any nicety, we may know from their rash exposure of themselves in overcrowded steamboats, and unsafe conveyances of all kinds'.(30) The horse-drawn trams, like the unsafe conveyance that killed Harry, were similar in some respects to the railways, for their wheels ran on rails, set in the road. Perhaps they appeared to be less dangerous than the railways, and were treated with even less care.

In the early railway days, writes the historian Christopher Hibbert, 'there were fifteen times as many fatal accidents in England as there were in Germany, many of them the fault of the passengers' and public themselves. The railway companies recorded them in their reports: 'Injured, jumping out after his hat'; 'fell off, riding on the side of a wagon'; 'skull broken, riding on the top of a carriage, came into collision with a bridge'; 'fell out of a third-class carriage while pushing and jostling'; 'guard's head struck against a bridge, attempting to remove a passenger who had improperly seated himself outside'. One year there were twenty-two 'serious accidents' caused by passengers jumping off 'when the carriages were going at speed, generally after their hats, and five persons were run over when lying either drunk or asleep upon the line'.[(31)]

That industrialists could set their workers' lives at naught is well attested and well recorded. But so it seems could people so regards their own lives. The situation on the roads was as bad. In the contemporary world of horse-drawn private carriages and commercial delivery vans, of public trams, Hansom cabs and four-wheeled 'growlers' the death rate in 1878, for example, was 237 with 3,961 injuries. And that was just on London streets. [(32)] Multiply that across the nation and deaths per head of population far outweigh the present national average of five deaths a day.

On the 'shocking street accident in Swansea' the south-Wales public would have read between the lines, as they were intended to do. 'He was employed off and on, and was not a constant hand. In Cardiff he was employed as a cab driver'. Well well, a sad case, they would have thought. But what do you expect? But, you might reply, he 'was not intoxicated and he was a steady, sober man'. Ah, but we don't speak ill of the dead here in Wales. He may have been a steady, sober man. He may have been. And he may not have been. Why was Harry working voluntarily? Perhaps he saw it as a way into the company's good books and regular employment. He was now twenty-eight and probably thinking to settle down. 'No blame being attributable to anyone' was a constant refrain of the newspapers. It was said of the most horrendous accidents where the apportioning of blame was quite obvious. The attitude permeated people's thinking. 'It is no one's fault' Harry had said. 'I slipped.' Well, there's safety of access to the vehicle for a start. It was late in the evening. Had he worked all day? Was he rested? What about job training? But these are modern approaches and were to be a long time in the making. As a footnote, Harry was of course not a native of Baschurch but of Coleshill, and Baschurch is nowhere near Weston-super-Mare.

~~~

Henry and Emily were lucky to be at Boreatton Park. It held a community that they'd lived and worked among, here and at Sandywell, for the last ten years. For Henry, there were many friends and contacts in the wider world too. So there was plenty of empathic sympathy to comfort them. What family had not lost a child to illness or accident? Victorian urbanisation had driven up infant mortality rates, reversing for a while a long steady decline during the Georgian era. In rural mainland Britain, death claimed at least one in ten of its infant children but in industrial and urbanised counties like Lancashire it was nearer one in five.[33] It was a pattern that mortality rates in general were lower in the villages than in the towns and cities. So it is not surprising, statistically speaking, that Harry, though by then a young man, had died in Swansea rather than in Shropshire. Provided you had warmth, shelter and the means to acquire enough suitable food, as Emily and Henry did, then countryside living held the advantage. It was the logical place to stay but logic has a way of placing itself at the periphery of our thoughts and plans.

There were no pansies or sweet peas of Henry's to report on that year. Possibly at Richard Dean's compassionate suggestion, Henry now took his mind off events by putting his spare time and energies into composing a long article for *The Gardeners' Chronicle*. It was to be on the 'improvements of peas and other plants'. Garden peas that is, and in 'other plants' Henry quietly excluded any reference to improvements that anyone but himself was making to sweet peas and pansies. He may have been distracted but he wasn't *that* distracted! An article is a rare event with Henry and is worth quoting in full. It appeared three months after Harry's death on Saturday 2nd October, 1886.

Improvement of peas and other plants.

> In the trial grounds of a well known firm of seedsmen there were growing in the summer 167 varieties of garden Peas which were sown side by side, so that the merits of different sorts might be tested by comparison. It is by such experiments as these that improvements are secured, and the Ringleaders, Alphas, and other Peas perfect of their kind, developed. A useful modification of this particular vegetable which is now in progress is the diminution of the haulm [stem] of the taller sorts. The value of a Pea depends on the character of its pods and the seed contained in them,
~~~

and the height of some of the tall growing sorts is obviously a disadvantage. The firm in question therefore have set themselves the task of reducing the height of the crop without diminishing the number of pods. They employ their own hybridisers in the prosecution of this task, but like all others in the trade who undertake improvements of any kind they avail themselves of the labours of amateurs. There are enthusiasts and experts in every branch of plant improvement who devote their attention to a few, or perhaps to only one subject, and who attain such success as the professional experts employed by the various firms could hardly expect to accomplish, their attention being generally too much divided. It has happened, therefore, that the most noted breeders of fruits, flowers, and vegetables have been amateurs, and in the case of the Pea the most skilful manipulator who has outstripped all rivals in the work of improvement is a lady. By the magic of the various methods known to those who practise cross-breeding and selection, she has altered the Pea at will, both in reducing the superabundance of the haulm and also by enlarging the pods and seeds.

It can easily be imagined that a breeder of plants will desire, like an author of books, wide publication or distribution, and as he cannot be his own advertiser, on account of the expense, he leaves this special business in the abler hands of one of the great firms of seedsmen. This explains the large number of Peas in the collection just referred to. The firm we have in view employ their own hybridisers in various departments, but they are always anxious to acquire the best of everything by whomsoever it may be originated, and from one source or another they have become the possessors of from thirty to forty sorts of cross-bred Peas of recent date. These have, in fact, been raised during the past three or four years, and all of them were grown last summer in the trial grounds, and subjected to a searching examination.

Among them is a sugary Pea of great merit, which grew side by side with the Duke of Albany, which is a good "all round" Pea, of excellent quality, and profitable for market. But the sugary Pea is not a productive kind. It is exquisite in flavour, but the haulm is deficient, and the pods and seeds are small, so that this incomparable Pea on the table is unfortunately a shy bearer and therefore unprofitable.

In the improvement of such a variety by the increase of the haulm and enlargement of the seeds, without loss of flavour, the breeder finds his opportunity, and when we consider that the marrowfat tribe have in some cases been reduced from 7 feet in stature to the much more convenient height of between 2 and 3 feet without any diminution of the size of the seeds and the yield of the crop, we may hope that the sugary dwarf will be enlarged and rendered more productive without loss of quality.

In passing along the rows, several sorts claimed notice, such as the little early Pea Bijou, which does not exceed 18 inches in stature, and the still more dwarfish American Wonder. By cross-breeding with such varieties as these the giants of the Pea tribe may be readily reduced to a moderate size, for if you cross a "first early" Pea with a tall late marrowfat, you will naturally be landed between two extremities both in regard to superabundant haulm and to the size of the pods; and perhaps the reduction of the size of the pods may cause them to be better filled and more numerous and productive as they are in some of the most prolific sorts whose pods hang invariably in pairs and almost break down the stems that bear them.

Many of the diversities we observe in different varieties of garden plants – both fruits, flowers, and vegetables – are entirely due to the skill of breeders and selectors in availing themselves of a universal law of Nature. They have observed what Mr. Darwin calls the "slight differences" between individuals and turned them to account in the artful moulding of a new variety. We need not confine ourselves to culinary plants. The history of the Fuchsia affords as good an example of the results of cross-breeding as that of the Pea. About half a century ago the only Fuchsia familiar to gardeners was F. coccinea, an old sort, which is still common in the gardens of farmhouses and cottages, in the southern counties. About sixty years ago new varieties of this flower were received from the seaboard of the Pacific, where the genus is very widely distributed.

The well-known Fuchsia fulgens of our greenhouses, with exquisite long tubed flowers, is one of these, and another is the splendid Fuchsia Riccartoni[i], a hardy specimen, whose great bushes flourish in the West of England and in Scotland. Many persons must be familiar with Fuchsia corymbiflora, introduced by Messrs. Standish from Peru, a conservatory plant which is sometimes moved out-of-doors in summer when its long and waxen corymb of flowers impart a tropical aspect to the garden.

These are varied sorts, but the Fuchsias of the present day are unnumbered, and they have been produced by the skill of the same class of experts who have modified most of the cultivated plants in turn, amateur and professional hybridisers and cross fertilisers.

The seed of a superior hybridised Fuchsia has been sold at the rate of 50 guineas an ounce, though no one ever possessed an ounce of such seed, the yield of an "improved" plant being in inverse ratio to the monstrosity of the blossoms, for artificial modification is usually the reverse of that natural modification which sustains a plant in its struggle for existence; and the more highly plants are bred, the greater the disturbance of the reproductive system, and the greater the tendency to modification in their seedlings.

Our modern flowers are not like the old ones, and the gardens of the last century wore raiments widely different from those of to-day. We have shown how the Fuchsia has been changed and its forms multiplied within a short period, and those who have not given much attention to this subject would be surprised at the number of flowers of which a similar story might be told. Most persons know the Antirrhinum or Snapdragon, but the "good old-fashioned" English flower known by that name will hardly be found now except in old-fashioned gardens. An eminent florist has preserved a plant of the old sort, for the sake of comparing it with the modern forms which owe their smart appearance to the arts practised on this and other flowers by himself. His unimproved specimen of this plant is a rampant straggler, with few blossoms – the Snapdragon of our boyhood, well adapted for large flower-beds, or a front position in shrubberies, where it might do battle with the Foxglove and the Phlox, or the Michaelmas

Daisy. Modern Antirrhinums, on the contrary, consist of several "sections" – dwarf, intermediate, and tall, ranging from 1 to 2 feet in height, all of them having a compact habit, with many distinct colours in each section.

It would be a long story to relate the modifications of each particular florist's flower, and it seems needless to do so, since the same process of selection, with or without cross-breeding, has been applied to each one of them. Among the much changed plants are China Asters. Some plants are far more stubborn than others, but it is no wonder if Asters readily become mutable, considering the predisposition to change which repeated acts of cross-breeding invariably introduced by occasioning a mixture of breed and such a varied parentage that none can tell what freak of reversion a seedling may display. The Aster is one of those popular flowers which has been coddled by florists with wonderful solicitude. We examined half an acre of them growing last summer in a series of beds. Among the varieties were the Asters, known as Chrysanthemum-flowered, Paeony-flowered, having the petals curved inwards, the pyramidal section, and some wonderful dwarfs not more than 7 inches high, and most profuse in the number of their blossoms.

As a rule, Aster seed is grown abroad, but an eminent firm of seedsmen are making an attempt to deprive foreign countries of the monopoly by producing it in England, their immediate object being the improvement of the flower, since the blossoms of home-grown seed have proved both brighter and more durable than those from German seed. The growth of Aster seed, even in our driest district, requires special care owing to the damp weather of autumn, which sometimes rots the blossom of this late ripening seed crop. The blossom may be saved, however, by protecting the beds with a light roof covered with some such material as Willesden paper, to prevent the ball of petals from becoming soaked by rain. With this reference to an incidental method of improvement, we conclude without having by any means completed what is in fact a "story without an end." H.E.[(34)]

Taking Henry's gardening skills with his writing ability, his knowledge of plant dispersion – 'the seaboard of the Pacific, where the genus [*Fuchsia coccinea*] is very widely distributed' – his interest in evolution – 'what Mr. Darwin calls the "slight differences" between individuals' – one can see that Henry is likely to have spent as much of his time reading and in written and personal contact with his peers as he spent on his day-to-day garden supervising and hybridising work. He is an advertisement for an education that fitted the aspiring and gifted Scots lad with an appetite to develop a life of depth and richness. That said, one wonders who this 'most skilful' and successful lady hybridist was. One possible answer to that is the passionate gardener Ellen Ann Willmott (1858-1934). She was known to have raised new plants generally and was, as we'll see, to have an Eckford sweet pea named for her. But Ellen Willmott seems to have been more drawn to flowers than vegetables. The reference remains teasing, there being so few female gardeners then (or so few spoken of).

A fortnight after Henry's '*Chronicle* article, a gardener wrote in to take him down a peg. 'I am much... disappointed with Eckford's Duke of Connaught, sent out by Mr. Bull. After such high recommendation, and high price too – 2s 6d per quarter pint – one expects to get something for the money; it is also a mixed sort. I cannot see any improvement on many other kinds... gardeners are often driven to save their own seed... I am beginning to think raising new Peas... must be a very profitable business and after so much disappointment shall now keep to older varieties... W. Divers, Ketton Hall'.[35] To be fair to Henry, Mr. Divers was having a general moan about all new pea varieties, not just Henry's. And complaining about hype! So the Victorians had it too.

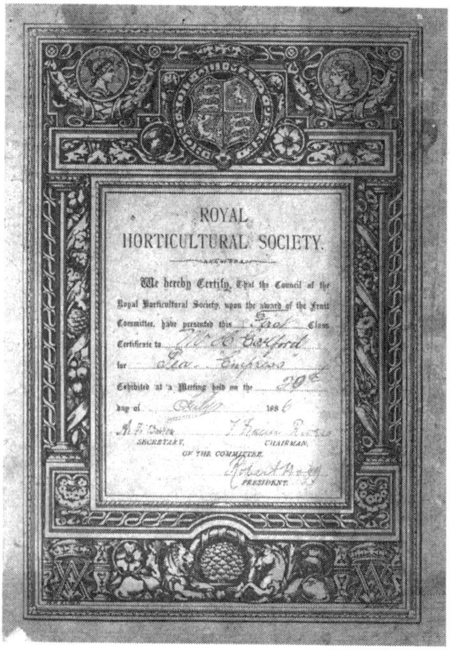

ROYAL
HORTICULTURAL SOCIETY.

We hereby Certify, That the Council of the Royal Horticultural Society, upon the award of the Fruit Committee, have presented this ... Class Certificate to ... for ... Exhibited at a Meeting held on the ... day of ... 1886

SECRETARY, CHAIRMAN,
OF THE COMMITTEE.
PRESIDENT.

RHS garden pea certificate, 1886, "Empress".

A month later, in November, further tragedy struck at our corner of Gloucestershire. Dr. Sankey's son William, a medical officer under his father at Boreatton, died at the early age of thirty-four.[36] 1886 became a year rent in grief at Boreatton Park.

~~~

Eighteen-eighty-seven was the year of Queen Victoria's Golden Jubilee. Addressing her from the horticultural perspective, 'Your majesty's subjects are eminently a garden-loving people' declared the Royal Horticultural Society, whose offices had 'directed its energies to the guidance and development of this national taste, which is as elevating and pure as it is practical and useful'.[37] Although the R.H.S. had come into existence (as The Horticultural Society) in 1804, it wasn't until 1891, when Henry was settled in trade, that he joined it.[38] He remained a full member until he died, and although he may well have been an R.H.S. judge, he never, it seems, became so involved as to become more than a temporary or co-opted member of the fruit and vegetable or floral committees. Unlike Richard Dean, who was regularly on both, or even Dr. Sankey, who had been on the floral committee back in the early 'sixties.

The R.H.S. has always had a rather patrician air about it. For most of the nineteenth-century, membership was denied to most working gardeners, as it was to most working people, by its from one to four guinea (one guinea = £1 – 1s) annual membership subscription. Women were thin in its ranks too, as they were in the profession generally. Although the first women members had been admitted in June, 1830, [39] it remained much as a well-to-do's masculine preserve until the 1890s saw a rise in women members, reflecting the rise of college-educated women gardeners, who were doing surprisingly well in the new colleges' examinations. Ordinary professional gardeners were encouraged to join then too, at a cheaper, patronising, 'associate' rate.

By the summer of Jubilee Year, life at Boreatton Park was getting back to normality. It was a hot summer that year but in spite of it, Henry's sweet peas had done well and Richard Dean had plenty of praise for them in the columns of *The Gardeners' Chronicle*.

> 'Mr. H.Eckford must be congratulated on the batch of charming new varieties of Sweet Peas he sent to a recent meeting of the Royal Horticultural Society. Like other raisers of new varieties of any flower, Mr. Eckford can see in his seedlings points of difference that are not so readily discovered by those who have a less intimate knowledge of a florist's particular fancy. But Mr. Eckford is raising
~~~

> new and distinct varieties that are gladly welcomed by those who take delight in these fragrant Sweet Peas. First-class Certificates of Merit were awarded to Mauve Queen, the standards and wings delicate mauve... Splendour... rose-coloured... and Primrose... the first real advance towards a yellow Sweet Pea... Two other varieties deserved this award, because of their disctinctiveness, viz., Apple Blossom... and Miss Hunt [probably named after a daughter of Boreatton Park's owners]... very pretty. Other good varieties were Mrs. Eckford... Delight... and Maggie Ewing...'.(40)

One attractive trait of an era less competitive than ours is that it lends itself more readily to an objective searching for the truth, rather than taking the 'fight your corner' approach. By the century's end there were some pretty impressive results in technology and medicine to show for it. Richard Dean was typical of his age. His friendship with Henry didn't mean that he'd defend and promote him above others. It was only right that Henry's strengths and weaknesses as a hybridist should be evaluated in comparison with his peers. Henry himself would have expected no less. His and others' garden peas were the subject of a lengthy review by Richard Dean in the '*Chronicle* that November.

> 'Peas New And Old. It is necessary occasionally to pass in review, and compare with old standard sorts, the new Peas that appear each season', it began. 'Some few are obtained by means of cross-fertilisation, the majority are but selections, and thus it is so many old friends are made to appear under new names. Peas that are obtained from *bonâ fide* crosses require to be carefully grown for some time, and very carefully selected, before they are named and sent out, as they are pretty certain to show a mixed character at first. All Peas, if they are to be maintained at a high standard of quality, need most persistent selection; some sorts deteriorate rapidly if neglected. During the present summer an admirable opportunity was afforded for comparing Peas new and old, as in the strong ground of Messrs. Hurst & Son's trial garden at East Croydon, they did well, and were seen to the best advantage...
>
> Eckford's Victor is very like [the variety] G.F.Wilson...
>
> Hurst and Son's New Wrinkled is a very fine Pea, raised by Eckford: in a dry state the seed is large, green and wrinkled, it has very fine

pale green pods like a large podded Champion of England, but earlier, appearing as if it would make a fine exhibition Pea; 5 feet...

Eckford's Magnificent is scarcely sufficiently distinct from Paragon, [cropping] a trifle later; it bears fine, well-filled, square-topped pods...

The grand old Jeye's Conqueror or Ne Plus Ultra appears under several new names. Fraser's Defiance is one of them, Eckford's Progress is simply a fine strain of it... It will take years to drive [this] Pea out of cultivation and it is difficult to match it for flavour'. (41) The present-day paragons, conquerors and victors of East Croydon's 'strong ground' are its skyscrapers. Perhaps a few peas escaped, and flower now unnoticed along the railway embankments out of East Croydon station.

~~~

The climate in Britain warmed throughout the second half of the century but the winters were still much colder than ours are today. Taking the central latitudes of the British Isles as a norm, a city like Edinburgh might see snow in September and be well into May before it had seen the last of it. (42) A few years before Henry's arrival in Boreatton, five degrees of frost on 18th October, 1877 merely brought forth the comment that it was a 'fine autumn day' and it snowed every day for five days at Christmas. In January 1880, temperatures fell frequently to 14°F (-10°C). Daytime temperatures were mostly below freezing too. (43) The winter wasn't noted as a particularly cold one.

The 'eighties saw a run of dry summers, but nothing lasts forever. The winter snows and the spring of 'eighty-eight gave way to the first wet summer for some years. It was a summer that, in the garden, rather suited plants like violets, Impatiens and Begonias and promoted the charms of pansies over Pelargoniums for the first time in a while. It was a quiet year for the garden pea and the sweet pea, with nothing of note in the press. Henry had, however, sent his sweet pea flowers to the R.H.S. and was 'Highly Commended' for the 'strain', in this case the quality of the flowers of the strain, compared to the same varieties grown elsewhere by others(44). The end of the dry summers was a change of routine. A change in one sphere can induce thoughts of change in another.

It was late autumn. In the kitchen garden, all the peas and pea-sticks had now been pulled-up. With brushwood easily available, the sticks were
~~~

probably used for kindling, though some may have been cleaned and put away in the dry for next year. The garden, the pleasure grounds, the park and all the Boreatton estate began to settle down for the winter. Poor Dr. Sankey, having only recently lost his son and, perhaps, lacking Henry's constitution, may just have been overwhelmed by it all for he was to die the following year. The signs may have been there for a while and Henry, in his self-interest, was wise to consider his position. Whichever of Dr. Sankey's other sons took over, allowing Boreatton Park's head gardener to continue with his passion for creating new plant varieties and neglecting, to some degree, his official duties was likely to be questioned. The problem was, who was going to employ, and indulge, a sixty-five year old gardener? To stay put or move was a dilemma to which there was no obvious answer.

Henry and Emily kept their cards close to their chests. Henry would appear as resident at Boreatton Gardens, Baschurch on the 1890 electoral register, compiled it would seem, when such things were taken more leisurely, a couple of years before.(45) As the smoke from the last of the autumn bonfires lifted and disappeared into the cool autumn air, Emily and Henry upped-sticks and left Baschurch.

In 1889, two obituaries appeared, one above the other, in the columns of the March 16th edition of *The Gardeners' Chronicle*. The first, Henry and Emily may not have known about. The second, the couple living not far from Baschurch, in Wem, they probably did.

> 'In the death of the Earl of Radnor... the poor [have lost] a kind and generous friend, and horticulture a liberal patron. He was a fellow of the Royal Horticultural and Royal Botanic Societies of London... [and] was a very good judge of all kinds of fruits and flowers...
>
> W.H.O. Sankey, M.D. – We regret to announce the death of William Henry Octavious Sankey, M.D., at Boreatton Park, Shrewsbury, on the 8th inst. The deceased gentleman was an enthusiastic horticulturist, much devoted to florists' flowers and the raising of seedlings'. (46)

Henry had been indirectly involved in commerce, having his plant creations sold through nurseries, but he'd never been in business on his own account. Now, the rise of the suburban garden, foreseen back in the 'thirties and 'forties by J.C. Loudon in his *Suburban Gardener and Villa Companion* and his *Suburban Horticulturist,* had combined with a general if erratic rise in living standards to create an expanding commercial sector for horticulture. Did Henry now recall St. Catherines, the mansion in Liberton that had been bought by an Edinburgh seedsman back in the 1840s? Whether or not Henry ever intended to tread the commercial path, his options had become limited and he took the plunge. Looking back from 1903, he recalled telling the postman 'that I wanted a place quite out of the way, where I could grow my Peas. He said "I think there's some land to be let at Wem, which would just suit you". So I came to look at it, and I took it'.(47)

The end, of which commerce was now to be the means, was in enabling Henry to continue with his garden and sweet pea work. With that goal in mind, a nursery would have to be a success, as indeed his so became. From being pitched uncertainly into commerce, success as an innovative nurseryman must have bemused him, as happens when our talents catch us pleasantly by surprise.

References to Chapter Eight

(1) 'New Garden Peas at Boreatton Park', *The Gardeners' Chronicle,* (18-8-1883), p.203.

(2) A.N. Wilson, *The Victorians*, (2002), p.456.

(3) Georgina Battiscombe, *Mrs Gladstone: The Portrait of a Marriage*, (1956), p.215.

(4) J.C.Loudon, *An Encyclopaedia of Gardening*, (1822), p.689; quote.

(5) William Cobbett, *The English Gardener*, 1829, (1996), p.123.

(6) J.C. Loudon, op.cit., p.1293.

(7) Lizzie Boyd, ed., *British Cookery*, (1988), pp.24, 88.

(8) Richard Sudell, ed., *The New Illustrated Gardening Encyclopaedia*, (1932), p.918.

(9) 'Home Correspondence', *The Gardeners' Chronicle*, (11-8-1866), p.756.

(10) 'Home Correspondence', *The Gardeners' Chronicle,* (22-9-1866), p.900.

(11) 'Notes on some changes and variations in the offspring of cross-fertilised peas', *Journal of the Royal Horticultural Society*, (1872), pp. 10-14.

(12) 'Eckford's New Sweet Peas', *The Gardeners' Chronicle,* (1-9-1883), p.264.

(13) 'Fancy Pansies in the Mass', *The Gardeners' Chronicle,* (15-9-1883), pp.338-339.

(14) 'Pansies', *The Gardeners' Magazine*, (30-9-1882), p.522.

(15) 'The Pansy and Viola', *The Gardeners' Chronicle*, (15-7-1899), p.44.

(16) Ruth Duthie, *Florists' Flowers and Societies*, (1988), p.87.

(17) See: G.W. Johnson, *The Gardeners' Dictionary*, (1877 edn.), p.814; H.J. Wright, W.P. Wright, *Beautiful Flowers and How to Grow Them*, (1922), p.118; Ruth Duthie, op.cit., p.87; Richard Mabey, *Flora Britannica*, (1996), pp.129-131.

(18) 'Mr. Eckford's Sweet Peas', The *Gardeners' Chronicle* (26-7-1884), p.114.

(19) 'Fancy Pansies from Boreatton', *The Gardeners' Chronicle*, (20-9-1884), p.371.

(20) 'Fancy Pansies from Boreatton Park', *The Gardeners' Chronicle*, (23-5-1885), p.670.

(21) 'Sweet Peas from Boreatton Park', *The Gardeners' Chronicle*, (1-8-1885), p.150.

(22) 'New Sweet Pea', *The Gardeners' Chronicle*, (14-11-1885), p.625.

(23) 'The People's Plants', *The Gardeners' Chronicle*, (26-7-1884), p.114.

(24) Molly Vivian Hughes, *A London Family, 1870-1900*, (1946), pp.5-6.

(25) G.Taylor, *The Victorian Flower Garden,* (1952), p.88.

(26) Christopher Hibbert, *The English*, (1988 edn.), p.655.

(27) Peter Laslett, *The World We Have Lost*, (1965), p.201.

(28) Letter from Derek Stainer Knight, great-grandson of Charlotte Stainer's sister Elizabeth, to John Good, 28-10-1992.

(29) 'Shocking Street Accident in Swansea', *Swansea Herald and Neath Gazette*, (14-7-1886), p.3.

(30) Charles Dickens, 'Lying Awake', in *Reprinted Pieces*, (1898 edn.), pp.64-65.

(31) C.Hibbert, op.cit., p.651.

(32) Jennifer Lang, *An Assemblage of 19th Century Horses & Carriages*, (1971), no paging.

(33) Pamela Horn, *Labouring Life in the Victorian Countryside,* (1976), p.184.

(34) 'Improvements of Peas and Other Plants', *The Gardeners' Chronicle*, (2-10-1886), pp.436-437.

(35) 'Home Correspondence', *The Gardeners' Chronicle*, (16-10-1886), pp.501-502.

(36) E.Scot Skirving, (compiler), *Cheltenham College Register*, 1841-1927, (1928), p.179.

(37) *Report of the Council of the Royal Horticultural Society,* (1887), p.27.

(38) *Book of Arrangements of the Royal Horticultural Society*, 1889-1897, p.71. Henry's name appears for the first time in the 'List of Fellows and Associates for 1891-92'.

(39) Brent Elliott, *The Royal Horticultural Society: A History*, 1804-2004, (2004), p.324. Serendipitously, Judith Anne, Lady Radnor, the wife of Henry's Coleshill employer, was the first woman ('the first lady Fellow') to join the then Horticultural Society in June, 1830.

(40) 'New Sweet Peas', *The Gardeners' Chronicle*, (6-8-1887), p.170.

(41) 'Peas New and Old', *The Gardeners' Chronicle*, (19-11-1887), pp.618-619.

(42) Robert C.Mossman, The Meteorology of Edinburgh, Part 2, *Transactions of the Royal Society of Edinburgh*, Vol.39, (1896-1899), p.169.

(43) 6001/4687, 'Notes on the Climate...', (1877-1880), Shropshire Record Office.

(44) 'Awards Made', *Journal of the Royal Horticultural Society*, (1889), p.xxxi.

(45) No.115, Register of Electors, West Division of Shropshire, (1890), Shropshire R.O.

(46) 'Obituary', *The Gardeners' Chronicle*, (16-3-1889), p.343.

(47) 'Famous Gardeners at Home', *Garden Life*, (26-9-1903), p.464.

Wem High Street in the Edwardian Era.

Chapter Nine

Remember how, in chapter one, we walked down to Liberton from Edinburgh? That landscape, the 'slow heaving sea' of this part of the Lothians, would have been farmland when Henry knew it. Now, travel up from Shrewsbury to Wem through the woods and fields of north Shropshire. The lie of the land, an undulating plain surrounded by hills to the south and west is uncannily familiar. Henry had been edging back northwards ever since leaving Coleshill. Wem was one final tack, to the northeast. It was as if, in some way, Henry was choosing to round off his life in the landscape of childhood.

One of Henry's later visitors spoke of 'the little village of Wem'. (1) He can't have looked around much. With a population of some four-thousand people, the same as Liberton's was when Henry was a boy, it was a small market town. Following Saxon settlement it acquired a Norman Castle and was big enough to survive disasters caused by the fifteenth-century War of the Roses and a seventeenth-century fire. It withstood a Royalist siege in The Civil War, the town eventually conquering its much bigger Royalist neighbour Shrewsbury. They're still edgy about Wem in Shrewsbury and angry at the King's fate. 'They're a funny lot in Wem' was the mildest comment I heard there. The Civil War is only eleven generations away.

Thomas Adams, a future lord mayor of London was born here in 1586 and founded the local grammar school. Edwardian additions to Wem included a town hall. An eighteenth-century artist, John Astley, was born here in Noble Street. A more notable resident of Noble Street was the, until recently, neglected essayist, critic and short-story writer William Hazlitt (1778-1830). He arrived here as a boy in 1787 and stayed, on and off, for the next few years. (2) Another eighteenth-century figure, born nearby, was John Ireland, friend and biographer of the artist William Hogarth. An unfortunate and undeserved association for Wem is with George Jeffreys,

'Hanging Judge Jeffreys' from Wrexham, whose name 'in an age of judicial barbarism... became a by-word for cruelty'[(3)] in the seventeenth-century. He became Baron Jeffreys of Wem, having purchased the Manor but he never seems to have lived in his house here, Lowe Hall, or even visited it.[(4)]

With no wages and no welfare state Henry and Emily had only their savings to bridge the gap between leaving Boreatton Gardens and starting a nursery. There were the costs of hired help in getting the nine-acre ground ready and the seed sown. All this and rent, food and heating costs... It's likely that they had to borrow some money. That the nursery was in production by the summer of 1889 is evidenced by the R.H.S's Floral Committee, meeting that July in Chiswick. 'Mr. H. Eckford, Wem, Salop, sent flowers of his well-known beautiful strain of Sweet Peas', they reported. Two weeks later the committee announced a unanimous decision to give Henry an Award of Merit for the strain, which was, they said, 'in excellent variety, and of clear, beautiful colours'.[(5)]

Wherever they lodged initially, (and at the Hazlitts' old house is one rumour) by 1891 we find Henry, Emily, their son John and possibly Isabella too as lodgers at 31 Noble Street, the home of the coal-merchant Issac Huxley.[(6)] This is the house, still standing, on the western corner of Noble

Henry Eckford's house, Wem

Street's junction with Market Street. Huxley had his yard out at the railway station on the eastern edge of town and lived here with his wife Laura and their six children. Although a large house, with two families and the odd staying visitor one can see that urban living lacked space and privacy. Peter, Henry's youngest was now twenty-two and still working up in Manchester as a tram driver. He met a girl there, Harriet Marsh and they married that November at Chorlton-on-Medlock.

From one new venture to another, Henry had forseen that he'd need help, so he persuaded his other son John to abandon his white-collar career and join him at Wem as a nurseryman. Which he did, once the business was up and running. (7) John Eckford recalled the event twenty years later. 'The suggestion did not appeal to me. I was born in a garden, and had the usual experience of a gardener's son – pulling weeds until my back ached. And I left home at an early age to pursue a business career in the City. It was with a great deal of reluctance that I yielded to my father's entreaties, but from that day to this I have never regretted my decision. I became fascinated with my father's work, and it was not long before I was as earnest as himself in the cultivation of the Sweet Pea in an effort to produce new forms and varieties'.(8) Beginning as his father's 'florist's assistant' John went on to run the business for decades after Henry died.

Business now was the growing and selling of seed. Would sweet peas prove commercial? In David Thomson's new 1887 edition of his *Handy Book of the Flower Garden,* no mention was made of sweet peas under his 'Select list of Hardy Annuals'; (9) in fact the book failed to mention sweet peas at all. A letter in *Gardening Illustrated* that summer would have maintained Henry's caution, confirming him in what he must have already decided, to concentrate on both the garden pea and the sweet pea. Of a list of 'sweet scented plants' there was no mention of sweet peas to rival the 'still very popular' mignonette, musks (musk roses), myrtles, rosemarys, Heliotrope (cherry pie) and the scented-leaved Pelargoniums. (10) Well, if sweet pea sales didn't take off they could fall back on garden pea sales... See what other traders were doing, too... The combination of the steady and reliable with risky higher returns makes such commercial sense that you wonder if Henry had had a nursery in mind from the moment he first began hybridising garden peas and sweet peas.

The land that had been on offer to the Eckfords was now nine acres of stock ground. It was sited on the north side of the Soulton Road; (11) an ideal spot, being quite near the station. With road and rail links, as we say nowadays. Henry may have been wanting to be 'away from it all' but he always tempered his inclinations with pragmatism. For irrigation, water was probably pumped into a horse-drawn tank from the brook that still runs there, the water then sprayed onto the plants from this mobile water-carrier. This was the method used by John when, after Henry died, he added some fields at nearby Tilley. [12] The nursery site was also within walking distance of home and a retail outlet that Henry had negotiated for in the High Street, [13] itself only a short way from home at the other end of Market Street.

A further venue was a bit further afield. The seed that was to be sold was grown away, mostly in Essex. Exactly where is unknown but it may have been in the north of the county. By 1910 there were a number of seed firms there. One was in the north-west; another, E.W. King & Co. was near Coggeshall. The old Edinburgh firm of Dobbies & Co. had their sweet pea grounds near there, at Marks Tey, near Colchester [14] with another, William Deal, at nearby Kelvedon.[15] These last two places had the advantage, as they do now, of road and rail links to London. *The Gardeners' Chronicle* could report, in June 1889, that the orchid growers Messrs. F.Horsman & Co. had been established 'in their healthy quarters at Marks Tey... for several years',[16] i.e., since soon after the arrival of the railways. An 1848 guide to Essex complained that, at Marks Tey, the new Eastern Counties railway line had inconveniently separated the Vicarage House from the church, [17] while at Kelvedon the railway 'has drawn off nearly all the great daily traffic, which formerly passed through the village, and was the chief support of the inns and shops'.[18] Funnily enough, loss of trade was an equally valid argument used against the railway *closures* of the 1960s. One person's loss is another's gain; in the small instance of the inconvenienced vicar can be witnessed the rising power of industry and commerce and the declining authority of the church.

This book on Essex, incidentally, belonged to my great-great grandfather Robert Wright, who lived at Lawford, on the Colchester-Ipswich road. His son, my great-grandfather George Powell Wright had ended his mostly farming life, spent around Colchester and Clacton, as an undergardener. It occurred to me, while using the book, that Robert, or more likely George, could have visited these seed-grounds here, perhaps even that of the Eckfords. In 1902 it was said that people drove (four-legged driving, mostly) for miles to see the seed fields like those of King & Co. whose grounds ran along the side of the Kelvedon Road.[19] 'They breathed into the summer air / Their fragrance, passing sweet; / And filled the rushing carriages...'.[20]

It hadn't crossed my mind until now that my ancestors might have met Henry Eckford.

The English south-east was an obvious choice for Henry. Being warmer and drier here than elsewhere made crop-failure less likely, for the plants could ride out a cold wet season better than they could in the north-west Midlands, once off the Essex clay. The light East-Anglian soils warm up more quickly in the spring, too, getting the crop off to a quick start, the region more closely resembling the flower's Mediterranean homeland than elsewhere. (If you live in Essex you may not believe this).

Noble Street, back in Wem, had obviously once been the posh place to be. It's physically superior too, for it is raised up above the fields northwards by being built, for part of its length at least, on the curved foundations of the old town wall.(21) By Henry's arrival it had fallen from grace sufficiently to accommodate the invasion of The Dickins' (now Dickin) Arms public house. Named for a local magistrate, one Captn. Thomas Dickin, this wasn't a name to find favour with the regulars, one suspects, becoming known instead as 'The Hole in the Wall'. It was now Henry's near neighbour. Worse was a tannery, the town's principal industry, sited further along Noble Street going westwards. What with the Gas Works west of that on Ellesmere Street it must have been pretty whiffy in Wem then, considering the prevalence of south-westerly winds. Only a malthouse was in a logical north-easterly spot on the outskirts of town, by the railway in Station Road.(22) You see this close mix of the posh and the plebean again in Shrewsbury. Birmingham and the Black Country, too, mixes the industrial and the residential like this and is, to southern-English eyes, one of the Midland's identity marks.

It is a pity, digressing a bit further, that the Noble Street residents Eckford and Hazlitt were a generation or two apart. (William Hazlitt died when Henry was seven years old). Both self-driven, both big in their own fields, the attraction of otherwise opposing temperaments could have made them agreeable neighbours, one feels. Horticulture was one of Hazlitt's many interests. 'If I have pleasure in a flower garden', he wrote in *Why Distant Objects Please* 'I have in a kitchen-garden too... If I see a row of cabbage-plants, or of peas or beans coming up, I immediately think of those which I used so carefully to water of an evening at Wem, when my day's tasks were done, and of the pain with which I saw them droop and hang down their leaves in the morning's sun'.(23) To which Henry, in his practical way, would have replied yes, wise to water in the evening if slugs and snails aren't a problem. But don't let plants droop – water them first. And Hazlitt would have heard more about peas, beans and cabbages, especially peas, than he wanted to. But they would both have gained from each other.

In August 1893 John would marry Alice Lucinda Sarah Boscombe from Corsham, across the county boundary in Wiltshire. She was twenty-four, he now twenty-eight. The ceremony took place in the Wesleyan Chapel in Chippenham. John, it seems, had now become a nonconformist, as his father had. Perhaps it was the Wesleyans' 'business-like other-worldliness'[(24)] that appealed to the businessman that John had become and would remain. After the marriage, Alice and John moved to Chapel Street on the other (south) side of the High Street. Here, the first of their seven children, Charlotte (Lotty) was born, on 21st July, 1894. The following year, John took the family back to Market Street, to number eight, a house half-way down and on the opposite side of the road to Henry. Here, between October 1895 and May 1901 were born Dorothy, Emily, Kathleen (Kitty) and Henry (Harry). Their last child, Joan (Queenie) was born at the nearby hamlet of Tilley, John and Alice's last move, in 1907.

'Lottie and I went to my grandfathers at an early age', Harry recalled, 'and were brought up by him. After he died my Aunt and Gran brought us up'.[(25)] The aunt was Henry's daughter Isa (Isabella). So, in the end, Isa did have children, two of her elder brother's. 'Gran', Emily, once again had small children to bring up; in both instances family but no blood-ties. At the age of eighty-five, Lotty also recalled the days. 'I do remember the lovely pinafores we wore, with a kind of large frill over the top of the arm. What a lot of washing and ironing in those days!'[(26)]

Between 1892 and 1909 Peter and Harriet had thirteen children, at least two of whom died in infancy. Peter doesn't seem to have cut his ties with his family, for many of his children had familiar Eckford and Stainer names; Isabella, Henry, Charlotte, Sarah, George... even an Emily for his step-mother and a John Stainer, his brother's exact name. Harriet was thirty-seven when their last child was born. In 1910 Peter left his wife and emigrated to Canada. Attempting to take six of their children with him, the port authorities prevented two of them from boarding the ship. Peter died a pauper in Ponoka, Alberta with, apparently, mental-health problems, shortly before the first world war. From the children, however, there are now descendants there.[(27)] These descendants' knowledge of John and the sweet pea business was probably as sketchy as John's descendants' view of Peter was to be. 'Peter William was never mentioned in our house', one of John's grand-daughters told Bernard Jones in a letter in 1989. It 'seems that he went ranching... He never did send for his wife and infant son'.[(28)]

~~~

By August, 1889, Henry, Emily and John had settled down together in their lodgings at Wem. To someone who, as Emily and John would have told you, had never had much time for ordinary garden plants, other than vegetables, the age was now championing Henry for his success with them! Well, one of them. 'All lovers of our hardy garden flowers will find in Mr. Eckford's grand new varieties of Sweet Peas something worth having in their gardens...' said the *Journal of Horticulture*. 'For a few years past [he] has, through the medium of the seedsmen of the country, been giving us new varieties. He was asked to send blooms of a few varieties to the Solihull Flower Show on the 24th July, to give the Midland people an opportunity of seeing what is being done with this favourite garden flower. A dozen varieties in bunches were sent...' They 'surprised and greatly pleased the large number of visitors to the show... As Sweet Peas are so easily cultivated and seed easily saved, Eckford's new Sweet Peas ought to be in every garden'. (29) That would have pleased Henry, apart from the bit about seed being easily saved, even though it was true!

In October, *The Gardeners' Chronicle* were publishing a paper on garden peas by Thomas Laxton.(30) With the garden pea and the sweet pea, Henry had a rival in Laxton reminiscent of his rivalry with Charles Perry and Verbenas. In August 1890, again on the garden pea, the *'Chronicle* wrote that 'as of making books, so of raising new Peas – there is no end...'. To Dr Maclean 'succeeded Laxton and Culverwell; and later... Eckford... men who went about their work of raising new varieties in an intelligent and scientific manner...'.(31) Intellect and scientific knowledge isn't quite the whole picture. Before they come into play something more like intuition is needed; one's unconscious preferences must be allowed to surface and be identified. The article rather gives the impression that Henry was the younger man but in fact he was Laxton's senior by some seven years. In contrast to mixed reviews for Henry's garden peas, the almost undiluted praise for his sweet peas continued in the following year's *'Chronicle*.

> 'Eckford's Sweet Peas. Every gardener knows well the value of a good strain of Sweet Peas, and where there is a large and constant demand for cut flowers, usually prepares two or more sowings from which basketfuls of flowers may frequently be gathered. In Mr. Eckford's novelties we have been favoured with something good and very far ahead of what was at one time thought
~~~

> excellent. Of the whites, Mrs Sankey might well be termed the queen, for it is a handsome thing. The individual flowers are large, with standards pure white, and borne on strong stalks, and the plant is exceedingly robust. I know of no white to equal it. Lemon Queen is very pretty...'. The writer also praised Henry's varieties 'Her Majesty', 'Purple Prince' and 'Ignea'. 'Countess of Radnor', 'Orange Prince' and 'Princess of Wales' he thought 'very peculiar'; 'Primrose', 'Apple Blossom' and 'Empress of India', as the real Queen Victoria was also now known, he all liked. 'These in their separate colours are very useful, and help, not only to beautify the borders, but when planted at intervals they may be kept to their colours better, taking care, should a "rogue" appear, to remove it at once. To give praise where it is due, I am sure there is ample room to congratulate Mr.Eckford on his success. H. Markham'.(32)

In a sort of mountain to Mohammed way, growers like Henry would take their flowers to the provincial summer flower shows. Having gone inland to Solihull in '89, it was to the coast and Liverpool that he took his flowers in the summer of '91. He handed out practical advice too, along with his blooms.

'At the recent Liverpool Show a charming stand of named varieties of Sweet Peas was exhibited by Mr. Henry Eckford. It is the first time Mr. Eckford has paid the Liverpool people a visit, but it is to be hoped it will not be the last, for the exhibit was greatly enjoyed... In a conversation I had with Mr. Eckford he stated great care is required to keep the varieties true to name and the limited quantity of seed some of them produced. He went on to state that with the choicer varieties the best way to succeed with them was to sow them in small pots from the middle to the end of January or a little later, according to situation, and then transplant. By doing so failures would be very few and the step greatly in advance of sowing them outside in drills'.(33)

~~~

On 25th November, 1891, Henry and Emily were 'admitted to membership' of Wem Baptist church. At first sight, this seems an odd move for Henry to make. He had been baptised at birth by the Church of Scotland, and there was no reason why, going south, he should not have transferred his allegiance to the Church of England. In fact he would have done, being
~~~

expected, as head gardener to the Lord of the Manor (as at Coleshill) to have attended Sunday Service in the parish church. Such popularity at the Baptists had then was due, in part, to the Non-Conformists' historical association with education in the form of Sunday-School classes.[(34)] Wem's latest Baptist Church, which had been built back in 1870 for an impressive £1,100, featured a schoolroom beneath it.[(35)] Henry knew, personally, the advantages of a good education, and this may have been a factor. The family story is that Henry fell out with the vicar of the parish church and became a Baptist by default. Being a Scotsman he may have never quite settled in the Church of England. Now, beholden to no-one, he could have used some disagreement as an excuse to leave.

The working-class were in the majority in the nineteenth-century, and they made up most of the congregations at church and chapel. Attendances overall however, were disproportionately middle-class. This was a situation that must have been particularly marked in the towns and cities, where newcomers found themselves free of the rural world's coercion towards religious observance. Through authority at church or chapel Henry could thus distance himself from the urban working-class, identify himself in his new trade status, enhance his position in local society and, incidentally, do no harm to his new business career. Genuine religious sentiment would have been enhanced, in his advancing years, by the reminder of his own mortality. The austere Baptists' slant on the nonconformists' argument with the established church may have rubbed off on him too: that it had become more show than substance, was ill at ease with those it professed to serve and lead, and was less concerned with saving souls than in saving and preserving the status quo in society.

The nonconformists' stress on individual effort and self-improvement would have appealed to Henry too, as it did for many. (Paradoxically, it's been noted, this attitude didn't square with their equal insistence on state-intervention to control the drink trade or to observe the Sabbath).[(36)] 'No doubt', the Rev. Canon Egerton of Wem had said when Henry had moved here in 1888, 'no doubt a great deal of the hostility to the Church arose from ignorance of the position, claims and history of it...', [(37)] as if people at large were likely to exchange an emotional perception for dispassionate analyses of claims, positions, etc., etc. The nonconformists attracted hostility in their turn when they denied proper equipment, even coal for heating, to church schools like those across the border in rural, Welsh-speaking mid-Wales.[(38)]

Emerson, in his *English Traits*, was having none of it. For him, religion could be summarised as 'the doing of all good, and for its sake the suffering of all evil.'

'...the religion of England, - is it the established church? no; is it the sects? no; they are only perpetuations of some private man's dissent, and are to the Established Church as cabs are to a coach, cheaper and more convenient, but really the same thing'.(39) It's likely that Henry had heard of this, if not read it. *English Traits* was popular, as affectionately critical studies of the British by outsiders often are, and it ran through many editions. Henry would have understood the practical advice, if not the sentiment. Anything convenient would do, then. And what was more convenient, for a busy nurseryman, than the Baptist chapel that just so happened to be in Market Street, five yards from his front door on the other side of the road? (Being on the corner of Market Street and Noble Street, Henry's house had entrances from both roads).

'As everywhere outside the established churches', wrote the historian Geoffrey Best, 'the men who were in demand to run things or who by some process of natural selection became the runners of things were the solid citizens, the pillars of the community, who could most efficiently raise funds and inspire confidence. A church not founded on rocks was all too liable to run *on to* them'.(40) Helping to keep the Wem chapel on (or off) the rocks found Henry taking on most positions of authority over the next few years. In 1893, he was on the finance committee; in May of 1896 he was made church treasurer and a Deacon. In October of that year he became, with the Pastor, a church delegate to the national Baptist Union Assembly. In January 1899 he was a delegate again in the wider sphere of the Free Church Federal Council. Emily 'took a leading role in the women's work', with John also later associated, together with his daughter Dorothy. (41) The Baptists later commemorated Henry with stained-glass windows in their Market Street chapel. In 1988, when they moved to bigger premises in Chapel Street they took these windows with them.

Among the Eckford artefacts collected from Shropshire by the late Bernard Jones is a copy of William Holman Hunt's *The Light of the World*. First exhibited in 1853, this night-time vision of Christ became the most popular of all Victorian paintings, destined to decorate millions of bedroom walls. Assuming it to be Henry's, it is not surprising that he should own a copy, indicating nothing more than that he shared the conventional religious piety of the age. It's interesting that it was so popular. Initially, that's understandable. As a Pre-Raphaelite painting it is attractive –

decorative and symbolic – and, painted just six years before the publication of Darwin's *Origin of Species* – a comforting icon in an age rocked by scientific discoveries. Yet it was uncomfortably honest, for Christ with his lantern is knocking on an ivy-covered, weed-surrounded door, the door to the human heart. And it seems as if we have all long denied Christ's light entry here and would continue to do so.

'Behold, I stand at the door, and knock: if any man hear my voice, and open the door, I will come in to him, and will sup with him, and he with me'.[(42)]

Along with the familiar message, that we are sinners needing to be saved, comes with the denial of his light an honest reflection of Victorian doubt. It would have spoken to Henry on both a conscious and unconscious level, the nocturnal setting adding a further touch of mystery. And, even if Henry hadn't known that the original models for Christ were two females, Christina Rossetti and Elizabeth Sidall, a touch of ambiguity; of which Holman Hunt, given the Pre-Raphaelites' more common use of boy models to represent girls, was perhaps unaware. Comfortable and decorous on the surface, raw and unsettling beneath, *The Light of the World* reflected the age for Henry and millions like him. Their moral meanings submerged, the visual attractiveness of the Pre-Raphaelite style saw it popular again a century later; icons now for the secular nineteen-sixties and seventies.

~~~

The summer of 1888 had marked a change to a period of cooler and wetter summers. This change in the weather favoured the newly fashionable pansy, which could now be taken up with success by gardeners in the drier south and south-east of England. Although Henry specialised in the garden and sweet pea, the nurseries grew a range of other plants too, and Henry's attention may have been momentarily distracted by this Scottish favourite's rise to prominence. Be that as it may, the first '*Chronicle* reference to Wem, in February 1892, shows that Henry's attention had remained focused.

> 'Growing Sweet Peas. The wonderfully improved varieties introduced lately, in which many lovely shades of colour are found, have given to the Sweet Pea a prominence amongst our garden flowers it had not previously enjoyed. Those, who like myself, have seen Mr. Eckford's fine varieties displayed by him at the Birmingham and Shrewsbury Shows, and elsewhere, can
~~~

testify to their beauty and to their appreciation by the visitors. As cut flowers they richly deserve attention, and they last so long in a cut state. For dinner-table decoration they are lovely objects, and a bouquet of Sweet Peas is an object the eye can rest upon with pleasure. The America florists have found out the value of the Sweet Pea as a commercial flower, and its adaptability to various kinds of decorative work, and the flower is becoming popular in the States.

...I went on from the great Shrewsbury show in August last to Wem to see the new varieties of Sweet Peas, and I found myself amongst 2 ½ acres of them in long rows of a sort, all growing free from each other, and in rows about 6 feet apart. Some plants from 24 inches to 3 feet apart, from 2 to 2 ½ feet through, branching and bushy from the bottom, and from 5 to 6 feet high... By giving the plants ample room, the flowers are much finer, and in Eckford's new varieties, it is not only in the very lovely new shades of colour that we see so much improvement, but it is especially apparent in the size, form and substance of the flower, and the increased number of blooms on each stem. I mention the bush form of Sweet Pea I saw at Wem to indicate what the Sweet Pea can do as a plant under favourable conditions; and as to general culture, I cannot do better than quote the cultural instructions found in Mr. Eckford's catalogue:-

"In preparing the ground, if not already tolerably rich, a liberal dressing of thoroughly decomposed stable manure should be dug in some time before the ground is wanted, leaving it rough and allowed to consolidate before sowing. When preparing for sowing, all that is necessary is to break the ground thoroughly with a hoe, and draw a drill about 3 inches deep, and sow the seed thinly, and cover with 2 inches of soil, leaving the drill hollow, and gently tread the row on either side, making the soil firm for a distance of 18 inches on either side. Sow early in February and for succession in March, April and May".

...It is wonderful what is being accomplished in colour, for in the new variety Primrose, we have a very near approach to a decided yellow; in Mrs. Eckford, a delicate tint of pale primrose; in Dorothy Tennant, a rosy-mauve colour; and in other varieties, very bright crimson-scarlet, rosy-claret and pale blue, pale mauve, bright purple-blue, orange-pink, very great improvements in whites, dark maroon-purple, and a host of intermediate shades of colour... When Messrs. Carter & Co. gave us Scarlet Invincible, we had a great advance, and this firm, as well as Mr. Eckford, has since devoted much attention to crossing, and with great results. Others are engaged in the same work, and the Sweet Pea will shortly be regarded as a popular flower, and deservedly so, for its long stems and fragrance alone are two great points in its favour. W.D.' (43)

The author this time was William Dean, Richard Dean's elder brother. The Deans were quite a family of gardeners. They were the four sons of a Southampton nurseryman, three at least of whom, William, Richard, and Alexander, had followed in their father's footsteps. A James Dean had died young, but as this was at Slough, and Richard had worked in the Royal Nurseries at Slough, it's likely that James had too and was a gardener also. Four gardeners of a gardener. And when the R.H.S. inaugurated the Victoria Medal of Honour in 1897 the two surviving Dean brothers, Richard and Alex, both received them. William, a pansy and *Viola* specialist, would have noted any new progress with the pansy, but there was none to report. It may be that Henry's interest in pansies had waned since William Sankey's death and his leaving Boreatton Park.

Before putting new varieties on the market, Henry would send them to his friends and associates around Britain to try out and report back on. Henry would pay visits too. One such recipient was Robert Brotherston, (1848-1923), gardener to the Earl of Haddington at Tynninghame, East Lothian. Another was William Gumbleton, an Irish aristocrat whose estate of Belgrove skirted Cork harbour, near Queenstown, (now Cóbh) in County Cork. Gumbleton was a strong-minded, dominant sort, one of those drawn to Henry's abilities by the attraction of near opposites; for Henry, though he knew his mind, was of an altogether quieter disposition. A water-colour illustration of Henry's sweet peas in *The Garden* in March was from flowers growing, as Gumbleton wrote, 'in my garden during the past autumn by my friend Miss Travers.

The four varieties... were... raised by Mr. Henry Eckford of Wem, Salop, and sent to me for trial about this time last year. All of them bloomed most abundantly and beautifully in my garden during the summer and autumn, and were much admired by all who saw them. Only one of them, Orange Prince, is yet in commerce. Mr. Eckford hoped to have been able to distribute seed [of 'H.M. Stanley' and 'Dorothy Tennant'] this spring, but owing to the cold and wet summer of last year, combined with a series of severe gales which literally blew away large numbers of the flowers, the yield of seed was so abnormally small, that these beautiful varieties cannot be sent out until 1893. Mr. Eckford has about 2 ½ acres of land under Sweet Peas, from which his usual harvest of seed amounts to about 25 bushels, but last year ...... he was able to save only about a bushel and a half, so Sweet Peas will be scarce this summer, I fear.

I consider H.M. Stanley to be quite the largest-flowered and finest dark Pea I have ever seen, and the deep mauve Dorothy Tennant is quite a new shade in these flowers, and I am sure will be much admired. Mrs. Eckford is also a fine, large flower, and a great improvement on Primrose, the so-called yellow Sweet Pea sent out some two years ago, but which is really only of a pale buff colour. Another beautiful group of four other [Eckford] varieties drawn by the same artist from my flowers, and embracing the striped variety Senator, the fine scarlet Cardinal, the deep purple Monarch, which comes next to H.M. Stanley, and the lovely delicate blush Venus appeared on January 15 in *L'Illustration Horticole* of Brussels'. (44)

The Garden expanded on Gumbleton's article with descriptions of Henry's newest varieties:

'The hardy annual Sweet Pea that twines its slender shoots round many a rustic porch and fills the air with a sweet fragrance is known to all, but the common variety of the cottage garden is different to the beautiful types that have, thanks to Mr. Eckford, ... sprung up of recent years. The flowers are larger, exhibiting a greater breadth of colour, and just as sweet as those of the old favourites, and the result of thoughtful and successful hybridising. At several of the shows last year a boxful of Sweet Peas from this raiser was a conspicuous and welcome exhibit, but, unfortunately, such fragile flowers are not seen in beauty on the exhibition table. They have lost their freshness, and therefore something of those delightful colours which are their great and enduring charm.

Painting of four Eckford Sweet peas, 'The Garden', 12-3-1892.

Painting of four different Eckford sweet peas, 'L'illustration Horticole', 15-1-1892.

There are many kinds, and the best known are Butterfly, Invincible Carmine, Invincible Purple, Purple Striped, and Fairy Queen; but some of Mr. Eckford's new introductions overshadow the older favourites. Captain of the Blues is the finest blue variety, the standards of a rich blue shade, the wings paler, and producing a rich contrast of two tints. Very distinct from this, but delightfully soft and delicate in colour, is Mrs. Gladstone, the standards pink and the wings blush. Primrose is, as its name suggests, primrose in colour... the nearest approach to a... yellow-flowered Sweet Pea. Purple Prince, maroon and purple-blue; Empress of India, rose-purple and white; and Countess of Radnor, the standards mauve, and the wings of a paler shade of the same colour, are all good varieties, and with just the same characteristics as the old kinds. In the raising of these distinctly improved forms Mr. Eckford has not relied upon chance seedlings, but endeavoured to fertilise the best varieties with pollen from others that have some distinctive characteristic likely to produce a novel and beautiful progeny...

Autumn sowing was once strongly recommended, but just as good results are to be had from sowing the seed in pots or boxes in the month of February...'.[(45)]

Gumbleton and Brotherston continued to give their opinions rather publicly in the pages of the horticultural press. 'I quite agree with Mr. Brotherston as to the great beauty of Mrs. Eckford and Her Majesty', Gumbleton wrote in a September edition of *The Journal of Horticulture*, 'both sent out for the first time this year, but am not inclined to give so high a position as he does to Countess of Radnor, as though undoubtedly a delicately pretty and uncommon shade of colour, it is unfortunately of a weak and delicate habit of growth... it is also an extremely shy seeder...'. As a substitute, Dorothy Tennant, 'sent out, I think, for the first time this year... might, I think, be found; it has... two entirely distinct and different shades of colour, one on opening, changing to the other and deeper shade after... a day or two.

Another very fine Pea is H.M. Stanley, which is, I believe, to be sent out next spring or this autumn for the first time [yes, as 'Stanley']. This is, in my opinion, by far the finest dark variety that has yet been seen, and has been at once noticed and remarked as such by

> all friends who have seen it in my garden during the summer... Another most beautiful Pea sent me this year by Mr. Eckford is named Gaiety, and is white, clearly streaked with rose colour... Another fine and most distinct variety, though by no means so new... is Orange Prince, also raised by Mr. Eckford, which I think should be in every collection'. (46) One unfavourable comment was that growers were bringing out many very similar varieties under different names. It was to become a common complaint.

Two of the visitors to Wem that summer put pen to paper. One, 'going on to Wem' from the Shrewsbury Show, and thus from the south, with the byline 'D' may have been Richard Dean's brother Alex or William, there to see the garden peas he 'had heard so much about'. Henry was and will remain noted for his work with the sweet pea but it is his work with the garden pea that is his forgotten success. His nurseries could never grow enough of them. The other visitor is an unknown Scotsman that Henry had known, or met, at Coleshill. This gentleman seemed to have some trouble in understanding the town!

> 'Some readers', he wrote, 'might have wondered what Wem could be; but they have learned that it is not a plant but a geographical expression. It is the name of an ancient looking village some ten miles from Shrewsbury, and the home of certain Peas... all raised by that diligent florist and experienced gardener, Mr. Henry Eckford. His Sweet Peas in their season of flowering are a sight to see, and the long rows of culinary varieties laden with huge pods are not likely to be soon forgotten by those visitors who... wend their way to Wem. Many do this not from various parts of this country only, but from the Continent and America; it is the Sweet Peas mainly that attract them.
>
> Mr Eckford has worked so perseveringly in improving these charming fragrant garden flowers that he may almost be said to have revolutionised them. The advances, however, have been made step by step, a few new colours and a slight increase in the size of the flowers having been obtained each year, so that we have to contrast the varieties of a generation ago with the new ones up to date to fully appreciate the change that has been brought about by the skill of the hybridiser and the selections of the florist. It is easy enough to raise new Peas, at least by those who know how;

but the work is not half done then, and rejection, retention, and fixation may be the work of years before pure and distinct stocks are established.

Mr. Eckford has, at an estimate, about 3 acres of Sweet Peas at Wem, and 2 acres of culinary varieties; these are of the newer varieties only, some of the latest in commerce, with others to follow, and the bulk of his seed is raised elsewhere, where better harvests are obtained than in their cold native parish. Peas grow luxuriantly at Wem, and flower profusely, the Sweet Peas apparently right into the autumn. They are grown in rows 5 or 6 feet apart, with Potatoes between them, of which Mr. Eckford's favourite early variety is Sharpe's Victor, because he finds it the first ready for digging, and one of the best for use. The Peas occupy the same ground year after year, without any loss of vigour; but it is possible their seeding may be prejudiced by the abstraction of potash, one of the chief essentials for a good harvest of seed. When a plant shows an improvement in its flowers, as is the case here and there in most of the rows, it is carefully bent from the row and secured to a stake, while any that may fall below the standard of merit are drawn out altogether. The rows, however, are singularly true, and the long lines of colour, from white through various tints of rose and pink to glowing carmine, from pale lilac through different hues of lilac and mauve to purplish black, have, when in full beauty, a delightful effect, and the air is laden with perfume'.

From Henry's sweet peas the writer picked out eight, which he described; 'Emily Eckford', 'Peach Blossom', 'Ovid', 'Royal Robin', 'Venus', 'Stanley', 'Blushing Beauty' and 'Lady Penzance'.

'Among culinary Peas Mr. Eckford has raised several very fine varieties by systematic crossing. With the view to obtain high quality he chose Ne Plus Ultra and British Queen for his base of operations; and with quality he seeks productiveness. To secure retention a variety must produce its pods in pairs... In use during the Shrewsbury Show week or the last half of August were the following' which the author describes: 'Ambassador', 'Superabundance', 'Colossus', 'Copious', 'Censor', 'Heroine' and

'Wem', the last three receiving the highest number of marks in the R.H.S.'s Chiswick Trials that year.

> 'Good work is being done on the outskirts of the primitive looking Shropshire village, with its laconic name; and as many gardens have been enriched already so will many more be in the future, through the skill and industry of a genuine British gardener, aided by his attentive and industrious son, who evidently means to uphold the reputation of the name he bears. Thirty-five years have come and gone since I first met the elder Mr. Eckford – a midnight meeting in a Mushroom house that he well remembers – and I am glad in having had the privilege to "tak a cup o'kindness" with him in these latter days of his busy life, now being happily spent among the Peas at Wem'.(47)

The visitor inspecting Henry's garden peas added four more to the above – 'Critic', 'Eckford's Gem', 'Shropshire Hero' and 'Eckford's Fame', describing them all.

> 'With a view to obtaining dwarf-growing earlier kinds' Henry had begun work with the famous old variety 'Ne Plus Ultra' as one of the parents. As we've seen, the other was 'British Queen'. 'Another point that has been regarded at Wem as a prominent one to be attained is, a dark green-coloured pod of good size, and well filled with dark green seeds... A visit to Wem teaches a lesson as to sowing and growing. Mr. Eckford grows probably 2 acres of Peas each year of his new sorts and seedlings, for the purpose of keeping his stocks quite true, by chasing out any "rogues", i.e., plants which are not of the true character. A great deal of his personal attention is devoted to this object, and when the character of the variety is fixed, bulks of seed are sent away to be grown by the acre in various parts of the country, and personal visits are made by Mr. Eckford during the podding season to ensure any further weeding out of "rogues", should there be any...
>
> At Wem the ground... is a good loam, and suits Peas and Potatoes well, and, in fact, almost everything else'.(48)

From both reports we can see that, of Henry's nine-acre site, three acres is given to sweet peas and two acres to garden peas. Most of the other four acres would have gone to vegetables, and to bedding plants like pansies and primulas. Both writers note potatoes, and from Henry's surviving catalogues we will see that every kind of vegetable was grown for seed. Even ornamental grasses were grown, to compliment sweet peas for table decoration.

So. Henry had chosen his site well, as you would expect. It is interesting that Henry travelled around the country to inspect his peas. He'd travelled a lot in his life, and as a hale and hearty sixty-nine-year-old he wasn't about to stop. Shows were good for business and he picked up some certificates for his sweet peas from that year's Shrewsbury Show.(49) Henry professed not to value such things in comparison to a prestigious R.H.S. First-Class Certificate, but he kept them all the same and some have survived to this day.

Information on family affairs is rare, but 1892 is thought to be the year that Isabella, Henry's one surviving daughter's attempt to leave the family failed. Engaged to be married, you'll recall, her fiancé died.

~~~

The first month of the new year, 1893, found William Dean writing in the amateur-gardening oriented *Gardeners' Magazine*. Promoting the sweet pea as an exhibition flower, he renewed the point, made in his '*Chronicle* article, that the United States was taking to the flower as we were, 'ahead of us', actually, 'in the adoption of indoor floral decorations, in which the sweet pea is brought into prominence, and Mr. Eckford has hitherto been unable to supply the demand for his new varieties to his transatlantic customers. If collections of sweet peas could be more generally seen at our flower shows their cultivation would be general, and give satisfaction to all the visitors, for, as a rule they are greatly admired, by ladies especially'.(50)

In anticipation of a 'back to nature' movement in the United States, Henry had been providing seed companies there with varieties of his own sweet pea seed. The United States was industrialising, as we were, but it had been interrupted by their Civil War in the 1860s. Now, there was a rush to make up for lost time. Headlong industrialisation and urbanisation could be expected to bring about a reaction, a nostalgia among white Americans especially for lost farmsteads and their sweet-smelling mostly European flora like the sweet-pea. (51) Mostly European because, surely, of the mass European migration to North America of the last few decades. Frequently, the initial reaction to an alien environment is to try and surround oneself with familiar things. It is only later that the confidence comes to develop an indigenous identity, and for society to produce an indigenous culture.
~~~

William Dean had noticed that women especially took to the sweet pea. Flora Thompson noted it too. Writing of the 'new generation of housewives' of her Oxfordshire hamlet in the 1890s, she tells a story of a vase of flowers one of the young women had 'placed on her table at mealtimes. Her father-in-law, it was said, being entertained to tea at the new home, exclaimed, "Hemmed if I've ever heard of eatin' flowers before!" and the mother-in-law passed the vase to her son, saying, "Here, Georgie. Have a mouthful of sweet peas". But the brides only laughed and tossed their heads at such ignorance. The old hamlet ways were all very well, some of them; but they had seen the world and knew how things were done. It was their day now'.(52) The sweet pea was a flower with a future, as well as a past.

There may have been some of James Carter and Co.'s varieties in that vase; or one of Thomas Laxton's, his 'Carmen Sylvia' perhaps, 'most attractive', it was said, 'its heliotrope blue wings and carmine standard give the flower a combination of colours which cannot fail to please, especially with the ladies....'(53) But Henry was the leader in the field and they were likely to have been all or mostly of his. The young women would have gone for the latest novelties. In 1893 a bunch of Henry's newest varieties would have included one called 'Novelty' itself, with 'orange-rose standards, the wings delicate mauve, and lightly margined with rose – very bright'. Others were 'Meteor', its 'standards bright deep orange - salmon, the wings delicate pink, with veins of purple'; 'Mrs Joe Chamberlain' was 'white striped, and heavily flaked with bright rose – very striking and pretty', and 'Venus' which was 'salmon-buff, and rosy-pink, very distinct'. (54) A feast but no, not an edible one.

Joseph Chamberlain, the prominent Liberal Unionist, was never seen out without an orchid in his button-hole. It was part of his identity, like Churchill and his cigars were to be. He had recently married a young American woman and Henry naming a sweet pea after her may have been calculating. Perhaps she would persuade her husband to forsake his orchid for the more up-to-date sweet pea! What a bit of free advertising that would be! It was, after all, the sort of thing the prime-minister's wife did. It was known that when Gladstone 'goes down to the House of Commons prepared to make an important speech, he is always well brushed, his hair is oiled, and he wears a flower in his button-hole. Mrs Gladstone always "revises" him before he leaves home on important occasions'.(55) Unfortunately, Mary Chamberlain was quite unassuming – 'I am not in the habit of making suggestions to my spouse'(56) – unfashionably plain, and simply dressed. The 'Puritan maid' she got called, and the orchids stayed put. Well, it was worth a try.

The 4th February, 1893, was something of a red-letter day. It matched Henry's first mention in *The Gardeners' Chronicle*, twenty-seven years earlier. It was the first advertisement in the '*Chronicle* for his new enterprise, and it was wholly typical of the man that it took up merely an inch or two in the classified ads. section. It is instructive, too, in that it was for his garden peas, not his sweet peas, that he advertised. 'Eckford's New Culinary Peas' it stated, in fairly large capitals, followed by their names, 'Epicure' and 'Censor', in smaller ones. The advert continued in very small letters, as if to apologise for all that ostentation.(57) Up until now, Henry had only supplied the trade, and the inclusion of mail-order retail selling marked a change, perhaps a gain in confidence. Mail-order itself was a relatively new concept. A mail-order catalogue had been produced by the new Army and Navy stores in London back in 1872. In the United States the giant Sears, Roebuck stores were offering anything and everything, from farming equipment to gravestones. It took a while for people to come round to this impersonal way of buying things but by 1893 Henry could be pretty confident of its general acceptance.

Two weeks later saw another article on the sweet pea in *The Gardeners' Magazine*. From 1891, there was never to be another year, to Henry's death and well beyond, when the flower did not figure prominently year by year in the horticultural press. 'Sweet Peas. The remarkable diversity of colour which sweet peas now afford is very largely due to the patient and persevering energy of Mr. Eckford, of Wem, Salop... Those who were fortunate enough to see the collection growing at Chiswick [the Royal Horticultural Society's gardens] during the past summer will remember how fine are the colours, how strong the constitution, and how distinctly the colours are being developed in those varieties emanating from Wem. Of the dozen varieties of Lathyrus odoratus which have received awards from the Royal Horticultural Society, eleven were Mr. Eckford's productions, and on three occasions his strain was certificated by the same society. Probably at no very distant date Mr. Eckford will be able to place a distinct yellow, or perhaps golden, sweet pea upon the market. The tendency is in that direction, at any rate...'. A probability that has remained distinctly elusive!

The writer went on to describe the varieties 'Mrs. Eckford', 'Lady Beaconsfield' and 'Her Majesty', and considered 'Countess of Radnor', 'Dorothy Tennant', 'Boreatton', 'Mrs. Gladstone', 'Captain of the Blues' and 'Mrs. Sankey' to 'make up a fine selection. We trust that Mr. Eckford may long continue to improve and popularise these lovely flowers. Mr. Thomas Laxton, of Bedford, is offering several new sweet peas of his own raising that are of fine form and colour...'.(58) Thomas Laxton seems to have been Henry's only serious competitor in the sweet and garden pea field, his efforts drawing both praise and criticism from the unnamed reporter (obviously William Dean again).

The next '*Chronicle* reference to sweet peas – and they come in thick and fast now – was in July. In reference to the coming Chiswick Show,

> '[visitors] should not miss the collection of Sweet Peas which will be then in full perfection. They are grown in rather an out-of-the-way corner, and might be overlooked. The varieties are chiefly from Mr. Eckford, who has done so much towards popularising the flower and adding to the store of good kinds already in the trade. The large number of varieties will be thoroughly inspected by the committee, and the best singled out. From what we can see, there is need of considerable sifting, but many very beautiful varieties are in the collection, the flowers ranging from pure white to scarlet. Considering the dryness of the season, they are in good health, and constitute a "trial" of more than ordinary interest'.(59) Alas, none were to be mentioned in the Show report, indicating that the weather had indeed been a factor in successful showing; as it still is. Henry may have had to write the Chiswick Show off but still made a splash of sorts at home, in the Shropshire Show. 'H.Eckford... staged thirty-two bunches of his charming varieties of the Sweet Pea, including the pretty blue, Emily Eckford...'.(60)

The following month, our old friend Richard Dean was writing again in *The Gardeners' Chronicle*.

> 'Mr. Henry Eckford is unwearied in his endeavours to augment our list of varieties of Sweet Peas, and in doing so he has placed in the hands of lovers of flowers a mass of beauty with which to beautify the garden, as well as ample material from which to cut; and it is interesting to note that while Mr. Eckford, by means of intelligent cross-fertilisation, has given us colours and combinations of colours undreamt of in Sweet Peas a quarter of a century ago, the exquisite perfume of this old favourite remains the same, and its habit of growth is not in the least degree changed. With new colours have come also increased size and substance in

the blossoms, which is the natural development of most flowers when they are taken in hand and improved.

Until Mr. Eckford commenced the progress of cross-fertilisation with Sweet Peas, any new introduction had been apparently the result of sports, as in the case of the Invincible Scarlet selection. Mr. Thomas Laxton has helped in the work; but to Mr. Eckford is mainly due the production of new and distinct varieties. As a matter of course, when new forms are produced in quick succession, there is always the danger of sameness, and though some varities have been sent out which appear at first sight nearly to resemble each other, yet on a close inspection they are found to be really distinct. The latest novelties shown by Mr. Eckford seem to point to decided improvements in our striped varieties, while some are densely spotted or punctured and such are welcome novelties. But in all the new introductions there are some that do not find the marked favour that others do; at the same time it should be noted that all are more or less admired, and that what one deems weak another considers beautiful, and so all are admired.

Ten new varieties were shown by Mr. Eckford at the last meeting of the Royal Horticultural Society...'.

Richard Dean named and described 'Eliza Eckford' and 'The Belle' (that gained Awards of Merit), 'Blanche Burpee', 'Countess of Aberdeen', 'Excelsior', 'Lottie Eckford', 'Mrs Chamberlain', 'Novelty' and 'Princess May'. Although Elizabeth was a family name, there were no Elizabeths among Henry's grandchildren. He had had an Aunt Eliza and a sister, who would be eighty-two if she were still alive. It was, however, probably named for Charlotte Elizabeth, his daughter who had died. She may have been called Eliza to avoid confusion with her mother, also Charlotte. Henry had three brothers so it's also just possible that the flower was named for one of their grandchildren. Henry seems to have kept in touch with his extended family, many possibly living at and around Traquair in the Scottish borders. 'Blanche Burpee' was named for the wife of an American seedsman who became a friend and frequent visitor to Wem; W. Atlee Burpee of Philadelphia. Lottie Eckford was Henry's four-year-old grand-daughter who lived with him.

Of recent varieties, Richard Dean highlighted 'Venus', 'Emily Eckford' – 'coerulean blue, and reddish-mauve, one of the very best of the blue Sweet Peas' – 'Gaiety', 'Fire-Fly', 'Duke of Clarence' and 'Blushing Beauty'. 'Older types' Dean still found attractive were 'Mrs. Eckford', 'Dorothy Tennant', 'Lemon Queen', 'Senator', 'Captain of the Blues' and 'Princess Victoria'. Dean finished his article by persuasively arguing the case for autumn as opposed to the spring sowing of sweet peas. All the garden writers did this and they were all mostly ignored. The main advantage was to gain earlier flowering, which would have appealed to those still in private service. (Early sowing would have long been practised by the commercial cut-flower trade). For amateur gardeners, it was just too much trouble for too little result.

However, let us give the floor to Richard Dean:

> 'We do not employ the Sweet Pea sufficiently for autumn sowing, either for blooming under glass, or for planting out in the open, to flower long before the spring-sown plants can do so. By sowing Sweet Peas at the beginning of September, having at most three plants in a 5 - inch pot, and keeping them in a cold frame all the winter, they can be had in bloom quite early in the year under glass, when they are found most useful to cut from. By sowing in autumn in the same way, and wintering in a cold frame, and the plants put out in the open as soon as it is safe to do so, and planted in rich soil, each one being allowed ample room, the Sweet Pea will grow to a large size, and bloom much more freely and finely than those raised from plants the seeds of which are sown in the open in spring. It is only when Sweet Peas are grown as single plants in good soil that anyone can be made to see how amply they branch, and what bushes they become'.(61)

Thomas Laxton died between the writing of this article and its publication. He was still only in his early sixties. When Charles Perry, Henry's *Verbena* competitor had died, it coincided with Henry's loss of interest in, and a general demise for, the *Verbena*. Now, with Laxton's death, others would come in to fill his place, while Henry would go on from strength to strength. Richard Dean's brother William visited Wem too this year, filing his report for the *Journal of Horticulture*. Unlike Richard, he commented not only on Henry's sweet peas but on his garden peas too, an especial interest of his. William pointed out that Henry began crossing the garden pea in 1879, the same year that he began working on sweet peas.

As before, Henry was happy to share the secrets of his garden pea breeding programme. His focus was on sweet peas, and here he kept his cards close to his chest.

> Mr. Eckford 'commenced crossing Ne Plus Ultra with Pride of the Market, Dr. MacLean, Champion of England, Veitch's Perfection, and others, producing the wrinkled type with a view to supersede the race of early round Peas so generally grown for the first crop... With one of the [new] seedlings 580 seeds were saved from one plant, and these were sufficient to plant five rows, each from 50 to 60 feet long, the plants 2 feet high...'.(62)

The last reference to Henry in '93 was in a November edition of the '*Chronicle*. 'I had heard so much about the great breadth of ground for Peas at Wem as to be somewhat sceptical' wrote a commercial grower, 'but I took a journey to Wem and found 5 acres devoted to the sweet and culinary Peas already sent out and to be sent out'. Meaning, as we've seen, seed to be sown elsewhere to produce seed in commercial quantities. 'Sent out' was a commonly used expression but one imagines that the author was already losing the attention of some of the '*Chronicle's* private-service readership. With Henry's garden peas, 'a variety named Wem will make a reputation' with the author describing it and 'Critic', 'Aston Gem', 'Rex', 'Armorial' and 'Chieftan'.

Of a new batch of seedlings under trial, 'Mr. Eckford has made Ne Plus Ultra and its progeny the pollen-bearing parents', meaning that its pollen was taken to the stigmas of the other plants of the cross, which become the seed or pea bearers, 'with a view to secure square-ended pods of large size, and of Ne Plus Ultra quality', and earlier cropping.

> 'Thin sowing is adopted here, and in the Wem peas a branching habit has been aimed at, together with obtaining pods in pairs instead of singly, and it was surprising to see the large numbers of full-sized pods on a single plant. Of varieties already sent out I specially noticed... Epicure, Censor, Superabundant, and Consummate.
>
> It was on a hot day that I had a run through the [Sweet] Peas, and the great variety seen there was somewhat bewildering, for the character of each variety was well shown in the long rows and masses of colour, and with these, as with culinary Peas, thin sowing, or rather planting, is the rule. Foremost amongst new

> kinds to be introduced are Duchess of York, white, striped with pink; Duke of York, bright rosy-pink', 'Meteor', 'Eliza Eckford' and 'The Belle'. All the flowers were of a 'good size and improved form; and in these, and the newer varieties sent out lately, the flowers are more numerous on the stalks than on the old kinds'. From Henry's recent introductions and his older varieties the writer was struck by a further twenty-one varieties.[63]

So, the end of another year, 1893, and it had all gone rather well. If Henry's family life had been littered with tragedies, his working life, where he had more control of events, had been anything but, and fame beckoned. Was it all going rather too well? Could it possibly last? Only time, he knew, would provide the answers.

Henry opened his new year campaign in typically quiet fashion with the smallest advertisement on the page, in *The Gardeners' Chronicle* of the 3rd February. 'Eckford's Sweet Peas and Culinary Peas', it ran, making it his first *'Chronicle* advertisement for sweet peas. (It had appeared two days earlier in *The Journal of Horticulture*).[64] When Henry worked for Dr. Sankey, he sold to a nursery before going into business on his own. With sweet peas, he may have continued in a similar way, supplying others to establish his name, before branching out on his own. If so, this may be the moment he changed to retailing. The month is very instructive. All the garden writers pushed the virtue of autumn sowing for sweet peas. A perusal of advertisements for the time, however, shows that the period for sweet pea seed sale was late January to early March. No-one advertised in the autumn. Gardeners read the articles but stayed with word-of-mouth, their own experience, and late winter sowing.

They were right to do so. In defence of the written word though, the age of the specialist book was only just arriving. The Reverend Samuel Reynolds Hole had written on our most popular flower in *All About Roses*, back in 1869, and one or two other plants had recently acquired their monographs. But there were none yet for the sweet pea, where knowledge was still being gained. The first of them, the first in English at any rate, was reviewed that April in the 'New books' column of *The Gardeners' Magazine*. 'All about Sweet Peas. By Rev. W.T. Hutchins. (W. Atlee Burpee and Co., Philadelphia). This handy little brochure fully justifies its title, for in the one hundred and thirty pages the author has given us the most complete information upon sweet peas that could be desired either by raisers or growers. Sweet peas appear to be not less popular in the United

States than they are in this country, the principal varieties grown being those introduced by Mr. H. Eckford, of Wem, whose labours receive just recognition from Mr. Hutchins' pen'.[(65)]

Hutchins, who we'll meet later, was an enthusiast who found time between his pastoral duties to buy, grow, sell, exhibit and now write a book on his favourite flower. Later on he visited Wem, which he located for the benefit of his American audience as being 'about fourty miles south of Liverpool, England'.[(66)]

In May, Richard Dean was once again promoting thinner sowing of sweet peas, this time in *The Gardeners' Magazine*.

> 'Mr. Molyneux is on the right tack when he states that sweet peas are sown much too thickly, and I can quite understand Messrs. Henderson, of New York, saying that one plant, if "well attended to, will produce something over one thousand blooms." Mr. Henry Eckford found that out long since, and when I used to visit him at Boreatton Park, and subsequently at Wem, I have seen individual plants of sweet peas, through being isolated and mulched during summer, so free branched and so floriferous as to appear to be almost miracles of culture. Anyone who has the opportunity of visiting Mr. Eckford at Wem will notice that his culinary, as well as his sweet peas, are sown very thinly indeed in the lines, and being also mulched with manure in summer, they make a remarkably branched growth and bloom freely and persistently, and it is mainly owing to his good culture, and allowing the individuals ample space in which to develop, that he produces the fine and striking bunches of blossoms he is in the habit of exhibiting in London and elsewhere...'.[(67)] Edwin Molyneux (1851 – 1921) was a well-known head gardener who had recently written books on Chrysanthemums and grapes and was now, like so many, turning his attention to the sweet pea.

> 'Never in my experience have I seen Sweet Peas do so well as this season', a gardener wrote in to *The Gardeners' Chronicle* as early in the season as July. 'Of the fifty or more varieties sown, none failed. Some are much more robust than others, under the same kind of treatment. The earliest to bloom were some that were self-sown, and which came up after rain in October; and these, owing to the mildness of the autumn, were very forward by Christmas... and began to bloom towards the end of May', a

> good month's gain in flower. He was surprised to see that all his sweet peas had already made seven, or in the case of 'Captain of the Blues' eight foot, and claimed his flowering stalks to be '1 to 1 ¼ feet' in length.

He mentioned seventeen varieties, sixteen of them Henry's, a fact he didn't state or need to, remarking only that the odd one out, an American variety called 'Emily Henderson' 'had done very well; and no doubt other gardeners, like myself, were induced to work up a stock of it from cuttings as soon as the plants were large enough'. He didn't realise that cuttings were unnecessary as the variety would (or should) have come true to seed. The beauty of sweet peas is their simplicity; simple flowers, simple to grow. It was understandable ignorance and interestingly he finished by displaying more: 'If three sowings are made at intervals of about one month, the first being early in February, to be planted out in March; another about the last week in March, to be planted out in April; and a third in the open ground in May, [it] will furnish blooms till frosts come. H.C. Prinsep, Buxted Park'.(68)

This is very good advice for hardy annuals and the sweet pea is a hardy annual. But the simple way to get sweet peas to bloom 'till frosts come' is to pick the flowers. There is nothing wrong with successional sowing. There is just no need for it. It seems almost inconceivable that no-one noticed that in its two hundred years of cultivation here, but if anyone did, they kept to the ancient oral tradition and never set pen to paper. Actually, one anonymously modest gardener had discovered it, six years before, and wrote to *The Journal of Horticulture* about it, but sweet peas weren't so popular then and, in the way of these things, it got overlooked. 'If the flowers are allowed to form seed the supply will soon cease, but if they are cut before they wither a long succession will be produced'.(69)

Henry's interest in hybridisation was his horticultural specialism. His interest in plants, however, was too wide to allow this to limit him further to the garden pea and the sweet pea. He was developing new strains of bedding plants too. At the Shropshire Horticultural Spring Show in April he had exhibited a collection of his Cinerarias, receiving a 'Certificate of Merit'(70) too. Which would have been nothing to him personally but was good for business. Henry hadn't lost his interest in pansies either, especially now that they were 'once again one of the more popular of florists' flowers with the general public',(71) and a new national Pansy and Viola Society had sprung up.

Appropriately enough, it was the pansy and *Viola* specialist William Dean who now drew '*Chronicle* readers' attention to the pansies growing at the new Soulton Road nurseries. Under the heading 'Bedding pansy, Eckford's Bronze Prince', he wrote:

> 'Not only do Sweet Peas occupy a foremost position at Wem, but Pansies are also cultivated there in large numbers for seed, and I was particularly struck with a quantity of bushy plants in full flower in the middle of August of the above variety. It is of compact habit, throws well above the foliage a quantity of well-formed bronzy orange-red flowers, with a dark central blotch. It is a particularly striking and acceptable colour. It is brighter in the bronzy-orange-red colouring than that popular old variety, Thomas Granger, esteemed so much for its bright orange-tinted red colour, and also Bronze Queen, a bedding Viola which was esteemed not long since. Mr. Eckford has a fine strain of richly-coloured striped Pansies of medium size and good form, and a selected strain of the distinct Peacock strain, so very rich in blue and violet tints, as well as other shades of colour each with a distinct wire margin of white'.[(72)]

William Dean appears to have taken in the Shropshire Show on his visit to Wem. Henry had been exhibiting locally in a small way for the last two years, picking up certificates for his sweet peas in the Shrewsbury Floral Fête in the summer of '92.[(73)] There seemed to be a revival of interest in the Show. The attendance on the first day numbered 15,000 and 40,000 on the second, giving the highest receipts so far. William thought he'd call by and see how Henry and the other west- midland growers were faring this year. It gave him enough for another little article, this time in *The Gardeners' Magazine*. 'Mr. Henry Eckford's collection of sweet peas... attracted much attention, especially from ladies, and deservedly so... More new colours with greater breadth and form of standard and wings are being obtained, and generally more flower on a stem. Two charming new varieties were especially noticeable, viz., Alice Eckford ...and Eliza Eckford...'. William noted down a further thirteen 'especially fine' varieties.[(74)]

The following month, on the 20th October, and under the heading 'Sweet Pea culture', *The Garden* magazine ran the following:

> 'Mr. Eckford, of Wem, Salop, now so well known for the many fine varieties of Sweet Peas he has raised, in replying to us as

to their good cultivation says: "I do not like the Celery trench fashion. If the ground is in a tolerably good state of cultivation, that is, has been fairly well dug, simply put on a fair coat of stable manure and dig deep now, leaving it rough. In the beginning of March when the soil is in good condition, thoroughly break with a fork, which will be sufficient preparation for the seed. To obtain the best results, clumps of two or three plants at 1 yard or 2 yards apart are better than continuous rows. In staking put three or four bushy stakes thus : : round the clump, but well away from the plants, which should have a few smaller sticks to lead them to the taller ones.

Round the whole put a string or bit of wire to keep them together, so that when the plants have grown up a sort of cone may be formed. The sticks should be if possible 8 feet or 10 feet high, as planted in this way the Peas will, if mulched with half-spent manure or any kind of refuse to protect the roots from hot sun, &c., grow very strong and tall, and if the flowers are cut close every morning, so that no seed can form, they will continue to bloom till the frost puts an end to them. Should the weather prove dry, a soaking of weak manure water two or three times during the season would be beneficial. Should they from excessive growth get untidy, take the hedge shears and clip them over neatly; they will in a few days throw out fresh growths and a profusion of flowers. If this way of growing Sweet Peas is adopted, it is a good plan to put the seed singly into small pots, and when the seedlings are strong enough to plant them out; in doing so make the ground very firm about them – they delight in firm ground. If the weather be dry tread well in. If this plan be followed we shall hear nothing of pink Mrs. Sankey, which I feel certain is the result of starvation"'.[75]

'*The Garden*' was another of William Robinson's magazines and this article, bar the last sentence, was inserted into the sweet pea entry for subsequent editions of Robinson's popular gardening encyclopaedia *The English Flower Garden*. The complaints of pink 'Mrs Sankey', supposed to be a white and perhaps unfixed, are dismissed by Henry as a result of inadequate feeding. One wonders at this fixation with pure whites. What's so wrong with a white tinged with pink? It sounds lovely. This article was followed by another on sweet peas from Robert Brotherston. Most of this article also found its way into *The English Flower Garden*. 'Mr. Eckford was here a few weeks ago, and he confessed to be unable to grow them so fine. He said that he had never previously seen the flowers of his own Peas grown to be so large a size or so fine in colour...'.[(76)] Nothing like blowing your own trumpet, is there! Visiting old haunts perhaps, Henry had taken the opportunity to call in at Tynninghame. He had been sending selected sweet pea seedlings for Robert Brotherston to trial,[(77)] and wanted to see how they were getting on. Quite well, apparently!

With less going on in the winter, the gardening press brought out the longer articles they'd been saving. Ones like 'Botanical exploration in Borneo', or the possibilities of women studying market gardening and other 'lighter aspects of agriculture' at college. These, together with articles reflecting on the past season all helped to fill the spaces between more topical items like winter work in the Orchid house and in the orchard. In December, Edwin Molyneux had three long articles on sweet peas in *The Gardeners' Magazine*, in the last of which he summarised his views on fifty-six varieties, mostly Henry's.[(78)] 'From their greater usefulness', he began, 'those varieties of sweet peas having white flowers deserve first place in the list'. By how far this is opinion dressed as fact is unclear. A previous letter to the '*Chronicle* had stated that 'white sweet peas are much liked' and that the white 'Mrs Sankey, one of Mr. Eckford's raising, should be grown... in fact, I know of no better...'.[(79)] The usefulness of white may have been its ability to harmonise with all colours in flower arrangements.

Edwin Molyneux wasn't so taken with 'Mrs. Sankey' however. It 'is classed as a pure white variety', he wrote, 'but the tinge of pink that it carries all through the year is somewhat detrimental to its position... Another objection... a serious one – is the tendency of the standards in this variety to curl inwards too much'. He preferred 'Emily Henderson', the pure white American variety. The idea that white means pure white went with some other ideas of the time, that notches in the petals or any tendency for the standard petal to curl forward – hooding – should be prevented. Followed to the letter, the result could be quite odd-looking. The imposition of quite arbitrary definitions of how a flower should look takes us back to the rules of the old florists' flower societies. These could be quite sensible of course. That the plant should have healthy foliage or that the flower stem should be strong enough to support the flower head[(80)]. But to shape the flower from a rule-book is to put the cart before the horse.

Ruth Duthie, in *Florists' Flowers and Societies*, reminds us of Mary Mitford's stories of Berkshire village life in the 1820s, and the farmer's wife who 'was a real genuine florist: valued pinks, tulips, and auriculas for certain qualities of shape and colour, with which beauty had nothing to do; preferred black ranunculuses ... Of all odd fashions, that of dark, gloomy, dingy flowers appears to me the oddest'.[(81)] Perhaps the gaudy Victorian bedding plants had been, in part, a reaction against this late Georgian fashion, and when the time came to react against *them* then cool whites were favoured, the wheel of fashion unable to return so soon to its former spot.

Edwin Molyneux was an able and articulate professional gardener. Through growing the plants himself through a season or two he was able to see things that others, popping down to a nursery or to a Show were not. Henry's 'Duchess of York', this critic thought 'one of the best varieties in existence', partly because he noticed that part of its colour 'deepens as the autumn advances'. With 'Lemon Queen' 'it is only in the unopened buds that any trace of lemon can be detected'. 'Captain of the Blues' colour lightened with age; in 'Lottie Eckford' 'the flowers curl up too much and are easily affected by rain, and except in dry weather the variety seems hardly worth growing'. 'Mrs Gladstone' he thought 'still one of the best varieties... Towards autumn the blooms assume a deeper tint of pink, and during a spell of wet weather the blooms are occasionally splashed with rich pink or pale rose'. 'Orange Prince''s growth was weak, its flowers 'showy in a mass, but easily affected both by rain and hot sun'. 'Lady Penzance''s flowers were 'quickly damaged by rain'. The author also regarded five pairs of Henry's varieties to be identical or almost so.

This generally reliable and objective observer's favourite sweet peas, in December 1894, seem to have been 'Emily Henderson' and seven of Henry's; 'Blanche Burpee', 'Mrs. Joseph Chamberlain', 'Mrs. Eckford' (similar to 'Primrose'), 'Duchess of York', 'Emily Eckford' and 'Mrs. Gladstone'. All eight, coincidentally, named after women. Of all the fifty-six varieties described, all still had only two or three blooms on a stem, apart from the American 'Emily Henderson', whose 'three blooms upon a stem is the rule, although I have found four on some'. Although Henry was clear of the field in Britain, the Americans might be edging ahead in blooms per stem. Well, it was nice to have some competition. It kept you on your toes. Heights were generally seven to eight feet, Molineux found, and scent was – well, he didn't actually mention scent. Perhaps he thought it too obvious, a bit like saying the leaves were green and it climbed.

What he might have said, and perhaps he hadn't noticed, was whether scent was strongest in certain weather conditions – after a shower for example. Was it like the *Verbena*, strongest in the whites? J.C.Loudon, naturally enough, had spoken about scent, as had Thomas Fairchild, James Justice and others in the eighteenth-century. But contemporary writers did not. It's odd. But then, you often don't notice something 'till it's gone. The scentless sweet pea – now there's a contradiction in terms! How could that possibly happen?

References to Chapter Nine

(1) 'A Visit to Wem', *The Gardeners' Magazine*, (29-2-1896), p.129.
(2) Geoffrey Keynes, ed., *Hazlitt's Selected Essays*, (1970), pp. xi-xii.
(3) Stevie Davies, *A Century of Troubles*, (2001), p.173.
(4) Marjory Ellis, 'Judge Jeffreys of evil memory', *The Shropshire Magazine*, (2-1956), p.30.
(5) 'Floral Committee', *Journal of the Royal Horticultural Society*, (1889), pp. xcv & xcix.
(6) Census return, Parish of Saints Peter and Paul, Wem, 1891.
(7) Harry Eckford, Correspondence to Bernard Jones, c. 1980-82.
(8) 'The Cult of the Sweet Pea', advertisement, *The Daily Mail*, (21-2-1911), p.10.
(9) David Thomson, *Handy Book of the Flower Garden*, (1893), pp. 149-150.
(10) 'Sweet Scented Plants', *Gardening Illustrated*, (15-6-1889), p.194.
(11) 'Famous Gardeners at Home', *Garden Life*, (26-9-1903), p.464.
(12) H. John Philips, lifelong Wem resident, in conversation with the author, 22-7-2001.
(13) E.R. Kelley, ed., *Directory of Shropshire*, (1891), p.464. The commercial high-street premises were short-lived, being unlisted in the 1895 directory.
(14) Thomas Stevenson, *The Modern Culture of Sweet Peas*, (1910), p.93.
(15) Ibid., p.97.
(16) 'Nursery Notes', *The Gardeners' Chronicle*, (22-6-1889), p.776.
(17) William White, *History, Gazetteer and Directory of the County of Essex*, (1848), p.153.
(18) Ibid., p.173.
(19) 'Sweet Peas in Essex', *The Garden*, (2-8-1902), p.75.
(20) 'Sweet Peas', *The Sweet Pea Annual*, (1906), p.45.
(21) 'Tracing the Growth of a Market Town', *The Shropshire Magazine*, (10-1958), p.23.
(22) Ordnance Survey map of Wem, 2nd. edn., 1902.
(23) G. Keynes, op.cit., pp. 130-131.
(24) R.C.K. Ensor, *England 1870-1914*, (1949), p.138n.
(25) Harry Eckford, letter to Bernard Jones, 12-5-1982.
(26) Charlotte (Lottie) Ryley, née Eckford, letter to Bernard Jones, 10-9-1979.

(27) I am indebted to Barry Eckford, Peter Eckford's grandson, for some of this information.

(28) Letter from Denise Tanner, née Barclay, to Bernard Jones, (8-12-1989).

(29) 'Eckford's New Sweet Peas', *The Journal of Horticulture*, (8-8-1889), p.111.

(30) 'Progress in Peas', *The Gardeners' Chronicle*, (5-10-1889), p.387.

(31) 'New Seedling Culinary Peas', *The Gardeners' Chronicle*, (23-8-1890), pp.211-212.

(32) 'Eckford's Sweet Peas', *The Gardeners' Chronicle*, (5-9-1891), p.284.

(33) 'Sweet Peas', *The Journal of Horticulture*, (20-8-1891), p.157.

(34) O. Chadwick, *The Victorian Church*, Part 2, 1860-1901, (1970), pp.227-228.

(35) Iris Woodward, *The Story of Wem and its Neighbourhood*, (1952), p.84.

(36) Mark A. Smith in C. Williams, ed., *A Companion to Nineteenth-Century Britain*, (2004), pp.343-344.

(37) 'Wem', *The Shrewsbury Chronicle*, (16-3-1888), p.7.

(38) Molly Vivian Hughes, *A London Family, 1870-1900*, (1946), p.586.

(39) Ralph Waldo Emerson, *English Traits*, 1856, (1902), p.134.

(40) Geoffrey Best, *Mid-Victorian Britain, 1851-75*, (1971), p.185.

(41) *The Henry Eckford Celebration*, Wem Baptist Church (pamphlet), 1988, passim.

(42) The Book of Revelation, 3:20-21, Authorised Version of the Bible.

(43) 'Growing Sweet Peas', *The Gardeners' Chronicle*, (13-2-1892), p.204.

(44) 'Lathyrus odoratus hybridus', (illustration), *L'llustration Horticole*, (15-1-1892), opp. p.9.

(45) 'Garden Flora', *The Garden*, (12-3-1892), pp.232-233.

(46) 'Sweet Peas', *The Journal of Horticulture*, (1-9-1892), p.185.

(47) 'Wem', *The Journal of Horticulture*, (1-9-1992), pp.188-189.

(48) 'New Culinary Peas', *The Gardeners' Chronicle*, (26-11-1892), p.651.

(49) 'Shrewsbury Floral Fete', *The Gardeners' Chronicle*, (27-8-1892), p.254.

(50) 'Sweet Peas at our Exhibitions', *The Gardeners' Magazine*, (21-1-1893), p.33.

(51) David Stuart, *The Garden Triumphant*, (1988), pp.275-276.

(52) Flora Thompson, *Lark Rise to Candleford*, (1948), p.161.

(53) 'Sweet Peas', *The Gardeners' Magazine*, (18-2-1893), p.93.

(54) 'Sweet Peas', *The Gardeners' Chronicle*, (12-8-1893), p.182.

(55) 'Current London Topics', *The New York Times*, (25-4-1880).

(56) Michael Balfour, *Britain and Joseph Chamberlain*, (1985), pp.201-202.

(57) 'Eckford's New Culinary Peas', advertisement, *The Gardeners' Chronicle*, (4-2-1893), p.125.

(58) 'Sweet Peas', *The Gardeners' Magazine,* (18-2-1893), p.93.

(59) 'Sweet Peas at Chiswick', *The Gardeners' Chronicle*, (1-7-1893), p.13.

(60) 'The Shropshire Horticultural', *The Gardeners' Chronicle*, (26-8-1893), p.249.

(61) 'Sweet Peas', *The Gardeners' Chronicle*, (12-8-1893), pp.182-183.

(62) 'Amongst the Wem Peas', *The Journal of Horticulture*, (4-10-1894), p.324.

(63) 'Eckford's New Peas', *The Gardeners' Chronicle*, (4-11-1893), p.566.

(64) 'Eckford's Sweet Peas', advertisement, *The Gardeners' Chronicle*, (3-2-1894), p.132 & *The Journal of Horticulture* (1-2-1894), no paging.

(65) 'New Books', *The Gardeners' Magazine*, (14-4-1894), p.208.

(66) Rev. W.T. *Hutchins, Sweet Pea Annual for 1897*, (1896), p.1.

(67) 'Notes of Observation', *The Gardeners' Magazine,* (5-5-1894), p.250.

(68) 'Home Correspondence', *The Gardeners' Chronicle*, (28-7-1894), pp.101-102.

(69) 'Early Sweet Peas', *The Journal of Horticulture*, (5-4-1888), p.285.

(70) 'Shropshire Horticultural Spring Show', *The Gardeners' Chronicle*, (21-4-1894), p.507.

(71) 'Florists' Flowers', *The Gardeners' Chronicle*, (8-9-1894), p.278.

(72) 'Home Correspondence', *The Gardeners' Chronicle*, (25-8-1894), p.222.

(73) 'Shrewsbury Floral Fête', *The Gardeners' Chronicle*, (27-8-1892), p.254.

(74) 'Notes of Observation', *The Gardeners' Magazine*, (1-9-1894), p.518.

(75) 'Sweet Pea culture', *The Garden,* (20-10-1894), p.363.

(76) 'Sweet Peas at Tynninghame', *The Garden*, (20-10-1894), p.363.

(77) 'Sweet Pea Lady G. Hamilton', *The Gardeners' Chronicle*, (12-2-1910), p.106.

(78) 'Sweet Peas and their Culture', *The Gardeners' Magazine*, (29-12-1894), pp.803-804.

(79) 'Sweet Peas', *The Gardeners' Chronicle*, (5-3-1892), p.310.

(80) Ruth Duthie, *Florists' Flowers and Societies,* (1988), p.33.

(81) Mary Mitford, *Our Village*, 1824-1832, (1997), p.30.

Chapter Ten

Well, what do you know. You wonder out loud when people are going to note that sweet peas are scented, or note more often that picking the bloom increases the flowering season and hey presto! the pages of the horticultural press are full of it. In February of the new year, 1895, *The Gardeners' Chronicle* ran a résumé of the Molyneux articles.(1) 'To Mr. Henry Eckford,' it began, 'we annually look for new, beautiful, and striking developments in sweet peas. Given a good strain of seed and good cultivation, no other annual can surpass the sweet peas either for beauty as plants, for fragrance, free and continuous flowering, or for cultivating from'. Well of course you knew they knew it was scented. All the same... The writer rubbed it in that the American variety 'Emily Henderson' was now known as 'the Queen of White Sweet Peas'. At least Henry could argue that 'Emily Henderson' was almost certainly derived from seed he'd sent there a few years earlier. Some people in fact considered it merely a carefully selected stock of Henry's 'Queen of England'. The paper had included an illustration of Henry's 'Blanche Burpee', 'Emily Henderson''s close rival (though monochrome drawings do nothing for flowers, especially peas).

In March the *'Chronicle* ran an article by Edwin Molyneux that was little more than another résumé of his *Gardeners' Magazine* articles of three months earlier. The *'Chronicle*, neither then nor since has catered for the amateur market. Horticulture was changing and society becoming too diverse for the *'Chronicle* to have held on to its premier position, but becoming a digest would be its own fault. Molineux did have one new announcement to make however. The article began familiarly enough: 'No one has done so much toward the improvement of Sweet Peas as Mr. Eckford, at his seed-ground at Wem, in Shropshire. His aim in raising new varieties is to obtain extra stout and long stems, and as many closely-set blossoms on each stem as possible. Some few sorts are furnished with as many as four blooms on a stem, and this 15 inches long. Concurrently with an improvement in the colour of the flower, a different method of culture has obtained, and in place of being grown in rows of mixed colours, they are now frequently grown in clumps of one colour only.' In the middle of the article he had this to say: 'Many persons make a mistake in not cutting the blooms sufficiently often. The more they are cut the longer and better they continue to flower'.[(2)]

In August, there was a letter on 'Colour and Perfume in Sweet Peas', too, from a reader in Ard Cairn, Cork.

> 'Have any of your readers observed how different are the perfumes of Sweet Peas? Some of them have that of the Rose, all more or less quite distinct from each other. The white varieties are particularly weak in odour, while the scarlets and pinks are very strong. To-day I ventured to suggest to a few friends, and my warehouse assistants, that I could blind-folded tell the names and colours of three very distinct varieties...' which he did. I 'venture to suggest that colour in flowers may have something to do with perfume... What say your readers?' [(3)]

It does; if true, sweet peas were, interestingly, contrary to the *Verbena*, whose scent in hybrids we saw came from the white night-scented *V.teucrioides*.

~~~

William Dean was seventy now and suffered from asthma and bronchitis. He died suddenly one March evening, at the end of the winter that became remembered as 'the great frost'. Like his brother, William had been born in Southampton and also like his brother had had a wide and active interest in horticulture, which he gained from the experience of living and working in various different places. This included an early start as foreman in a Belfast nursery, where his youth and country, neither being favourably regarded by his workforce, saw him move to London. If William, his brother and Henry were at all typical of successful Victorian gardeners, it was a success achieved by travelling around a great deal within their profession, gaining experience both of it and their nation. The Victorian age was known for its energy and these three, driven by that and their endless curiosity, shared it. It eventually brought them to the heads of their fields, though you wonder how happily this fame sat with them.
~~~

Living in Birmingham, William had often travelled the three miles from his Sparkbrook home in the Dolphin Road to the Botanic Gardens at Edgbaston. Here, his '*Chronicle* obituary stated, 'he took a leading part in the exhibitions... sometimes acting as secretary, particularly in connection with the Midland Carnation and Picotee Society, as assistant to his old friend, Mr. Robert Sydenham', [4] a very famous name to be in sweet pea circles. William Dean would have introduced Henry, his fellow adopted midlander, to Robert Sydenham and to the Gardens. The Gardens are still, in my view, the best reason to visit Birmingham or to live there. You would hardly know, in this delightful spot, that you were in the confines of a city. Henry once won a first-class certificate from 'The Midland Carnation and Picotee Society, Botanical Gardens, Edgbaston, Birmingham'. It's on my desk as I write. It was for his sweet pea 'Prince Edward of York', which came out in 1896, the year after William died. So Henry retained his connection to the Gardens.

Henry, two years older than William, still travelled around a great deal through his work which he loved, especially when it brought him into the company of like-minded enthusiasts like William Dean. Like-minded

Certificate for Sweet Pea 'Prince Edward of York' c.1896.

enthusiasts could be found at the provincial flower shows too. The railways were making travel easy and affordable. Nurseries like Henry's could take advantage of them to enter or at least have a display in our numerous summer shows. One such, in Berkshire, brought forth the following letter.

> Windsor Castle.
>
> July 2/95.
>
> Lord Edward Pelham Clinton, Master of the Household, is desired by the Queen to convey Her Majesty's thanks to Messrs. Eckford for the beautiful bouquet of Sweet Peas sent by them from the Flower Show held at Windsor on June 29th.[(5)]

Later in July, Henry featured in another article, in *The Gardeners' Magazine*.

> 'The collection of sweet peas, exhibited by Mr. H. Eckford the well-known raiser of Wem, Salop, at the meeting of the Royal Horticultural Society on Tuesday, was noteworthy not less for the taste with which the flowers were arranged than for the evidence it gave of the remarkable development these flowers have undergone as the result of the manipulative skill of Mr. Eckford. Instead of staging the thirty odd varieties represented in the collections in close bunches, on the flat orthodox stands, each variety was lightly arranged in a simple yet elegant vase about fifteen inches high. Staged in this tasteful manner, they not only contributed very largely to the attractions of the meeting, but demonstrated the utility of the sweet peas for the adornment of the house'.[(6)]

The old claret and blue sweet peas had long been sold as cut-flower, destined to decorate a room in a container of some sort. But it was only now, it seems, having become fashionable, that they were placed in a vase as a feature. It would have explained the surprise of Flora Thompson's new bride's in-laws at seeing them so displayed on the dining table at mealtime.

The '*Chronicle* was now reporting a surprising new development from the United States. 'A sweet pea which doesn't climb but attains a height of only a few inches'.[(7)] It had originated as a sport with Mr. C.C. Morse of Santa Clara, California, and was being sent out by Atlee Burpee & Co. of Philadelphia.[(8)] They were known as 'cupids', and made a British appearance at an R.H.S. committee meeting on 25th June. They were found to be bushy, scentless, white-flowering plants of some 5" or 6" in height. A photo appeared four days later in the *'Chronicle.*[9] Henry was to stock them, possibly derived from seed brought over by the American W.T. Hutchins, who visited Henry this year.[10] Henry, like Hutchins, wasn't too enamoured with them. They were a long way for a tall seventy-two year old to bend down to! These smaller forms have acquired a niche value now as attractive plants for window-boxes, containers and hanging baskets.

Small sweet peas making a splash – and small garden peas too. 'So far as the tables of the rich are concerned', said *The Garden*, large [podded] Peas are each year becoming more and more unpopular...'.[11] The larger ones looked good in an exhibition but people were finding, not for the first or last time, that size wasn't everything and that smaller pods contained tastier peas. Who, we wonder, already knew that? The following week, a reader wrote in to agree 'that many small-podded varieties of Peas have much to recommend them... I have this year grown several of Eckford's newer varieties, and find that many of them are short-podded sorts. He must have had this idea in his head when working to obtain new varieties' and the writer went on to praise them.[12] That summer's hot weather may have benefited them too, in comparison with the larger sorts.

Four months after his brother's death, Richard Dean was visiting Lincolnshire and the seed-trial grounds of a Boston seed merchant, to look at the garden pea. With much of the country labouring under a summer drought, 'living green' Lincolnshire, in Dean's phrase, had escaped it. All the new varieties obtainable were there and, having grown well, were fit for comparison. Richard Dean duly summarised the results for a long *'Chronicle* article to be published in the autumn.[13] Henry had seven new pea varieties in the trial but with his focus elsewhere they were but few among dozens of others. Henry had none among the early varieties. Among the main-crops, his 'Critic', five-foot high, produced 'very fine blunt tipped pods, and as seen here well deserved the three marks given to it in the Chiswick trials'. 'Censor' 'is a pea growing to the height of 2 ½ feet, having fine full dark green pods, slightly incurved, certainly a very fine garden variety'. His 'Consummate' was very like another grower's variety here, the three foot 'robust grower' 'Sharp's Queen'.

Henry had three of the four peas in the 'taller section', perhaps unsurprisingly for a tall man who wouldn't want to bend down too far. His 'Epicure' was an illustration, six feet tall, with large slightly curved pods like the well-known variety 'Telephone', and 'a great cropper... Wem is a very tall grower in Lincolnshire, producing very large whitish pods of excellent quality in the seeds; it is a free cropper. Memorial is a deep green wrinkled Marrow, growing quite tall, with very long, slightly-pointed pods, promising to be very fine for exhibition'. In unstaked open field cultivation 'Superabundant was very fine... it is a flat-podded variety, and very prolific... Critic and Censor were also very good' here. 'Invincible' was another in this section for market gardeners to 'keep an eye on'. 'Invincible' was the name of an old variety of Henry's. Dean doesn't say whether a new variety had taken its name or not.

Of all the peas on trial there were many good ones, with an early, 'English Wonder', seeming to catch Dean's eye the most. What did catch his eye was the quantity. 'That it may be fairly said we have too many Peas', he concluded, 'there can be no doubt. Messrs. Hurst & Sons' catalogue for the present year contains 122 names, and of these a few may be accepted as synonyms; perhaps the growth in numbers is not very remarkable after all, for their catalogue issued in 1852 contained forty-nine varieties. It is perhaps well to have an ample choice to meet individual tastes, and adapted to certain localities; still there is an amount of bewilderment in the contemplation of so many, and the differences between some are only very slight. Perhaps the old adage, that it is possible to have too much of a good thing, may be accepted in the case of Peas; and it does appear as if the list is likely to be extended rather than reduced. Certain it is, that no one who values Peas as a cooked vegetable can fail to have his taste gratified among so many claimants to his favour.'

A hundred years on, very few pea varieties had survived to be grown by the BBC in their wonderful television series on 'The Victorian Kitchen Garden'.(14) One of them from here had though: 'Gradus, announced as a large-podded first early wrinkled variety, produces fine-pointed pods... but the work of selection is by no means complete, judging from what was seen at Boston. But with rigid selecting, a very good type of early Pea is likely to be produced; in height it is about 2 feet'. Two years later, rigid selection had worked, but whether the BBC had *it*, an earlier form or even one completely different would be hard to say. In Henry's day old varieties would appear under new names, as much by accident as design, and one supposes the practise continued. What is certain is that the days of being bewildered by the contemplation of so many claimants to our palate are long past.

With his focus more on sweet peas, Henry found unequivocal success again at Shrewsbury, where he won a gold medal for 'an extensive collection of Sweet Peas, tastefully displayed…'.[15] The firm also won six silver medals at Summer Shows in Edinburgh, Trentham and London.[16]

The new year, 1896, was eighteen days old when the following article appeared in the columns of *The Gardeners' Chronicle*: 'One of the most puzzling things about variations is the simultaneous appearance in widely different localities of the same "sport". A little while since we published a figure of a dwarf Sweet Pea, which reached us from America. Now we learn, that at the same time Mr. Eckford had the same dwarf variety in his grounds at Wem in Shropshire'.[17] Although Henry's cupids may have been American, such a mutation was likely to appear anywhere. A fortnight later it was reported that the seedsmen Cooper, Taber & Co. had it, while the Ernst Benary seed firm at Erfurt in Germany already had one in its catalogue.[18]

A problem soon identified itself. The new form was rather delicate, here and in Germany. It was okay under glass but not outside. A grower in Cork however found that they did do well outdoors if treated like bedding plants, hardened off and planted out in May.[19] The abundant artificial cross-fertilisation now taking place appeared to be throwing up these odd forms, almost impossible to occur naturally. An even more dramatic change, however, was yet to occur.

~~~

It is easily possible to be dominant in your own field without having any wider recognition, even within your own profession. Anonymity within horticulture stopped for Henry on 29th February, 1896. This was the day that William Robinson published Henry's portrait and a full-page article on him for his *Gardeners' Magazine* readership. 'A visit to Wem' it was called, and became the first of a number of features on what was fast becoming one of horticulture's iconic figures.[20]

The article treads familiar ground, familiar to us anyway, but there were some interesting comments on Henry's garden peas – all Mr. Eckford's best sorts are wrinkled marrowfats of the highest flavour, and a great point is that the pods are produced in pairs – and there were some new sweet pea varieties to report on. Among them 'Little Dorrit took my fancy [above the others] as it produces immense flowers with a broad and bold rosy cerise standard and white wings; it is a lovely flower, and cannot fail to rapidly become popular'. The name would have been recognised as the child born in a debtors' prison from Dickens's novel of the same name. Another new one, appearing in the 1896 catalogue, was 'Prima Donna', a soft pink, while
~~~

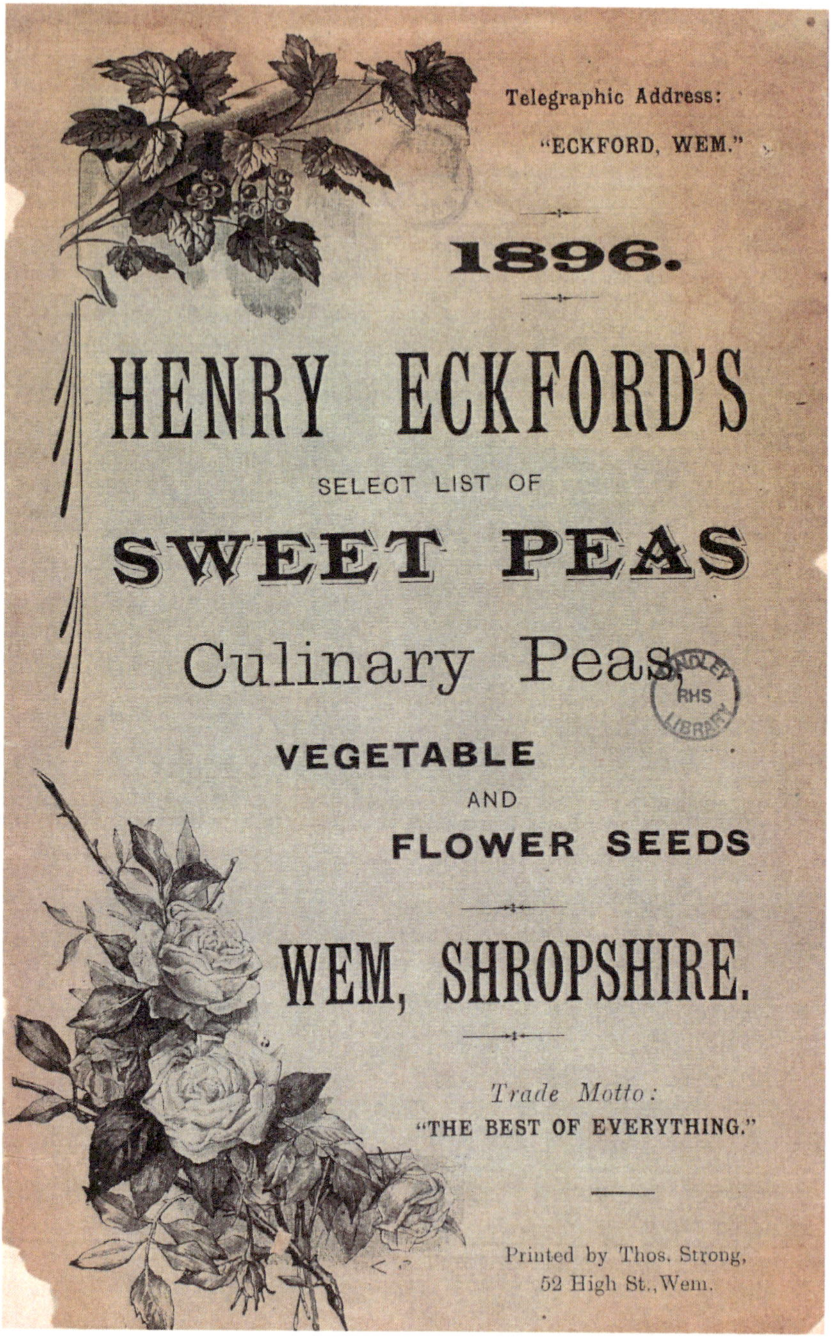
Telegraphic Address:
"ECKFORD, WEM."

1896.

HENRY ECKFORD'S

SELECT LIST OF

SWEET PEAS

Culinary Peas,

VEGETABLE
AND
FLOWER SEEDS

WEM, SHROPSHIRE.

Trade Motto:
"THE BEST OF EVERYTHING."

Printed by Thos. Strong,
52 High St., Wem.

Henry Eckford Seed Catalogue for 1896

Grisell Baillie (Lady Hamilton)

another, 'a delicate and lovely blue form', not yet in commerce Henry had named 'Lady Grisel Hamilton'.

The late, philanthropic Lady Grisell Hamilton (1822-1891) had been the sister of Robert Brotherston's employer at Tynninghame, the Earl of Haddington. As we've seen, Brotherston was one of Henry's sweet pea trialists. It appears that Henry had known Lady Hamilton for some time. Henry 'had previously raised and lost a variety of much the same colour', Brotherston recalled years later, 'and on the occasion of his exhibiting [it] for the first time in London, he wrote me of his intentions, and, Lady... Hamilton being in London at the time, I wrote to tell her of Mr. Eckford's forthcoming exhibit'.(21) Lady Hamilton's biographer, her sister, only mentions two London visits, both out of season for sweet peas, so there must have been others. Early in 1890, visiting her sister there they attended a performance of Handel's Messiah.(22) One can imagine Henry accompanying them on a trip out to Sydenham and the Crystal Palace there, a popular venue for choral music. Similar in age to Henry, Lady Hamilton was however pious and unworldly. Being from the Scottish borders too, a day away from which is considered as a day wasted, (or is that just Hawick?) she would have welcomed the more travelled man's companionship. She may have helped Henry financially in the past: when he first went to London, for the Hamiltons had some connections to the Thrieplands of Fingask, one of Henry's last Scottish employments before moving south. But this is all speculation, like so much in this story.

A few pages on from 'A Visit to Wem' (a copy of which Henry kept) brings us to one of the new Eckford adverts (which he didn't). Either Henry realised that advertising wasn't his thing, or John told him, because the style set now is that continued by John long after his father's death. A quarter-page spread is headed: 15 Gold and Silver Medals. Eckford's giant sweet peas. Their beauty is altogether indescribable. I am convinced none of us really knows what Sweet Peas are until we have seen yours, says Mr. Percy Chipp (a former Kenwood colleague of Henry's?) of Caen Terrace, Highgate. Etc, etc. But this was really no worse than anyone else's advertising. A free descriptive catalogue came with Queen Victoria's 'special commendation'. And if that was the Windsor Castle letter, (which it was, on the back page) then it's a bit of a cheek because such a note is easily obtained. But that, we're told, is business. It's interesting to see that the Eckfords were already diversifying with strains of Cinerarias, *Primula sinensis* and hybrid primroses, garden pea varieties and other vegetable seeds for sale too.(23) It's evidence of their earlier decision that, with the fickleness of fashion, sweet peas could drop from favour. They could rely on garden pea sales and hedged their bets further with a broader range of subjects.

The 1896 catalogue is the first to come down to us and was, perhaps, their first retail one. The green-coloured cover, oddly enough, features illustrations of grapes and roses, two plants that they didn't grow. It begins with garden peas, indicating that they were its biggest seller. With vegetables, the Eckfords don't seem to have their own varieties but they were, perhaps, their own strains. Items other than sweet pea seed included lawn and sports grass seed mixtures, with or without clover; bulbs, corms and 'garden sundries' like fertilisers, tying materials, gloves and so on, all of which Henry had used all his life. A wide-ranging collection but lacking even a nod, in a wild, natural and native Robinsonian climate, towards hardy ornamentals. Henry's impressionable years were far behind him and he probably couldn't have changed, even if he'd wanted to. The catalogue invited customers to win 'special prizes' of from £2 to 10/- for exhibiting Eckford sweet peas (and garden peas at Shrewsbury) at seven regional Shows from Dundee and Edinburgh to East Anglia and the Midlands.

If Henry had wondered if interest in his sweet peas would wane, there was no sign of it. Quite the reverse in fact. *The Gardeners' Magazine* had another article on them in their editorial pages in March. Following a brief history it anticipated, following the 'cupid' breaks, moderately dwarf varieties to about 3′ that would do away with the expensive necessity (for the urban and suburban gardener presumably) of hazel sticks and the like to grow them up.(24) The Eckfords needed around two-thousand bundles of hazel twigs and birch tops yearly for their Soulton Road nursery alone. But the dwarfer types never really caught on. Neither did another 'break' that July, when James Carter & Co. announced the arrival of sweet pea flowers with double and triple standard petals, a feature which, they said, 'makes the blossom far more attractive than has hitherto been the case when only one has appeared'.(25) This wasn't the general view however, as it negates the flower's basic attraction of charm and simplicity.

No aspect of the sweet pea could now escape scrutiny. How about the right way to exhibit them? There was no better way than in tall 'glass vases about fifteen inches high, and with slender stems such as those used by Mr. Eckford, of Wem', wrote *The Gardeners' Magazine* that June. (26) A correspondent to the 'Chronicle thought that a 'striking artistic effect' could be obtained by growing the Eckford sweet peas alternately with *Tropaeolum speciosum*, the nasturtium. (27) Further evidence of the sweet pea's growing popularity was its inclusion in the popular school-children's 'object lessons'. On the list for the school year 1896-97 for Standard Three pupils at St Paul's Church of England Combined Schools, Lemington Spa were 'snail, butterfly, silkworm, coal, daffodil, sweet pea...'. Where possible, the teacher would place an example of each item on the pupils' desk. Examples from further down the list, however, would have tested the teachers' ingenuity in this regard – icebergs, glaciers, streams, camels and coal gas...(28)

There was an interesting entry at the Shropshire Horticultural Show in August. 'The class for Sweet Peas was a strong one, Mr. Sankey, Baschurch, Salop, being first, showing the best and newest kinds; the arrangement being tasteful and natural'.(29) The new Dr. Sankey's head gardener, Bert Ballinger (30) would have heard all about Henry and would soon meet him if he hadn't already. He had absorbed Henry's enthusiasm for the sweet pea with, obviously, good results. It was nice to see that, with Henry's departure, the flower was still blooming well at Boreatton Park.

Eighteen ninety-six was the year that seventy-three-year-old Henry Eckford got a bike.(31) So too that year did the fifty-five-year-old Prince of Wales, the future Edward VII, or 'Prince of Wheels' as *Punch* magazine dubbed him.(32) It seemed that nearly everyone else was on a bike too, for those that

Henry Eckford, at 73 years of age, with his first bicycle

couldn't afford to buy them could hire one for sixpence an hour.[(33)] The bicycle had been in development throughout the century. From a steerable contraption in 1817, to a crank-driven sort of hobby-horse in 1839, to the more popular but still aptly-named bone-shakers from 1861 on and the frightening penny-farthings of the 1870s. These were all heavy solid-tyred machines, great for going downhill on.

Then came the invention of a pedal-driven rear-wheel drive using chain-transmission. With the general arrival of the long invented pneumatic tyre in 1888 followed by gears and the free wheel in the early 1890s the scene was set for an explosion in bike-riding. Following his boyhood experiences of two day's travel, in winter, on the outside of the old Defiance stage-coach, Henry had had particular reason to be circumspect of all forms of vehicles. So it was rather late in the day that he took a risk with the bicycle. The rural Shropshire pedestrian though, who had as elsewhere in rural Britain been used to having the road to him or herself, or at most sharing it with a slow-moving horse and cart, was now at more risk than Henry was!

After education, the bicycle was the great woman's liberator. The railways also brought travel but you had to get to a station and for a long time fares were too high for most village people.[(34)] Many men were strangely envious of women cyclists and at first tried to stop them. It was as if they had a premonition that this was the beginning of the end of their captive and dominant role. Cycling necessitated women dressing practically and in a masculine manner. You couldn't look or act in an overly feminine fashion on a bike. A year later, in 1897, a dummy of a young woman on a bike, so dressed, was hung above a shop in Cambridge as an example of why women should be prevented from gaining Degrees at their university. (They were, too, until 1947).

The independence that cycling brought was now there for the taking to anyone who lived above the poverty-line. Daddy 'had had all the fun hitherto', wrote Flora Thompson; 'now it was his wife's and daughter's turn. The knell of the selfish, much-waited upon, old-fashioned father of the family was sounded by the bicycle bell'.[(35)] With hindsight, the 1890s were probably also the safest time ever to be a cyclist or pedestrian. The 1901 census saw 'motor-makers' mentioned for the first time. The twentieth-century's reverence for the motor-car, which Henry managed to avoid, had begun.

~~~
~~~

William Robinson had been editor of *The Garden* magazine for twenty-five years. During this time he had taken to arranging his weekly paper into six-monthly 'volumes', dedicating each volume to some distinguished horticulturist. Someone special would now be needed to honour the occasion of the magazine's fiftieth, now complete. Two days into the new year, his readership found out who it was, for on the whole of the first page was a portrait of Henry. On a subsequent page, beneath a pair of cherubs, was written 'To Mr. Henry Eckford (of Wem) the fiftieth volume of "The Garden" is dedicated. W.R., January, 1897'. This was followed by a page with a picture of sweet peas in a vase, followed by a résumé of Henry's life and work.(36)

It was quite an honour for Henry. The go-getting, pugnacious Irishman had quite taken to the mild-mannered, tenacious Scotsman. Robinson had nothing to fear from Henry as a journalistic rival, so he was able to admire a successful fellow professional. '...The Sweet Pea is the most valuable of all annual flowers of the present day', he wrote. 'Its delicious perfume, its diversity of lovely colours, its lengthened succession of bloom, and its value for cutting entitle it to a place in every garden... The work of Mr. Eckford... shows how much may be done with simple and often neglected things in our gardens. The Sweet Pea certainly was always one of the most valued of flowers, but now with so many delicate and lovely hues, Sweet Peas are a garden of beauty. Who knows how many other things in our gardens may not have in them the germs of like improvement? Even some of the shrubs that now only have one aspect for us may someday show us a like variety. In any case we owe many charming things for our open-air gardens to Mr. Eckford, and wish him many happy years more of his charming and useful work'.(37)

Later in the month, a new column in the '*Chronicle*, 'American Notes', led with 'New Sweet Peas'. A few years before, Henry had sent some of his sweet pea seed to the United States and 1897 was the year they seemed to take off there. 'Everything goes to show that the Sweet Peas will retain this year the remarkable favour which they have been developing for several seasons past. The methods and the material foundation furnished by Mr. Eckford have been rapidly utilised and adapted in this country. There are many Sweet-Pea amateurs and professional experts in the Eastern States; but California is now the scene of most of the commercial variety-breeding and seed-growing. Immense tracts of land are there given up to the production of Sweet Pea seeds in a specially favourable climate, and in the charge of notable specialists. A very promising list of new varieties is offered for 1897'.[38]

The American *Florists' Exchange* journal for 2nd January gave a series of outlines of twelve of Henry's newer sweet pea varieties. The largest diameter of the central, standard petal was thirty-eight millimetres. 'In form the standard is flat or involute on one or other or on both margins. The notch at the apex varies in depth, and the short stalk at the base is equally variable'.[39] The Victorians were great ones for collecting, measuring, identifying and cataloguing. There was an urge to nail things down, both literally and metaphorically. So it's surprising that flower measurements don't appear more often. In an age of scientific discovery the jargon of the medium was used with some abandon. Other than in a botanical journal, we would probably now just say that the flowers varied in size and shape.

American news continued. 'Sweet Peas up to date – Messrs. W.Atlee Burpee & Co., of Philadelphia, have published a little pamphlet under this title. From the interesting statements and historical data it includes, it is something more than a trade catalogue. The author... is the Rev. W.T. Hutchins, who gives details as to culture, and a descriptive catalogue of 105 varieties, mostly raised by Messrs. Burpee & Co., or by our renowned champion H. Eckford, to whose efforts it is that the present fancy for these sweet flowers has attained such large proportions. It is to be hoped that the florists will not be too severe in their code of properties; already there is a tendency to flatten petals, remove notches, and render the flower formal. The flower is naturally irregular, and to make it anything else may be curious, as showing what may be done; but it is a mistaken taste, and wipes away all the history and meaning of the flower'. [40] Let's hope that someone was listening.

Mr Henry Eckford

The Rev. Hutchins of Indian Orchard, Massachusetts, was, as we've seen, also the author of *All About Sweet Peas*. Between that and *Sweet Peas up to Date* he had brought out his *Sweet Pea Annual for 1897*.[41] Hutchins wasn't a seedsman but an exhibitor and seems to have set up a mail-order business, selling sweet pea seed to his fellow Americans. 'I work to get together a mail stock of the very best of every variety in this world', he wrote, a laudable aim he had, as he said, yet to achieve. Of the 121 varieties he lists, the majority are Henry's. With the new or newish varieties, Henry (from Britain) had 44, Germany 2 (C. Lorenz's 'Celestial' and 'Striped Celestial'), and the U.S. 21. Of 54 older sorts, 25 are Henry's, with the remaining 29 being mostly British and American. 'My prices will show that I am giving away my work and experience in this flower'. If Hutchins' sermons were half as good as his patter they would have been worth a detour to hear.

At first glance, Hutchins appears generous to Henry, who features in most of the catalogue. But Hutchins was, it seems, Henry's agent in the U.S.A. There are two photos of Henry on the front cover, a recent one and one of him 'at fifty', the second one known to have been taken before he was famous. The photo would probably have been taken for his fiftieth birthday. Early portrait photography was like portrait painting. You waited for an important occasion to celebrate; a twenty-first, a wedding, a fiftieth birthday, and you went along to the local studio to record it. So the earlier photo would have most likely been taken on or close to the 17th May, 1873. The two photos are similar except that his long beard, now he was in his seventies, has turned white. Another photo inside, of 'John S. Eckford's son' is actually of one of John's daughters, Charlotte (Lotty) or Dorothy.

> 'We greet you once more in the name of our most popular summer flowers' the booklet begins. 'The year 1897 will show no abatement of the widespread interest in Sweet Peas. Probably no circular, setting forth this flower, has ever approached in completeness and novelty this which we now offer you. The list of varieties includes everything in America, England and Germany of which we have been able to learn. Especially will the list of novelties be of interest. We are really just entering upon Mr. Eckford's finest work, and our American growers now offer us beautiful novelties whose additional merit is their splendid germinating quality and abundant bloom. The directions for growing have been so revised and improved by experience that we are confident they will give success to all who study them.
>
> We show you this year Mr. Eckford as he looked twenty years ago when he began to work on the Sweet Pea and also how he now looks. The old town of Wem, about forty miles south of

> Liverpool, England, is Mr. Eckford's home. It has historic marks of Cromwell's days, but its narrow streets seem like lanes to an American. Here this old florist has his comfortable residence and just outside the town is his floral workshop, a bower of beauty of several acres in extent, where he has made the wonderful novelties in Sweet Peas that have given him fame the world over. He is a Fellow in the Royal Horticultural Society, where he is a familiar figure, and in the fifty years of his floral work he has won a "cart load" of certificates. Seventy of the varieties on our Sweet Pea list are of his production, a large number having been awarded certificates of merit by the Royal Horticultural Society. Mr. Eckford has now in partnership with him his son, John Stainer Eckford, who will carry forward the business'.

'Read carefully', Hutchins continued, 'the sections of the list of novelties showing two notable departures made by Mr. Eckford from his long time rule, favoring his American patrons. He has reduced the price of the novelties from 60 cents per packet to 25 cents, and just as this catalogue was going to the printer he wrote to Mr. Hutchins that he might have a stock of his very latest novelties which we had not expected here till 1898... Mr. Eckford leads the world in the work of improving this flower'. Hutchins went on to warn that Henry's seed 'must in our vigorous northern climate be planted in boxes under glass, or in the house and transplanted'. He is thinking of States like his own, with cold winters, for this wouldn't make sense to anyone from the south.

Of the American sweet pea growers listed, W. Atlee Burpee & Co. of Philadelphia heads the list. They are followed by C.C. Morse & Co. of Santa Clara, California; the Sunset Seed and Plant Co., of San Francisco ('the pioneer growers of Sweet Pea Seed in this country'); Mr. M. Lynch of Menlo Park, California; 'my Oregon grower' unnamed, and D.M. Ferry & Co. Of seven drawings of sweet peas, five of them are Burpee's, one's Ferry's and one Henry's – 'Blanche Burpee', unsurprisingly. Despite his eminence here, Atlee Burpee's Company appears to just market the seed, relying on Waldo Rohnert, a young agricultural school graduate employed by Morse & Co., to do the development and hybridising work. Hutchins later stated that 'while we always give Mr. Eckford the first place in the history of the modern Sweet Pea I do not know of any worthier name to stand second than Henry Ohn's, the Chinese superintendent at the Morse ranches'. (42) From where, incidentally, came the dwarf 'cupid' sweet peas, that only did well in California. Despite sending cupid seed out, Burpee was quite ingenuous about it. 'Let us damn it with faint praise' was his verdict.(43)

One unusual feature of the catalogue is the reprint of Henry's Windsor Castle letter under the heading 'Her Majesty's Letter to Mr. Eckford'. After all, her majesty wasn't *their* majesty. It seems an odd thing to do, in this era of mass emigration from Europe to America. Such Europeans were mostly keen to shake the old monarchical know-your-place cobwebs from their shoes. Irish Catholic immigrants, the potato famine of the 1840s a living memory, would have held no nostalgia for Britain's head of state, nor, come to that, romantic memories of the narrow streets of old English towns. Even the mention of Cromwell in the introduction seems calculated to insult them. Odd really, to annoy potential customers.

The point to note about the Rev. Hutchins' home state of Massachusetts is that the typical recent immigrant there was an Irish Catholic. At peak immigration periods, like the present one of the 1890s, it gave rise to anti-Catholic and anti-foreign sentiment from the older Protestant settlers. These recent immigrants, the Germans with their lager beer gardens and the Irish with their St. Patrick's Day revelries 'seemed to be at wide variance with what was considered moral in America', (44) to quote the American historian Carl Degler. Through the civilised pleasures of the cultivated garden the old families could keep their distance and maintain their moral superiority from the newcomers. The only problem with this is that plants are altogether too democratic and promiscuous, and will happily grace anyone's flower bed or boudoir, anywhere. As someone said, somewhere else, the sweet pea 'has a keel that seeks all shores, wings that fly across all continents, a standard which is friendly to all nations'.(45) The Rev. W.T. Hutchins, actually.

~~~

Henry was now coming up to his seventy-fourth birthday. With John and the Company's labour force relieving him of much of the work, Henry began to look around for diversion, preferably something within cycling distance. Major changes in local government had seen a new County Council give Wem a Rural Parish Council in 1895. In April 1897 Henry became a member and was re-elected in 1898 and 1899.

Henry's first Council appearance was at the meeting held at the local non-conformist run 'British School' premises on 15thApril, 1897. He got himself onto the Fire Brigades Committee, avoiding membership of the
~~~

Allotments Committee. Not such an unusual choice; as we've seen, on the big estates like Colehill's, fire engines were the remit of the head gardener and Henry was familiar with them. Henry rarely proposed anything at council meetings and never anything of substance, though he often seconded motions. Unlike Gregor Mendel, who had become Abbott or head of his monastery in Brünn, Henry would never become leader of the council. Mendel, bogged down with leadership issues, became a heavy cigar smoker and died at the age of sixty-two. Henry's relations with the council were, by contrast, dutiful but slight. In that way he retained his health and his focus remained with his plants. We know little of Henry's social life, but if he had ever taken up Mendel's hobby of chess he would surely have been the better player of the two.

At council meetings, Henry teamed up with Councillor Reece Morris, regularly seconding Morris's continual proposals for a 'supply of pure water to certain cottages on Barkers Green in the parish of Wem. The County and District Councils having refused to take action...'(46) It turned out that Morris had a particular interest in a Barkers Green cottage owned by a Mrs. Morris, whoever she might be! Provision of clean mains water, under public ownership, had arrived in Wem with the Water Works built in 1884. Pollution of the water-supply by sewage was a problem that all British cities and towns, swollen by decades of population growth, were having to solve.(47) An outbreak of diphtheria in Barkers Green led to no action by higher authorities, leading the more perceptive councillors to see that without greater powers, their little council would remain a mere talking-shop. The question of sanitary provision was made worse by the uncertainty over whose exact responsibility it was to provide them, itself 'the cause of much complaint'.(48)

Better luck with the railways, perhaps. Henry and Reece Morris teamed up again to ask for a tunnel or footbridge by Wem Station, and that trains slow to 4m.p.h. at the level crossing there, 'the danger and inconvenience to the public' at present 'being very great'.(49) The station still lacks the convenience of a footbridge although there are now automatic gates at the crossing. Needless to say, the councillors' requests fell on deaf ears. Realising their lack of authority, Henry and Morris jointly proposed the formation of an Urban District Council for Wem, an idea it seems they were also canvassing for in the town. At the Council meeting on 17th October 1898, members found themselves evenly split on the issue. More authority, yes, but a leap into the dark nevertheless. The motion was won by the casting vote of the chairman. Eighteen months later, Wem Urban District Council would come into being.

Social history, the stories of the lives, times and concerns of ordinary people, can be read, among other places, in the journals of parish and urban councils. For horticultural history, we turn elsewhere. In October, 1897, *The Gardeners' Chronicle* made a good attempt to trace the history of the sweet pea since its arrival at these shores.(50) There was another attempt in the American *Florists' Journal* too, on 17th July. This has since become quite a popular pastime, with no-one's facts or speculation quite matching up with anyone else's. Much could be written (and starts have been made) on just a few of our common garden plants; on their original natural distribution and how, since arriving here, their individual journeys have resulted in their presence and appearance in today's gardens. There are traps, however, for the unwary author. Some early hybridised or otherwise changed garden plants, like forms of our native primrose *Primula vulgaris* journeyed abroad, becoming known on the continent as 'English' flowers before the end of the sixteenth century, (51) though their wild form was as common in France as in Britain.(52)

Henry was showing his Primulas again in the Shropshire Spring Show. They weren't *Primula vulgaris* though, probably a strain of *Primula sinensis*, whose flowering he would have had to retard for the Show. It was too early yet for sweet peas, even for autumn-sown ones. He didn't need to win prizes now, and in common with other growers would put non-competitive displays into the provincial shows. In the Shropshire Horticultural Show in August he went one better. As well as a sweet pea display, 'six lots of [garden] peas, consisting of three dishes each were put up for Mr. Eckford's Prizes. Mr. W. Pope, gr. [gardener] to Hon. Mrs. E. Kenyon, Melspen, Whitchurch, was 1st showing Eckford's Rex, Eckford's Prior, and Eckford's Magic...'.(53) Henry's garden pea varieties had won Henry's first prize. There were no flies on Mr. Pope, were there!

There was a gentle, *fin de siècle* dig in the columns of *The Journal of Horticulture* for 20th May, three days after Henry's seventy-fourth birthday. Sweet peas 'do not appear any more beautiful or useful because they are labelled "Messrs. Somebody's Extra Special Gigantic Rainbow Sweet Peas... If anyone has an inclination for a collection of plants representing the nobility and the celebrities of the present times, they can easily satisfy themselves with the Sweet Peas, for they have been honoured with names from Her Majesty downwards, including a liberal allowance of princesses and duchesses'. Henry may have smiled but here, and in other ways, he wasn't about to change. Interestingly, the article ends by saying: 'If grown simply for cutting purposes a few rows may be sown like garden Peas and duly staked, but they are seen to much better advantage if they can be allowed to trail over old tree stems or roots, up trellises or round summer houses. The habit of the plant is naturally so graceful that it seems almost deplorable to see it confined to rigid limits or formal stakes'.(54) This is a very modern idea (see Graham Rice's *The Sweet Pea book*) and interesting to see that it was being promoted back in 1897.

Henry had been growing his garden and sweet peas alongside each other for some years. He now noticed that the two appeared to cross-fertilise, with no help from him, for plants seemingly intermediate between the two would occasionally appear. Henry may have begun reflecting on such a possibility and now saw what he expected would be the result. Separate genera shouldn't interbreed but sometimes they aren't so separate. When their compatability leads to the occasional production of viable seeds, a cross like X *Fatshedera* (*Fatsia* x *Hedera*) can occur. It was reported on the following year in the columns of *The Gardeners' Magazine*.

> 'A distinct hybrid pea has been obtained by Mr. Henry Eckford at his nurseries, ... being the result of a natural cross between a culinary pea and a sweet pea. The newcomer is thus a bi-gener, for the culinary pea has been selected from Pisum sativum, and the sweet pea from Lathyrus odoratus. In growth the new plant is robust, but does not possess the stout stems and leafage of Mr. Eckford's new culinary peas; it is much branched, has white flowers, and produces small, full pods of round peas that are pale, buff-coloured, and of a waxy appearance.
>
> The peas are very sweet to the taste when raw, but we have not tried them cooked. The crop is a very heavy one. Whether this new pea will ever be of any commercial value is at present an open question, but the plant is very interesting, both horticulturally and botanically. The wonder is that the hybrid has been so slow in coming, especially when we consider the very many acres of both culinary and sweet peas that Mr. Eckford grows side by side at Wem...'.(55)

Until recently, *Lathyrus odoratus* has never crossed successfully with even another *Lathyrus* species, let alone with another genus. For the horticulturist E.R. Janes, writing in the 1950s, this raised the possibility that the original sweet pea introduction had been 'a natural hybrid and as such refused to breed with any other pea or pea-like plant'.(56)

Crosses between any *Lathyrus* species are extremely rare and, artificially, range from the difficult to the impossible. The few reported so far have come about by laboratory crosses; *L. annuus* x *L. hierosolymitanus*(57) and two hybrids from *L.tuberosus* x *L. rotundifolius*:- *L.*x *tubero* and *L.* 'Tillyperone'.(58) Another exception has also been achieved scientifically, through crossing Dobbie's 1906 sweet pea variety 'Mrs Collier' with *L.belinensis*, a species discovered growing in Turkey as recently as 1988. This is especially interesting as *L.belinensis* has flowers with yellow pigmentation. *L.odoratus* does not, and this may be one way of introducing the colour into the sweet pea. *L.belinensis* is also, like the sweet pea but unusually for *Lathyrus* species, distinctly scented. As is another, even more recently discovered pea from Chile: *Lathyrus cabrarianus*. In another attempt at introducing yellow into sweet peas, a successful cross, again by scientists, with *L.chloranthus* has been reported but they were unable to progress with the yellow. (59)

There are no recorded instances of hybrids occurring in the wild among native British *Lathyrus* species, our wild peas and vetches.(60) This refusal of *Lathyrus* species to hybridise naturally is probably universal. Because others since Henry have also claimed to have crossed the sweet pea with the garden pea, scientists working in the United States in the 1930s took seventeen species of *Lathyrus* (including *L.odoratus*) and *Pisum sativum* (the garden pea) and used them in 458 attempts at interspecific crosses (crosses between species, crossing *Lathyrus* species with each other) and intergeneric crosses (crosses between genera, here crossing *Lathyrus* with *Pisum sativum*). 'In most instances', they reported, 'the pollinated flowers merely dropped off a few days after pollination' and no seed at all was set for any of the *Pisum sativum* crosses.(61)

So, it's rather unlikely that Henry had a hybrid between the garden pea and the sweet pea. What a pity! The view now is that he and others mistook the occasional odd-looking sweet pea and garden pea for each other. The garden pea / sweet pea seeds that Henry's reporter ate with such fascination were what they must surely have suspected they were at the time – garden peas. It's unlikely that the self-fertile sweet pea ever even cross-pollinates itself. However, as Darwin observed, 'in England the varieties of Sweet Peas never or very rarely inter-cross. But is does not follow from this that they would not be crossed by the aid of other and larger insects in their native country'.(62) A more expected cross, and ones that actually occurred, would be between the new dwarf cupid sweet pea and the normal sorts, in the hope of bringing about a semi-dwarf or at least some variation in height. An early cross, reported on in that autumn of 1897, was with the American variety 'Emily Henderson' and resulted in a plant about 2'6" tall[63].

~~~

We now come to the 'Victoria Medallists'. Let *The Gardeners' Magazine* elucidate. '...the Royal Horticultural Society, with the gracious permission of her Majesty, resolved to commemorate the sixtieth year of the Queen's reign by founding a medal to be known as the Victoria Medal of Honour in Horticulture, and to be awarded *honoris causa* in the domain of horticulture. Under the scheme prepared by the council, the medal was awarded to sixty horticulturists who have rendered distinguished service, and it is intended that this shall be the maximum number' at any one time. 'In selecting the names for distinction the claims of all branches of horticulture were fully considered, and the medallists... comprise botanists, collectors, hybridists, nursery and seedsmen, market growers, journalists, cultivators, and general workers'. [64]

They were really too spoilt for choice. Sir Joseph Hooker's name was the most prominent. Henry wasn't among them, yet, but many of his friends and colleagues over the years were. There was the nurseryman William Bull from years back; Richard Dean, happily; Edwin Molyneux... David Thomson 'the foremost gardener in Scotland' was there as were just two women: Ellen Willmott, whom John Eckford would name a sweet pea for, and perceptively, considering the lasting influence she was to have on British horticulture, Gertrude Jekyll. Nothing for William Robinson, who would have to be content with his 1866 Fellowship of the Linnean Society.

Eighteen ninety-seven, the year that had begun so spectacularly for Henry in Robinson's *The Garden* ended on another high note. In *The Gardeners' Magazine's* 'Answers to Correspondents' column in December, a reader's query on 'forcing sweet peas' drew the following response:

> 'Sweet Peas will not stand much forcing, that is, they detest strong heat; then can, however, be induced to flower early if sown in early autumn. It is too late now to make much progress, but the following information, supplied to us by Mr. Henry Eckford, will be serviceable to you in future: "Sweet Peas for early spring blooming should be sown in boxes early in September, and as soon as the plants are fit to handle they should be potted singly
~~~

> into sixties, and kept in a cold frame close to the glass, with abundance of air, to prevent weakly growth.
>
> During the winter months they should never be subjected to a temperature higher than from 45 to 50 Degrees [Fahrenheit]. Artificial heat should only be used as a means to protect them from the frost. On no account must they be pushed during the dark months of November, December and January. About the end of November the plants should be fit for a shift into forty-eights, and again about the end of January, if their progress has been satisfactory, they may be put into their flowering pots, which should be a ten or twelve inch. In repotting, care should be taken not to disturb the roots more than necessary. The plants should be nicely trained up suitable sticks – hazel is the best – and if properly attended to will make elegant specimens for furnishing the greenhouse, as well as providing a wealth of cut bloom for table decoration. The sweet pea will succeed in almost any soil. The best compost is good well-decomposed old turf, such as most gardeners covet, with a moderate quantity of well-rotted horse or cow manure, with sufficient sand to make it friable. The thing to aim at is good, healthy, but not over-luxurious growth"'.[(65)]

Getting sweet peas to flower in pots, out of season, would have been aimed for by the great estate gardeners, with their need to place fruit, flowers and vegetables on their lordship's table week in, week out. The days of such men weren't finished but their era was disappearing. It was to be the amateur gardeners day now; day, year, and century to come. In what had recently been a mostly rural existence, people commonly found space to grow some flowers and, of necessity, fruit and vegetables. What little evidence from cottage gardens there is shows that seed was rarely bought, it being considered a wasteful expense, and that plants were traded, propagated from and generally passed around among villagers. With rising incomes, people had money for their gardens and commercial horticulture, in all its manifestations, rose to meet the demand. But amateurs had always been there. And not so amateur, either, when most work was in agriculture and its related trades and when farming, like gardening, followed and felt the seasons closely and the two shared a closer, simpler, kinship. That said, farmers weren't known for their gardens!

~~~
~~~

1898

Telegraphic Address:
"ECKFORD, WEM."

ENTERED AT
STATIONERS' HALL

Catalogue
OF
GIANT
SWEET
PEAS,
CULINARY
PEAS,
VEGETABLE
SEEDS,
FLOWER
SEEDS,
ETC.

TRADE MOTTO:
"THE BEST OF EVERYTHING."

HENRY ECKFORD

WEM

SHROPSHIRE.

Printed by Thos. Strong, 52 High Street, Wem, Shropshire.

Henry Eckford Seed Catalogue for 1898

1898 kicked off with Henry getting one of his tiny adverts past his son and into *The Gardeners' Magazine's* small ads. 'Eckford's Giant Sweet Peas, direct from raiser...'[66] could be read by any catalogue hunter with a magnifying glass. Henry, like other seedsmen, advertised in the winter. Sweet peas naturally germinate in the autumn in southern Italy but here in our colder British climate they're best held back to late winter or early spring.

There were one or two new features in the new pink-covered Henry Eckford seed catalogue. Besides new varieties of sweet peas and garden peas there were mushroom spores for sale, along with what may be Henry's own varieties of *Primula sinensis* – 'Blushing Beauty' and 'Aurea Magnifica'. The main novelty was ornamental grasses, with a list of twenty-two genera and species, an incursion into hardy ornamentals beyond sweet peas and pansies. (Who said Henry couldn't change?). 'Ornamental Grasses', he wrote, 'are not sufficiently known, and this is a wonder, when their beauty and elegance combined with their easy culture is considered... For mixing with Sweet Peas the Grasses are eminently suitable'. Without an editor, his grammar gets a bit slipshod, although if he wrote as he spoke (assuming it's Henry writing) you can catch his speech rhythms.

'During the past summer extensive trials of grasses have been made in my grounds, and indeed for some years past I have paid considerable attention to this class of plants, and am in a position to recommend them to my numerous customers; much admiration has been excited at the numerous exhibitions where I have staged a collection, either separate or mixed with the Sweet Peas. The Grasses that are intended for winter decoration should be cut when in full bloom as only thus will they remain bright and handsome, cut the stems as low down as possible, then place the cut ends in dry sand in boxes placed in a cool shed away from draughts, let each spike or plume be free, in this way they will soon dry and retain all their natural grace'.[67]

There was an interesting item in *The Gardeners' Chronicle's* 'American Notes' column in February. 'The enthusiastic Sweet Pea connoisseurs, Messrs. W. Atlee Burpee & Co., of Philadelphia, are introducing this year their new Dwarf Sweet Pea, Pink Cupid. This is a sport from Cupid, the original dwarf white variety, and is said to be somewhat stronger in constitution and easier to grow. The same company promises two more dwarf varieties for 1899, Primrose Cupid, and Eliza Eckford Cupid. This energetic development of such a striking line of novelties is all very delightful, and no one would have the heart to criticise it in a firm which has done so much for Sweet Pea culture in America; but after all, the Dwarf

Sweet Peas have not proved to be of any great interest to the ordinary grower of garden flowers, and they do not seem to affect favourably or adversely the immense popularity which the Sweet Peas in general are now enjoying'.[68]

One notes that a conversational tone has crept into the *'Chronicle's* columns. It was the smallest of nods in the direction of *fin-de-siècle* and would have raised eyebrows in an earlier readership. Atlee Burpee was already a good friend of Henry's. Unless there was a new niece somewhere, Burpee's small new 'Eliza Eckford Cupid' would have been named in memory of Henry's little Charlotte Eliza, born thirty years before but who had died aged just fourteen. That 'immense popularity' of the sweet pea was of course shared by us. Among many to join in the excitement was Silas Cole, the head gardener at Althorp Park, situated where the east midlands gently descends into the flat lands leading into East Anglia. One of his crosses that summer was between two of Henry's varieties, 'Lovely' and 'Triumph'; of more anon.

On the other side of the midlands that summer was the annual Shropshire Show. It was held, then as now, at Shrewsbury's twenty-nine acre Quarry Park, situated between the old town wall and the River Severn. It was a major national event and has since become, it's believed, the world's longest-running horticultural show; thought it soon became more than an exhibition of fruits, flowers and vegetables. You get an idea of what Victorian bedding might have been like when you visit the sunken 'Dingle' gardens there, a feature then as now. It would be nice if someone would take bedding out of the nineteenth-century and bring it into the twenty-first. Planting bedding was the first interesting job I had in my gardening career, and some new ideas would be welcome.

In Henry's day the Show was reported on in great detail in the national horticultural press. Henry's nursery was giving prizes for sweet peas this year, a Mr. Bessel of Ludlow taking top honours. 'Mr. Eckford's Challenge Cup, for thirty-six varieties of Sweet Peas was splendidly won by Dr. W.H.C. Sankey of Boreatton Park'. Well, it was splendidly won by Sankey's gardener Bert Ballinger, who'd taken over from Henry at Boreatton. All very family. 'Mr. Henry Eckford, of Wem, Salop, as might be expected so close to his home, showed a collection of his beautiful varieties of sweet peas, and a Gold Medal was awarded him for them'.[69]

There is no mention of any of Henry's garden peas in a July review of the subject in *The Journal of Horticulture*. It was left to *The Gardeners' Magazine* to include them in their report on a visit to Wem in the summer.

Among much else, it shows that Henry appears to have given up on growing sweet peas (and garden peas) for seed in Essex and had acquired some land in Clive, a hamlet near Wem. He was seventy-five, after all, and could cycle the short gentle journey to Clive. He couldn't cycle to Essex! Let *The Gardeners' Magazine* explain, as they did in December. The visiting reporter is probably the magazine's editor, George Gordon (1841-1914), an early president, in 1903, of the soon to be formed National Sweet Pea Society.

PEAS IN PLENTY

> In the month of August last it was my good fortune to revisit Wem, a little Shropshire village that has become well known by name throughout the civilised world as the home of that grand old florist, Mr. Henry Eckford, who has given us such a host of beautiful sweet peas which delight us for months together with their elegance and fragrance, and who also has raised and distributed many of the finest culinary peas in cultivation. It is about ten or eleven years since Mr. Eckford settled at Wem, and during that period his business has increased greatly, and his seed grounds have proportionately extended to keep pace with it. For many years a large proportion of the seed peas needed were grown for Mr. Eckford in Essex, but the yield per acre was not comparable to that obtained at Wem, while the quality was not so good as that of the home-grown sample. Then the distance of Essex from Wem is considerable. It proved a costly business to pay for frequent visits to the seed farms to ensure the varieties being kept true and free from rogues. Now, however, all this has changed, and besides having under his immediate control all his new varieties of sweet and culinary peas, trials, crosses, sports, and what not, Mr. Eckford can convey a visitor to his seed farms in half an hour, as all three of his establishments lie within a mile or so of each other. Altogether, Mr. Eckford has about thirty acres of land, and not less than nine-tenths of this area is devoted to peas that are beautiful to look upon or good for food. The latest addition is a fourteen acre lot at the little hamlet of Clive, about two miles from Wem on the way to Shrewsbury, an open spot set in the midst of bracing and charming country.

SWEET PEAS AT MR. H. ECKFORD'S CLIVE SEED GROUNDS.

The Clive grounds, Gardeners' Magazine, 10-12-1898.

Culinary Varieties

It was at once pleasant and provoking to see magnificent crops of culinary peas at Wem at a time when in many a small southern garden this succulent vegetable was but a memory, and in larger ones was proving a vexation of spirit to the gardener who had none too much of labour or water. No matter what the season, one may depend upon seeing famous crops of peas at Wem, and when it is remembered that the pods are left to ripen –for seed, not green peas, is the end in view – then we are compelled to admire the wonderful constitution and productiveness of the Eckford varieties. In the matter of cultivation Mr. Eckford's methods do not differ materially from those adopted in good gardens; good dressings of manure are essential, and these are provided with an occasional year of rest, when lime in some form or another is given. The land is deeply ploughed as soon as possible after the crop has been harvested, threshed, and housed ready for being picked over. The soil is allowed to remain rough during winter and is worked down during suitable weather early in the year. Sowing is a big business here, and it is highly probable that no other seedsman sows and produces under his own immediate supervision such quantities of peas as does Mr. Eckford. The staff is not a big one but it is capable, and its members are well aware they have a good employer; they know all about sowing peas (previously red-leaded), the labels are ready, and with Mr. H. Eckford, his genial son Mr. J. Eckford, and their capable foreman Mr. Jones to see that there is no mixing or misplacement of tallies the work is done in a wonderfully short time in spite of the fact that besides the catalogued varieties of sweet and culinary peas there are several hundreds of crosses and trials to be sown each year. Last year all the peas were sown within three days, but, of course, the weather was fine and open and the workers did not worry about keeping an eight-hour day.

My note-book contains notes of some fifty varieties of culinary peas, these including only sorts that are catalogued, or are named ready for distribution, or have been on trial sufficiently long to justify large breadths being sown... A very good idea of the Eckford culinary peas is to be obtained from Mr. Eckford's 1898 catalogue, for, putting aside the raiser's descriptions, it is seen that more than half of the twenty-eight varieties listed have gained awards from the Royal Horticultural Society, and most of these have been tested and proved at Chiswick. Even the latest issue of the society's journal testifies to the excellence of the Wem seedlings, for to the trial of thirty-nine varieties conducted at Chiswick this summer Mr. Eckford sent two, Bruce and Prior, and both these splendid forms gained awards of merit. A great feature of the Eckford race of peas is that the pods are produced in pairs, and each pod is packed as full as possible with large, highly flavoured, wrinkled, marrow peas. There is no waste of energy in these varieties – that is to say, the peas are not clustered in the centre of the pod, leaving vacant ends. So far as number of peas per pod is concerned, the girls who in former days were anxious to find a pod containing nine peas, so that they might lay it on the lintel of the kitchen door, and thus ensure to themselves as husband the first single man who afterwards entered, would now have no difficulty whatever in finding such pods at Wem during the summer, or even with ten, eleven, twelve, aye, and thirteen peas per pod; indeed, the greatest difficulty at the Wem seed ground would be to find pods with fewer than nine peas.

A few varieties must be mentioned, and perhaps a selection of twelve will answer every useful purpose; it shall be a baker's dozen, and arranged in the order of use: Eckford's N°. I, a grand early pea, five feet high, and bearing profusely the whole length of the haulm; Shropshire Hero, two and a half feet, and The Don, four feet, first-rate second earlies; then follow Fame and Juno, the former four feet and the latter under two feet in height; Consummate and Critic, superb main crop forms, two and a-half and four and a-half feet high respectively; the latter is one of the very finest of peas no matter how considered; Colossus and Memorial, rising five feet each, both very prolific, and the former carrying immense pods full of a dozen or so of large, splendidly-flavoured peas; Superabundant and Superiority will follow, the former two and a-half feet and the latter five to six feet, and of the pair my inclinations are toward the latter, as it stands dry weather as readily as Critic; Heroine and Censor will bring up the rear, and are two and a half and three feet high respectively. The varieties are paired, so that a smaller selection can be made if necessary. If you want later peas you must start again with the early varieties. If you desire the newest of the new, the finest of the best, and have no objection to paying a rather higher price, then you cannot do better than try Ideal or Pioneer as early varieties; Prior as second early; Bruce and Royalty for main crop supplies. Still later productions that will appear in the 1899 catalogue are Wem Wonder, a beautiful dwarf early that is wonderfully prolific; The Clive, a useful second early that fills up speedily; and Diamond Jubilee, a good late that succeeded splendidly under field culture at Clive and had no sticks to support it.

Each year the Eckford peas are becoming more and more popular, more extensively cultivated, and they are being annually improved; as fast as any variety previously sent out is superseded it is dropped out and its successor takes its place. This is the right method, for one does not want bewildering lists – but we do want the best varieties.

Sweet Peas

A great debt of gratitude is due to Mr. Eckford, for to his energy, patience, and persistence the development of the sweet pea from a comparatively insignificant flower to one of exquisite beauty, grace, and fragrance is mainly due. Then the intensely beautiful and varied shades of colour, or combinations of colours, now to be found in sweet peas suffice to please the most fastidious and exacting, for among them are some that will harmonise with the most subtle art shades in which my lady's boudoir can be decorated. The evolution of the sweet pea is one of the wonders of the age, horticulturally, at any rate, and, if ever a history of floriculture is written, the name of Henry Eckford will occupy a prominent place, for to few is it given to enjoy such a long and honourable career, and few, very few, have achieved so much as he, and single-handed.

Charming as are sweet peas as we see them in our own gardens, in our homes, and at exhibitions, it is only when seen by the acre, as at Wem and Clive, that they become thoroughly impressive. The grand clumps or bushes of a variety that Mr. Molyneux annually produces at the pretty Swanmore Park Gardens are, in their way, the finest I have seen, but these imposing displays are utterly eclipsed during July at Wem and Clive, when the sweet peas are in their greatest beauty. You climb an elevated platform and overlook the acres of sweet peas in the home nursery, all in long orderly rows, neatly staked, and with the flowers so closely set that each row seems to be a dense swarm of some exquisite butterfly, and the motion provided by the gentle summer zephr accentuates this idea. Some of the more intensely coloured varieties gleam like breadths of silk or velvet in the sunshine, and at each motion of the breeze or turn of one's head new shades of colour appear in those forms that combine several hues. Such a sight is worth going a long way to see; indeed, several American enthusiasts – and Eckford's sweet peas are as well known to Brother Jonathan as John Bull – have crossed the Atlantic on purpose to see this beautiful sight. At the Clive grounds both culinary and sweet peas are grown without sticks, and here, just before the haulm begins to bend over by reason of the weight of flowers and early-formed pods of seed, an indescribably beautiful effect is produced. The long rows of sweet peas extend for several hundreds of yards across the wide field, and with a number of rows side by side – as shown in the accompanying view specially taken for the *Gardeners' Magazine* – the effect is better imagined than described.

Mr Eckford is a broad-minded florist, and recognises that there are a variety of tastes with regard to the form of even such a generally popular flower as the sweet pea; consequently we find that, in addition to those bold forms that have big broad standards, there are others less militant, and yet others, known as hooded, that, like a coy maiden, partly hide their charms by infolding their segments, this very peculiarity forming, paradoxical as it may seem, one of their greatest charms. Some eighty varieties are kept true at Wem, and constitute the trade collection, but year by year finer forms and distinct shades are introduced, or a great improvement is effected upon the depth of colour in existing and old forms, so that the forward movement is as noticeable in this quiet village as in the larger centres of industry. A glance at the R.H.S. list of awards shows that up to 1893 sixteen certificates and awards had been granted for sweet peas, and of this number thirteen were made to Mr. Eckford. Since then Mr. Eckford has fully maintained his position as a raiser, and during the past summer several beautiful forms gained the highest honour the R.H.S. could confer upon them. Just as in the case of roses, so with sweet peas, it is a very difficult matter to make a selection of even a dozen varieties that will exactly meet the tastes of all persons; fortunately, Mr. Eckford exhibits largely in summer time, and so growers have an opportunity of noting what best pleases them. For my own part a suitable dozen, excluding the new varieties that will be soon offered, would be Blanche Burpee, white; Queen Victoria, very pale yellow; Shahzada, deep and rich maroon; Salopian, brilliant red and orange; Prince Edward of York, a real beauty, rose and scarlet; Lady Nina Balfour, greyish mauve; Lady Grisel Hamilton, heliotrope; Prince of Wales, deep clear rose; Colonist, lilac and soft rose; Black Knight, maroon, with a metallic lustre; Lady Mary Currie, a lovely shade of pink; and Lady Beaconsfield, salmon-rose and yellow. Mrs. Sankey would form a fine additional white.

Some beautiful and entirely new forms that are to be distributed shortly are as follows: Sadie Burpee, a splendid white flower, but a black-seeded variety. For size, substance, freedom, and grace it cannot be surpassed; it is a hooded flower. Sadie Burpee quite captured the R.H.S. Floral Committee, and gained an award of merit during the present year. Lady Skelmersdale was seen at several exhibitions during the summer, and is a charming hooded variety in which white and rosy-lilac are sweetly combined. Duke of Westminster is a magnificent form both in size, shape, and colour, the latter being a rosy-maroon tinged in a unique way with violet; this and Sadie Burpee are, to my mind, the two finest novelties of the set. Countess Cadogan has slightly hooded flowers of a bright violet-blue hue. Mrs. Dugdale bears finely-expanded blooms with rose and primrose standards, and primrose wings that are splashed with very pale rose. Hon. F. Bouverie is a variety that cannot fail to receive a cordial welcome, especially from the ladies, by reason of its exquisite coral-pink colour; in size and freedom it is also one of the best. Mr. Eckford has a fine rich crimson-purple form named Othello, but whether the stock is sufficient to justify its distribution in 1899 has not yet been decided. The new set gives us new and pleasing shades of colour; they are robust, free flowering, and as beautiful as only first-rate sweet peas can be. The flowers are of wonderful substance, as many have had opportunities of noting at the Drill Hall, Crystal Palace, Hanley, Leicester, Cardiff, and elsewhere during the past summer; they also are carried on long, stout stems, and almost in every instance three flowers on a stem.

Pansies and Grasses

In addition in the masses of peas, both drawing and dining-room varieties, Mr. Eckford and his son devote a good deal of energy to the improvement of pansies, and a wonderfully fine strain they possess, especially in bronze and yellow forms; while the breadths of Bronze Prince are something to admire and remember. Ornamental grasses are largely grown, and these are eminently useful for association with sweet pea flowers in vases. Chinese primulas are carefully seeded and selected, so that the strain is of a high order of merit. Verbenas are good, and this reminds me that there was a time when each year found floriculturists as eager

> to discover what new verbenas Mr. Eckford had raised as to-day they are to find the newest and biggest chrysanthemums. Dahlias are not wanting, for Mr. Eckford does not like to be entirely "off" with his old loves, though he is well on with the new.
>
> There are few greater pleasures in the life of a busy horticultural scribe than that of a quiet day with Mr. Eckford at his seed grounds. Fresh, clear, and bracing air, peace and quiet, open country, beautiful flowers, and a companion whose wealth of knowledge, gained by experience, is only exceeded by his intense love of all that is beautiful in flowers; what more could one want? The stately old Scotsman has long since passed his allotted span of threescore years and ten, but he is hale and hearty, firm of foot, and keeps his seventy-odd inches as erect as most of us less than half his age. If further proof of the vigour of this grand old florist is needed, it may perchance be found in the fact that two years ago he learned to ride a bicycle, and now, week in and week out, he may be seen on most days merrily wheeling out or in between his comfortable home and the grounds at Clive or Wem; but even at Wem, unfortunately, grey hairs are not exempt from the camera demon, and an evil-hearted snapshottist caught Mr. H. Eckford "in the act" last summer, and gave me the result, which I shall keep as a souvenir of my brief sojourn where, even in 1898, there were Peas in Plenty.[(70)]

It's slightly odd that Henry had given up on Essex, especially as other growers would be happy to go there. It is most probably a matter of long journeys and old age. For the workers in Essex, sub-contracting as it were, it was just a job of work and their hearts wouldn't have been in the production of the best possible yield and quality of seed. So Henry may well have been right about the lower standards. After Henry's death, his son John, while keeping and growing sweet peas locally, later found it advantageous to grown them largely in France. It's surprising that the French, with their better climate, didn't work much with sweet peas in Henry's day, at least not so as to be noted abroad. Mons. H. de Villmorin, of Messrs. Villmorin et Cie, Paris, won the occasional mention.[(71)] When you think of all the old French fruit and flower varieties, you realise that their gardeners were well versed in hybridisation.

In a rare sharing of knowledge, the annual *R.H.S. Journal* touched on hybridisation in a history of the garden pea. 'It very likely happens that the best type obtained from a particular cross is found in selections made in the fourth and fifth year after the cross was made. By the same process of cross-fertilisation, the Sweet Pea has been vastly improved, and the varieties largely increased'.(72)

Of Henry's new sweet peas mentioned in *The Gardeners' Magazine* article, the 'Hon. F. Bouverie' – a bit of a mouthful that, it should have been named for an edible pea – was named after one of the Radnors. The hamlet of Clive, incidentally, is centred around Clive Hall, the birth-place of the Restoration dramatist William Wycherley. By coincidence, William Wycherley had married the widowed Countess of Drogheda, daughter of the first Earl of Radnor. The 'Earl of Radnor' title of Henry's former employer only seems, however, to have begun in 1765, well after Wycherley's death. Some informal connection, placing the grounds in the ownership of the present Lord Radnor might explain Henry's presence here now. A further possibility, another past employer, is the Duke of Sutherland. As well as owning over a million acres in Sutherland, he owned 17,500 acres in Shropshire.(73) The Shropshire acres being mainly coal-fields, however, Henry's need for some nearby land was probably met by a local farmer, glad of the income after some hard years. Society was tugging its collective forelock in the direction of industry and commerce now, and successful commercial gardeners like Henry could bask in the warmth of new found respect.

A pretty conceit for a pretty pass!

References to Chapter Ten

(1) 'Sweet Peas', *The Gardeners' Chronicle*, (2-2-1895), pp.66-67.
(2) 'Sweet Peas', *The Gardeners' Chronicle*, (30-3-1895), p.392.
(3) 'Home Correspondence', *The Gardeners' Chronicle*, (24-8-1895), p.217.
(4) 'Obituary', *The Gardeners' Chronicle*, (30-3-1895), p.404.
(5) Letter from the Royal Household, Windsor, to Messrs. Eckford, 2-7-1895.
(6) 'The Collection of Sweet Peas', *The Gardeners' Magazine*, (13-7-1895), p.416.
(7) 'Dwarf Sweet Pea', *The Gardeners' Chronicle*, (13-4-1895), p.462.
(8) 'Sweet Peas', *The Gardeners' Magazine*, (3-8-1895), pp.478-479.
(9) 'Sweet Pea "Cupid"', *The Gardeners' Chronicle*, (29-6-1895), p.793.
(10) 'Sweet Pea Reminiscences', *The Sweet Pea Annual*, (1918), p.14.
(11) 'Peas', *The Garden*, (27-7-1895), p.65.
(12) 'Small Peas', *The Garden*, (3-8-1895), pp.90-91.
(13) 'Pea Trials in Lincolnshire', *The Gardeners' Chronicle*, (12-10-1895), pp.423-424.
(14) Jennifer Davies, *The Victorian Kitchen Garden*, (1987), p.71.
(15) 'Shrewsbury Horticultural Show', *The Journal of Horticulture*, (29-8-1895), p.203.
(16) Henry Eckford's seed catalogue for 1896, p.17.
(17) 'The Synchronism of Variation', *The Gardeners' Chronicle*, (18-1-1896), p.83.
(18) 'The Sweet Pea Cupid', *The Gardeners' Chronicle*, (8-2-1896), p.177.
(19) 'Sweet Pea Cupid', *The Gardeners' Chronicle*, (12-12-1986), p.728.
(20) 'A visit to Wem', *The Gardeners' Magazine*, (29-2-1896), p.129.
(21) 'Sweet Pea Lady G. Hamilton', *The Gardeners' Chronicle*, (12-2-1910), p.106.
(22) Countess of Ashburnham, *Lady Grisell Baillie: A sketch of her life*, (1892), p.Liii, n.
(23) Advertisement, *The Gardeners' Magazine*, (29-2-1896), p.142.
(24) 'Evolution of the Sweet Pea', *The Gardeners' Magazine*, (7-3-1896), p.143.
(25) 'Home Correspondence', *The Gardeners' Chronicle*, (18-7-1896), p.73.
(26) 'Societies', *The Gardeners' Magazine*, (27-6-1896), p.791.
(27) 'The Sweet Pea', *The Gardeners' Chronicle*, (23-5-1896), p.657.
(28) Amanda Clarke, *Finding out about Victorian Schools*, (1983), p.38.

(29) 'Shropshire Horticultural: Cut Flowers', *The Gardeners' Chronicle*, (22-8-1896), p.225.

(30) Yoland Brown, *Boreatton Park*, (1989), p.33.

(31) 'Peas in Plenty', *The Gardeners' Magazine*, (10-12-1898), p.800.

(32) Frank E. Huggett, *Victorian England as seen by Punch*, (1978), p.168.

(33) Flora Thompson, *Lark Rise to Candleford*, 1945, (1948 edn.), p.455.

(34) G.E. Mingay, *Rural Life in Victorian England*, (1976), p.113.

(35) Flora Thompson, op.cit., p.455.

(36) *The Garden*, (2-1-1897), pp.ii, iv, xi.

(37) 'Mr. Henry Eckford', *The Garden*, (2-1-1897), frontispiece.

(38) 'American Notes', *The Gardeners' Chronicle*, (30-1-1897), p.73.

(39) 'Sweet Peas', *The Gardeners' Chronicle*, (30-1-1897), p.78.

(40) 'Sweet Peas up to Date', *The Gardeners' Chronicle*, (27-2-1897), p.141.

(41) Rev. W. T. Hutchins, *Sweet Pea Annual for 1897*, (1896), no paging.

(42) 'Sweet Pea Reminiscences', *The Sweet Pea Annual*, (1918), p.14.

(43) Ibid., p.12.

(44) Carl N. Degler, *Out of Our Past: The Forces That Shaped Modern America*, (1984), p.325.

(45) Often quoted: see for example Rev. D. Denholm Fraser, *Sweet Peas: How to grow the perfect flower*, (1913), p.ii.

(46) CP325/1/2/1, Minute Book, Wem Rural Parish Council, (18-10-1897), p.166, Shropshire Record Office.

(47) E.C. Midwinter, *Victorian Social Reform*, (1971), p.54.

(48) CP325/1/2/1, Minute Book, Wem Rural Parish Council, (12-6-1899), p.225, Shrops. R.O.

(49) CP325/1/2/1, Minute Book, Wem Rural Parish Council, (19-12-1898), p.207, Shrops. R.O.

(50) 'Sweet Peas', *The Gardeners' Chronicle*, (6-3-1897), p.160.

(51) Alice M. Coats, *Flowers and their Histories*, (1956), p.209.

(52) Alastair Fitter, *An Atlas of the Wild Flowers of Britain and Northern Europe*, (1978), p.148.

(53) 'Shropshire Horticultural: Miscellaneous exhibits', *The Gardeners' Chronicle*, (21-8-1897), pp.134-136.

(54) 'Our Hardy Plant Border', *The Journal of Horticulture*, (20-5-1897), pp.426-427.

(55) 'A Distinct Hybrid Pea', *The Gardeners' Magazine*, (27-3-1898), p.553.

(56) E.R. Janes, *Sweet Peas*, (1961), p.164.

(57) 'New interspecific hybrids in Lathyrus (Leguminosae): Lathyrus annuus x L. hierosolymitanus', *Botanical Journal of the Linnean*

Society, (1996), p.89.

(58) Greg Kenicer, Unpublished doctoral thesis on the genus Lathyrus, (2007), Royal Botanic Gardens, Edinburgh.

(59) 'Interspecific hybridisation between Lathyrus odoratus and L.belinensis', *International Journal of Plant Sciences*, (1994), p.763.

(60) C.A. Stace, (ed.), *Hybridisation and the Flora of the British Isles*, (1975). No Lathyrus hybrids recorded.

(61) 'Experimental data for a revision of the genus Lathyrus', *American Journal of Botany*, (2-1938), pp.67, 69.

(62) 'National Sweet Pea', *The Gardeners' Chronicle*, (18-12-1909), p.426.

(63) 'A dwarf sweet pea', *The Gardeners' Chronicle*, (9-10-1897), p.256.

(64) 'The Victoria Medallists', *The Gardeners' Chronicle*, (20-11-1897), p.725.

(65) 'Answers to Correspondents', *The Gardeners' Magazine*, (18-12-1897), p.818.

(66) 'Eckford's Giant Sweet Peas', (advertisement), *The Gardeners' Magazine*, (15-1-1898), no paging.

(67) Henry Eckford's seed catalogue for 1898, p.68.

(68) 'American Notes', *The Gardeners' Chronicle*, (12-2-1898), p.95.

(69) 'Shropshire Horticultural Show: Cut Flowers, Decorative Classes', *The Gardeners' Chronicle*, (20-8-1898), p.151.

(70) 'Peas in Plenty', *The Gardeners' Magazine*, (10-12-1898), pp.799-800.

(71) See for example 'Sweet Peas, 1894', *The Royal Horticultural Society Journal*, (1895), p.ccxxii.

(72) 'Garden Peas', *The Royal Horticultural Society Journal*, (1898), p.259.

(73) John Bateman, *The Great Landowners of Great Britain and Ireland*, 1883, (1971), p.431.

Chapter Eleven

Eighteen ninety-nine was, for the botanically-minded, the year of the Hybridisation Conference, an event eagerly awaited by the horticultural press. With hindsight, it would have been better if it had waited until Mendel's work had been discovered and acknowledged. But then, it had been the growing interest in the subject, typified by the Conference and, earlier, by Mendel's conclusions, his work gathering dust on shelves since the mid-1860s. 1899 might also have been the year of a bicentennial sweet pea celebration. The plant was growing in Robert Uvedale's Enfield garden in 1700, but the seed may well have been sent in 1699 and sown that autumn under glass, Uvedale having the rare luxury of owning a greenhouse. It would, however, be decided to have a big 'do' in 1900. One conference at a time was their sensible British solution, if possibly lacking in logic.

So, the sweet pea celebration was a year away but there was to be no let-up in interest. First off was *The Journal of Horticulture*, in reflective mood. 'I can well remember', it recalled, 'twenty-five years ago hearing the cottagers talk a lot about their "Posy Peas". These were short-stemmed, small leaved, and small flowered varieties, about 3' high, and the flowers were mostly dark purple...'.(1) Although there were already a number of sweet pea varieties around before Henry began working on them, all the evidence shows that they weren't much known about, and little grown. Henry's acquaintance, R.P. Brotherston, had moved from East Lothian to Dumfries, and despite the south-westwards direction had discovered unfavourable conditions. 'In this district in Scotland', he wrote in the '*Chronicle,* 'it is impossible to bloom sweet peas earlier than July; and in a late season I have known it to be August before the plants were in full bloom. By sowing seeds in pots during January, however, and encouraging germination by a little warmth, I have known flowers to be produced in May...'. Wise advice and perhaps the professional gardeners it was aimed at took it up. Sowing much before February though must have seemed a lot of bother and expense to amateurs. Amateurs south of about Perth, excluding Dumfries!

Brotherston's practical advice continued with the first reference to stopping the leading shoots to check growth and gain earlier flowering. Recommending it to another grower 'it was received as quite a new idea'. He ended with a long list of his favourite sweet peas. 'For my own part, I consider American introductions a long way in the rear of the best of Mr. Eckord's varieties, and so leave them out'. The list, mostly of Henry's plants, naturally included 'Lady Grisel Hamilton', 'by far the finest of the light blue varieties', though Brotherston admitted to finding its seeds 'difficult to ripen'. It may well have been that the American varieties did badly, due presumably to the climate. This was also the argument used against British seed in North America. Considering the variable climate occurring within both Britain and the United States, one feels that factors other than objective analyses were in play on both sides. Robert Brotherston, imagining he has rejected the inferior American varieties, included in his list 'Emily Henderson', introduced some five years earlier by, not surprisingly, Messrs. Henderson and Co. of New York[2] but offered through a British seed firm. 'Among white varieties', he wrote, 'Emily Henderson is a favourite flower, the tint of green for which it is remarkable being so pretty'.[3] That this and other American varieties had their origin in seed that Henry had sent there in around 1890 happily muddies the waters all round.

The new Eckford catalogue for 1899 featured a front-cover advertisement for *The Horticultural Trade Journal* Co., a paper recently founded by Henry's nephew John S. Brunton in Padiham, Lancashire. The new catalogue also shows that Henry and his son had thought of another way of drawing in business. 'I usually have several capable and reliable Gardeners on my register', they wrote, 'and shall be pleased to render assistance to ladies and gentlemen who are seeking reliable men'.[4] *Men*, note. There was also mention of two new *Primula sinensis* varieties, 'Splendour' and 'Lady of the Lake', among the usual long list of current sweet pea varieties.

Now, the sweet pea, fashionably as it was fast becoming, would have to share the limelight at the 'International Conference on Hybridisation' about to take place at Chiswick and London on Tuesday 11th and Wednesday 12th July. It was hoped that members of the Government and other 'public men' would meet socially with all present. *The Gardeners' Chronicle*, in a leading article looking forward to the conference, noted that 'plant breeding has been left almost entirely to the commercial and amateur horticulturists;

Eckford seed catalogue for 1899

it is indeed only recently that a plant of garden origin has been allowed to have a scientific interest. Hence the fewness of hybrids raised in botanical establishments compared with those raised elsewhere'.[5] There probably always will be this divide between those interested in plants from a horticultural or a botanical viewpoint. The article's author, 'W.W.', would have been aware of it, as secretary of the R.H.S., and regretted it. The Rev. William Wilks (1843-1923) is better known to us as the originator of the hybrid poppies he raised in Surrey, in his garden at Shirley, near Croydon.

Although Mendel's work had been referred to in a few papers in mainland Europe, his breakthrough hadn't been realised. His name wasn't mentioned here, nor, it seems, did it crop up at the conference. With three days to go, the *'Chronicle* was hoping for the best. It 'will, we believe, deal with hybridisation in a broad sense and under many aspects, scientific as well as practical...'.[6] Short notes on some European hybridists included the now elderly Henry Eckford, together with his portrait.[7] 'Mr. Eckford is a veteran worker in the field of cross-breeding. Of late years he has made a name for himself in raising by cross-breeding as well as by selection, superb varieties of Sweet Peas, and is famed also for the new culinary varieties, which owe their origin to his judgement and skill'.[8]

The Gardeners' Magazine were a little slow off the mark this time, leaving it until the conference had ended to portray those, like Henry, who had been invited to take part.

> 'Mr. Henry Eckford has made his name famous the world over by reason of the wondrous improvements he has made in sweet peas. Had he done nothing else for horticulture Mr. Eckford would deserve our thanks and a foremost place among raisers for his persistence and eminently successful efforts in increasing the productiveness and vigour of the plant, and the size, number, and colours of its flowers. But during the seventy-six years he has lived Mr. Henry Eckford has been equally successful in the improvement of verbenas, dahlias, pansies, and sweet peas. Of the former he was the champion raiser when verbenas were at the height of their popularity, and scarcely a box of two dozen show dahlias is put up at our autumn shows that does not contain some one or more examples of his skill as a raiser. Culinary peas have been greatly improved by Mr. Eckford both in flavour and productiveness, and curiously enough a hybrid between his two great specialities, sweet and culinary peas, was secured at his

establishment at Wem, Salop, last year'.[9] (The less said about that the better).

Eminent hybridists and raisers who were invited to the Royal Horticultural Society's conference on hybridisation, illustrated in a feature supplement to The Gardeners' Magazine, July 1899

'Great pains were taken to make this gathering a success', began the *'Chronicle's* conference report, '...but a larger gathering of botanists to meet the foreign visitors would have been more graceful'.[10] The *'Chronicle's* editor was the botanist Dr. Maxwell Masters (1833-1907). He had begun this critique of his fellow botanists while giving the conference's opening address. 'It is indeed altogether surprising that the botanists should have objected to the inconvenience and confusion introduced into their systems of classification by the introduction of hybrids and mongrels, and that they should object to hybrid species, and much more to hybrid genera; but it would be very unscientific to prefer the interests of our systems to the extension of the truth...

The days when "species" were deemed sacrosanct, and "systems" were considered "natural" have passed, and Darwin, just as [William] Herbert did in another way, has taught us to welcome hybridisation as one means of ascertaining the true relationships of plants and the limitations of species and genera'.[11] His frustration with narrow thinking was aimed as much at the absent British botanists as at those who were there. Gone were the days when, in the seventeenth-century, Britain led the world with as many botanists as the whole of the rest of Europe.[12] Masters could see that despite British advances in practical horticultural hybridisation, breakthroughs in scientific understanding were likely to come from abroad. The garden historian Miles Hadfield summed up the Conference and its on-going importance in his *A History of British Gardening*.

> 'It was attended both by practical plant breeders and scientists. In the preceding years the [Royal Horticultural] Society's Scientific Committee, urged on by Wilks and Maxwell Masters, had been co-operating with scientists such as William Bateson who had been studying hybridisation scientifically and by means of statistics. The Conference was the first ever held to consider the problems of plant breeding, and its consequences went far beyond the rather limited agenda. Wilks followed up its success energetically, and arranged for Bateson to speak on heredity to the Society on 8th May 1900. In the train on his way to attend the meeting Bateson read for the first time an hitherto unknown paper sent to him by Hugo de Vries, (1848-1935), the eminent Dutch botanist and scientist. The Paper was Mendel's, in which he stated his now famous laws…'.[13]

The theory of evolution by natural selection and the discovery of genetics were the two ground-breaking events of the age, as they would have been for any age. Their resonances went far beyond plants and gardens, yet it was from the enjoyment of the growing and study of plants that many of the insights were gained. For gardeners, that's quite a nice thought.

Richard Dean had a horticultural paper on 'The Pansy and Viola' at the Conference. He spoke of the now much hybridised Show Pansy's failure in the south, 'due to a disease, which carried off the plants by the hundred'. Dean doesn't relate cause and effect; environmental factors like pollution and a run of long dry summers; hybridists' focus on the flower to the neglect of health; they surely all played a part. Henry's successes aren't mentioned here, though there may be some editing. 'What is wanted', he concluded, 'is a race of early-flowering varieties for use as carpet plants in the spring garden in March, April and May. Such a race would be very welcome to gardeners, who find it difficult to carpet their bulb-beds with plants which will be in flower simultaneously with the Hyacinths, Tulips, Daffodils, &c., and furnish blossoms when these have gone out of flower'.[14]

Winter flowering was a problem that we had seen exercising George Fleming's mind at Trentham. Now, thanks to the Dean brothers, it was highlighted again. Richard's brother Alexander had been considering it two years earlier in the *R.H.S. Journal*. He had considered hardy annuals like *Silene pendula* (catchfly) and *Limnanthes douglasii* (the poached-egg plant) but saw that, sown in late summer, they mostly waited 'till early spring to flower and failed to answer the question of flower in winter. We have since developed spring bedding, so called became the plants, though planted in the autumn, still flower more in the spring than the winter. It has become a familiar and rather hackneyed staple of local authorities' parks departments, to which wallflowers, forget-me-nots, *Primula polyanthus*, *Bellis*, tulips and, sometimes, daffodils have been joined in recent years by ornamental cabbage and a strain of winter-flowering pansies.

There was one reported contribution from Henry. 'Hybrid between the sweet pea and the field pea. Mr. Eckford brought up to the Conference some accidentally-produced seedlings, apparently intermediate in floral characters between the two plants mentioned, and which attracted much attention'.[15] If Henry had hoped for some botanical explanation, none was reported. Henry would have known the physical near-impossibility of natural cross-fertilisation occurring, as would the reporter, but let his hopes cloud his judgment.

The field pea, *Pisum arvense*, is a natural pink-flowering variety of the garden pea, a probable mutation in either seeming to have caused confusion. All plants are liable to mutate, especially highly hybridised ones. There can be changes to shape, form and colour, and when this occurs in the flower, and can be maintained, is one way of developing a new variety. When changes occur in the stipules it can cause problems of identification, especially in legumes. It had doubtless happened before and would do so again. 'At different times, 'wrote a sceptical E.R. Janes in the 1950s, 'we have been told that crosses have been made with various species of peas and leguminous plants, from vetches to cytisus, but no one has ever seen the progeny. In this connection even gorse has been mentioned...'.[16] These are all good examples of the limits of empiricism.

With hybridisation and sweet peas much to the fore, Silas Cole, who we met last year, was continuing his work in Northamptonshire. Having crossed Henry's two varieties, 'Lovely' and 'Triumph', he now crossed the best of the resulting plants with Henry's 'Prima Donna'. What would happen? Who knows, that's the fun of it. He would only have to wait another year to find out. Cole was to find his gardener's natural patience desert him as events began to get exciting – as we shall see.

The Hybridisation Conference was hardly over before a similar sort of thing for sweet peas was being mooted. 'It has been suggested', a '*Chronicle* reader reported, 'that in view of the marvellous advance made in the production of beautiful varieties of these flowers, that a conference and special exhibition might well be arranged for at Chiswick next July. The proposition is an attractive one, and might even go further and include every section, and so far as possible known, of the great Pea family... Sweet Peas are especially in need of such selection as a conference and exhibition of them could furnish... Raisers should work hard to give us stout, erect standards, with large flowers and beautiful colours. The raisers and admirers of edible Peas might find enough to talk about also, and then there would still be the numerous beautiful hardy perennial section which so much merits greater development. One special feature... might be prizes for the most tasteful method of exhibiting sweet peas as art flowers. A.D.'(17)

The correspondent was, without doubt, Richard Dean's brother Alexander (1832-1912), keen to push his vegetable interests by promoting the garden pea's inclusion. The pea family is vast, and Alexander's argument could, logically, have seen beans included. It might have been an idea to have added other ornamental *Lathyrus* species but the organisers would decide, perhaps wisely, to limit themselves to the fashionable sweet pea. From a proposed event, attention turned to a traditional, established one.

> 'There are few readers of *The Gardeners' Chronicle* but have some interest in the annual shows held at Shrewsbury. Nothing commands attention sooner than success, and for this reason the Shrewsbury Show... excites at least a passing interest in many to whom flower shows as a rule do not appeal'.(18)

Much the same attitude prevails today, for most of us have heard of shows like those at Chelsea and Hampton Court. Their annual arrival excites our passing interest but few of us will visit. With sixteen million of us gardening regularly in Britain it's just as well! At Shrewsbury that August the '*Chronicle* reported that the sweet peas 'produced such a display, that a tent might well have been devoted to their special benefit. For Mr. Eckford's Special Prizes, Thos. Aldersey, Esq., was well to the fore with eighteen varieties [or rather, his gardener was] which might well be considered as typical of the best kinds of the present day. The bunches were light, and tastefully arranged, the individual flowers very fine, and the colours well blended...'. There followed a list of fourteen varieties 'given as indication of the best kinds to grow', of which eleven were Henry's. Henry himself put in a trade stand of sweet peas for which he won a gold medal.[19]

On the thirteenth of the following month, the (still flourishing) Royal Caledonian Horticultural Society opened its autumn Show in Edinburgh. On the same day, at the Royal British Hotel there, 'a goodly number of English and Scottish horticulturists'[20] met to formally decide on a bicentennial celebration of the sweet pea in London next year. A nineteen-strong committee of eighteen men and one woman – Ellen Willmott – was announced. It was a good mixture of youth and experience that gathered there, if almost entirely representative of what A.N. Wilson calls the era's 'unashamed and undiluted masculinity'[21]. They weren't all sweet pea specialists but they were all eminent, or were to rise to eminence, in horticulture.

COUNTY BOROUGH OF HANLEY
Horticultural Show and Floral Fete,
HELD IN THE HANLEY PARK,
JULY 5th & 6th 1899
Certificate of Merit
Awarded to Eckford & Sons.
for new Sweet Peas
Lord Kenyon General Gordon Fascination
Countess Lathom Duchess Westminster
O Thomas
Mr Fitzgerald
Alfred Outram
Judges
Joseph Kent
Secretary

Hanley certificate for sweet peas, 1899.

Henry promised a three guinea (£3-3s.) donation but, at seventy-six, let his son John take his place on the committee. Other notables included Owen Thomas, the Queen's gardener at Windsor. George Gordon presided and Richard Dean (who else!) became secretary and treasurer. Over the ensuing months nine people left and six were added. Additions included the Rev. W.T. Hutchins, who must have been living in Britain for a while. Horace Wright was joined by his brother Walter Page Wright (1864-1940), a figure prominent not only with sweet peas but who was to find fame as the founder and editor of *Popular Gardening* and author of numerous books such as *Beautiful Gardens* (1909) and *Beautiful Flowers* in 1922. (And beautiful books they are too).

The meeting was reported with delayed enthusiasm in the horticultural press, for, not being topical, they waited until December to run it. 'It has been thought advisable', said *The Gardeners' Magazine*, 'that some attempt should be made to celebrate the bi-centenary of the flower, not only as a matter of sentiment, but with a view to recognise its importance in commercial horticulture, and to bring into special prominence the finest of the varieties, of which there are now such large numbers in cultivation'. The meeting's 'proposals included the holding of a conference for the purpose of discussing questions relating to the sweet pea, the classification of varieties into their several colours, and the selection of the best varieties from each colour. An exhibition of a comprehensive character' is proposed with visitors 'from the United States and several Continental countries...'. However, 'to celebrate the bi-centenary in a manner becoming its importance and to hold an exhibition worthy of the occasion will involve a considerable outlay...'.[22]

Richard Dean's defence for such a grand undertaking was that sweet pea varieties were 'multiplying with singular rapidity; second, that it is now so generally employed for garden and floral decorations; third, that it has become a very prominent subject at horticultural exhibitions; and, fourth, that its culture for cut flower and seed purposes has become a great commercial industry...'.[23] Richard's brother Alexander was, for some brotherly reason, all against this sweet pea thing of which Richard was the secretary. 'The hero of the Sweet Pea today is Mr. Eckford, the patriarch of Wem', he fired back five days later in the columns of the rival *Journal of Horticulture*. 'Not ten-thousand celebrations could accomplish one tithe for the Sweet Pea that he has done...'.[24]

While being anti-Richard, Alex's fulsome praise for Henry may have been fuelled by guilt. He was simultaneously campaigning to exclude nurserymen like Henry from association with the various R.H.S. committees of which he was, and Henry had been, members. This rearguard action against commerce, successful for a while in its narrow achievement, could do nothing in the bigger picture. Sweet peas were becoming more commercialised, as was horticulture itself.

> 'The garden is indeed in a highly satisfactory condition at the present time', commented the leader-writer of *The Gardeners' Magazine's* Christmas Number, 'and as one result the commercial industries associated with horticulture are in a flourishing state'.

> 'Flowers are more freely and tastefully employed in the decoration of the home and as the result of the enterprise and skill of the commercial growers many flowers of great beauty are placed on the market at prices which place them within the reach of households that two or three decades ago had to be content with the commonest kinds in extremely limited quantities. Much the same conditions obtain in the case of fruits and vegetables...'.

> 'So satisfactory is the garden that we would gladly confine out attention wholly to it, for immediately we pass beyond its boundaries and then towards South Africa we see much that must depress the mind and greatly detract from the enjoyment of a season with which peace and good-will is so intimately associated...'.

This was two months after the start of the second Boer War and a month away from the British disaster to come at Spion Kop. It's not quite the jingoism evident elsewhere, for horticulture's role in war, if it is a role, is the non-aggressive part of providing gardens for rest and recuperation, wholesome food, floral comfort for the wounded, dying and grief-stricken, wreaths for the dead, plantings for cemeteries and Gardens of Remembrance.

> 'That those who have lost husband, father, or son in the struggle will have the deepest sympathy of horticulturists there can be no doubt, for they are essentially men of peace and imbued with a keen sense of the sufferings of others, nor are they likely to forget their duty in doing their utmost in helping to cast a ray of comfort into those houses of mourning which have to bear the burden of poverty as well as the loss of a dear one...'.[25]

Nothing stands alone. Henry, an ordinary man living a fairly ordinary life, shaped by every twist and turn of the nineteenth-century, leaves it linked in to warfare. The graphic war reports in the national newspapers may have reminded Henry again of his mortality, for he now contacted Wem solicitor Edward Bygott about making his will. Henry now owned 31 Noble Street, originally rented, as well as the adjacent 9 Market Street together with number eight, half-way down on the other side where John and Alice and half their family lived. Henry decided to leave not only his personal belongings but all three houses to his wife Emily. After her death

John would have 9 Market Street and 31 Noble Street but not 8 Market Street, where he lived. This would go to John's sister Isabella.

So; three houses for Emily and none (yet) for John and Isabella. Peter, Henry's last surviving child who lived away was already under a cloud. 'I bequeath to my son Peter William Eckford the sum of three hundred pounds but... in such instalments and proportions weekly monthly or otherwise as my... executors shall... think fit...'. Should he 'at any time do or permit any act or default whether voluntary or involuntary whereby the said three hundred pounds or any part thereof should not be payable to him then the whole of his interest in the said sum shall cease and the amount for the time being remaining unpaid to him shall go to and belong to all his children in equal shares...'.

Henry left the seed business in the joint hands of Emily, John and Isabella. Isabella was to be paid 25/- a week (£65 per year), Emily £100 per year and John also £100 per year – not less, 'so long as he shall assist in the management of the said business...'.[26] You get the feeling that although Henry had little faith in Peter his hopes for John weren't sky high either. There are so many variables in all this – that the business continues and continues profitably for one – that makes such Wills a wish-list as much as anything. What if the bottom fell out of the sweet pea market?

There were no such worries for the near future. That *Gardeners' Magazine's* article on South Africa was followed immediately by one on sweet peas, the flower of the moment. 'As the year on which we are entering is the bi-centenary of the introduction to this country of the wondrously beautiful and delightfully fragrant sweet pea', it began, 'we have considered it advisable to mark the event by selecting some of the finest varieties for illustration on the almanack present with this issue... .[27] no one who has a knowledge of the sweet pea as it existed some thirty years ago will fail to be in some degree surprised by the immense improvement that has during this comparatively short period been accomplished as the result of well-directed effort... they will be filled with admiration for the energy and perseverance by which the development of the sweet pea from a comparatively unattractive flower to one of exquisite grace and beauty has been effected...'.[28]

The Gardeners' Magazine produced some quite beautiful calendars in colour, and this one for 1900 was one of the best. The Eckfords cashed in on the publicity with a full-page advertisement.[29] It focused on the garden and sweet pea but there was mention, too, of new *Primula sinensis* varieties of Henry's. A strain of his *Verbena* seed could be purchased too, bringing

Gardeners' Magazine Calender for 1900

(or attempting to bring) the plant back into the limelight. So, for gardeners, the year (and for the so-called 'zeroists'[30] the century too) was rung out by the promise of peas to come, and whatever wild skies the twentieth-century had in store were, Henry knew, for others to discover. But he'd make what he could of them.

During the 'nineties, Henry had acquired number 8 Market Street for his son John's family. He also acquired number 9, which must have been where his seed could be sorted and stored in the dry, although it's not listed in the commercial directories. He also, of course, had his own house in Noble Street. Other financial outlays included the fields of peas in the Soulton Road and at Clive, probably rented; and a domestic servant. The Eckford household would not have considered this last a luxury. It defined their status, as it did for many; an absolutely indispensable part of the household until, one day, societal changes showed them that they were, actually, completely dispensable.

~~~

'Unquestionably, the popular flower for 1900 will be the Sweet Pea, and this is not to be wondered at'[31] said *The Gardeners' Magazine* with confidence. Already popular, the flower was to become even more so. Sweet pea enthusiasts nowadays might think of 1901 instead, when a 'sport' or mutation of one of Henry's varieties (with the possible involvement of two more) took the flower to super-stardom for the next fifty-years. They would be correct, too, but it's important to remember that its later popularity came from an already high esteem. Such interest ensured that the change, when it came, would not be overlooked by gardeners. As, similarly, the previous year's hybridisation conference had paved the way for Mendel's rediscovery. The Sweet Pea Bicentenary Committee, meeting in January, decided to hold the show at the Crystal Palace, Sydenham, in July. For the exhibition part of the conference, 'Mr. Henry Eckford's generous offer of £15' in prize-money was a gratefully accepted 'offer made by the Wem veteran to commemorate the fact that this is the twenty-first year of his work in selecting and cross-fertilising Sweet Peas'.[32]

The Eckfords had advertisements out now in the horticultural press, as did other seed companies. John blew an even louder trumpet in them than their competitors did, unmindful perhaps that noise both attracts and repels. 'SWEET PEAS! THE BIGGEST SUCCESS of all is ECKFORD'S GIANT SWEET PEAS. THE PURITY and STAMINA of these Home Grown Stocks are so conserved by SPECIAL METHODS of CULTURE
~~~

that the seeds are PHENOMENAL for UNRIVALLED STRENGTH and VITALITY...

CULINARY PEAS! ECKFORD'S CULINARY PEAS are positively UNRIVALLED by any on the market for FLAVOUR and CROPPING CAPACITY! ENORMOUS YIELDS...' etc., etc., with Cinerarias, *Primula sinensis*, pansies, Verbenas and vegetable seed collections getting similar treatment.(33)

These advertisements appeared in the February 3rd edition of the '*Chronicle*, and in the same issue 'E.M.', Edwin Molyneux, was stating that 'the first week in February is a suitable time to make sowings of Sweet Peas... The plant does not make much headway before April, and I fail to see the advantage of autumn sowing...'.(34) Molyneux was the first commentator to advocate what was already happening on the ground. Most of his recommended varieties were Henry's. A reader wrote in to agree, he planting them out in April. 'Owing to lack of space', he wrote, 'I can only find room to grow the following varieties, which, however, give a good range of colour:- Blanche Burpee (white), Lady Nina Balfour (mauve), Mrs. Eckford (pale yellow), Salopian (crimson), Shahzada (purple), Countess of Powis (orange-pink), Princess of Wales (striped-blue), Emily Eckford (blue), and Aurora (striped-pink)'.(35) All Henry's varieties too, bar the last.

As well as raising new sweet pea varieties, Henry was, in a lesser way, also raising points of interest on the parish council. Thanks largely to him and Reece Morris, Wem now had an Urban Council to look forward to. Elections for another layer of local bureaucracy however, fixed for Friday 23rd April, don't seem to have greatly fired the interest of Wem's electorate. Or that of *The Shrewsbury Chronicle*, who lethargically reported the results a fortnight later. 'It will be remembered', they observed, 'that some time ago, on the petition of numerous ratepayers, a committee of the County Council sat in the Town Hall to hear evidence for and against a proposal to constitute the town an urban district... . Such a radical change in the government of the town might have been expected to excite great interest, but it is a fact that only half of the electors voted on Monday. There were 13 candidates for the 11 seats, and the following was the result of the poll:- C.H. Kynaston, maltster, 190 votes; J. Kynaston, ironmonger, 177;... R. Morris, draper and outfitter, 124; H. Eckford, seedsman, 121... The figures were announced shortly before 10 o'clock but caused no excitement'.(36)

The Kynastons, father and son, had both served on the parish council. This influential family had recently purchased the Wem tan yard with a view to reopening it. It had been the town's principal industry and the electorate possibly thought that if the Kynastons wanted to be councillors again it might be wise to elect them. A fifty per-cent turn-out at local elections would nowadays seem quite respectable. Someone less of the hail-and-well-met political campaigner than Henry would be hard to imagine, and with his age one would think his re-election doubtful. In the end, Henry was returned fairly comfortably. Elections, where ordinary people could vote, and in private, were still a novelty, and the voting is instructive. Those with the most votes were from industrial occupations rather than from the professions. Those who ran affairs prior to the widened and secret ballot, men like one of the candidates here who was described as a 'gentleman' and not needing to work, came last in the poll with thirty-three votes. It was from such expressions of the people's will that David Lloyd George, as Chancellor of the Exchequer, would rely on in his attack on the House of Lords and the landed classes in general in his 'people's budget' of 1909.

As before, Henry was a diligent councillor, attending the majority of the meetings. These began as weekly affairs again at the local 'British School'. As enthusiasm waned however, they became fortnightly and then monthly, never quite settling to any regular pattern. A 3 o'clock start was unsuccessfully objected to by Henry who reasonably wanted it at 6pm when the working day was over. Henry Kynaston was made chairman. As he and his son often disagreed over issues there was no 'status quo' to defer to and no encouragement for cliques to form in opposition. A councillor was therefore almost as likely to oppose a motion as to support one.

A number of interesting items stand out from that first year. On the 11th June it was 'Proposed... that a Joint Committee of the Parish and Urban District Councils be formed for the management of Fire Engine + matters arising thereto + that Mr Eckford be one of the committee. Carried unanimously'.[37] Henry was to also serve on the General Committee. Some items have modern resonances. 'That the attention of the County Council be called to the large amount of traffic on the said roads and that they be respectfully asked to take this into consideration when fixing the grant. Carried unanimously'.[38] (The grant given by the County Council to the Urban Councils). This was in 1900. What would Henry and his colleagues make of present conditions? Juggernauts turning into the High Street from Mill Street mount the pavements to make the turn – as closely witnessed by your author, who stepped smartly out of the way of one or you wouldn't be reading this. So it is now in countless small (and not so small) settlements.

19th June: 'Resolved... that Lord Barnard's offer to give to the Council the Town Hall & his rights connected with Markets, Fairs & Tolls and land... be accepted...'.(39) was a further example, like the Sutherlands' offer of Trentham Hall to their Council, of aristocrats' belt-tightening seeing a shift of some of their land and property into public ownership. Hob-nobbing with the likes of Lord Barnard was all very well, Emily may have reasoned, but what was the point of your husband being a councillor if he allowed the pavements outside your home to fall into disrepair? So it was that on 21st November 'the surveyor be instructed to at once put the footpaths in Noble Street + Market Street into repair...'.(40) Quite right too!

Back in the gardening world, Richard Dean's brother Alexander had just turned sixty-eight but had long turned into, well, a bit of a misery really. He was sounding off in the *'Chronicle's* columns again and from this distance is fun to read, in small doses. He's having a crack at his brother's refusal to include garden peas at a sweet pea celebration. 'Possibly some enthusiastic admirer of that more prosaic and useful section of the Pea family – those that are edible – will be disposed to promote a conference in relation to them...'. Unfortunately, however, 'there is no getting up of steam in relation to the floral feature of these useful garden Peas... whilst I do not quarrel with those who think the Sweet Pea worthy of a conference, I do think if conferences have in them any elements of practical value, that the podding Pea is more worthy of such honour and notice. But... what of benefit for the garden pea can a conference accomplish?' Alex continued, negating his argument. 'Hardly can the Pea be more widely grown than it is; hardly can it help to produce better varieties...'. But should such a one occur, then it 'should originate with the Royal Horticultural Society, and not with scratch committees',(41) meaning his brother's, the brother who had recently received the Victoria Medal of Honour in preference to him, Alex. Ouch! as my mother would say.

This was a continuation of his campaign, under the pseudonym 'alpha', in the pages of the amateur-oriented journal *The Garden*. He built up a head of steam in his favour here which the editors, Ernest Cook and Gertrude Jekyll, eventually halted due to its vindictiveness. This misbehaviour was also being felt within the Royal Horticultural Society, where Alex Dean was, as we've seen, campaigning to reduce such influence as nurserymen like Henry had with the R.H.S.(42) Henry had been on their Floral Committee, before he became a nurseryman, in the late 1870s; probably co-opted because his name doesn't appear in the records. Alex Dean's father had been a nurseryman, as Alex himself was in his early days. The effect of Alex's insecurities with his forceful and saturnine disposition was that nursery traders would be prevented, by a supine Royal Horticultural Society, from holding committee secretaryships until after the First World War. This slap in the face to commercial horticulture was at one with the Society's traditional antipathy towards having women and practising garden professionals as members. Henry would have felt the hurt, as would important long-established nurseries like those of Veitch, Bull and Low.

Continual sniping from Alex Dean over the coming months brought forth one of, if not *the* longest letter of sustained anger and exasperation yet to appear in the *'Chronicle's* columns.(43) Letters of such temper must have been written before but were rarely if ever published. That this one from a nurseryman was, meant that it likely had the support of the editor, Maxwell Masters. From the paper's professional readership, such support as there was for Alex Dean's anti-Sweet Pea Conference stand was muted. Yes, thought one contributor in June,

> 'the Sweet Pea is at present such a popular flower, and so universally cultivated, that it hardly requires the metropolitan celebration of its bicentenary to intensify its popularity. That it is amply deserving of such a commemoration' however 'must be the opinion of all those who, like myself, have grown it assiduously and lovingly for years... If the Rose is the noblest... of garden flowers, the Sweet Pea is assuredly one of the most exquisite', its fragrance refined.
>
> 'For many of the finest forms of this fair flower we are indebted to Mr. Henry Eckford, of Wem, in Shropshire; who, though he lives there, is a Scotchman, and a native of Midlothian. I learned this fact some years ago (when writing to him about his graceful creations) on the highest authority, viz., his own. In his work he has had no formidable rivals, though the Messrs. Burpee, of America, have achieved, through the medium of cross-breeding, some gratifying results. But of Sweet Pea raisers, Mr. Eckford is the unquestionable king; among his finest productions are the following varieties', and the reader described nineteen of them. 'Navy Blue', he added, 'an American introduction, is worthy of cultural association with these'.(44) You can't help feeling that some more objective comparisons with the admittedly few American varieties around would have given them more credit.

A week before the Bicentenary Conference, *The Gardeners' Chronicle* gave, in a near full-page 'slight sketch' of sweet peas' history as cultivated plants, a better review than any appearing before and as thorough an account of any made since. It ended by saying that 'what Mr. Eckford and his followers have done for it may be seen on comparing the figures of a century ago with the flowers now produced'.(45) By the following week, the Messrs. Carter seed firm had reminded them that as James Carter it was their firm that sold the bulk of the few sweet pea varieties that were around before Henry set to work, and listed some of them. 'No doubt other firms could add to the list', said the '*Chronicle*, a little dryly, but none wrote in. For the '*Chronicle*, 'the development of the Sweet Pea in modern times is especially due to Mr. Eckford, at one time gardener to Dr. Sankey, a great florist, and the results of his labours are now everywhere manifest. The dwarf forms, of which Cupid was the first, hail from America. They are good for pot use, otherwise they have little or no advantage over the taller-growing sorts'.(46)

Of 'the pale-green elmy undulations of North Essex', (47) to quote the poet John Betjeman, the part that lies in a triangle shape midway between Braintree and Colchester was, as we've seen, very popular with seed-growers. This year, Hurst & Son's seed firm at Kelvedon were running a trial of over two hundred varieties of sweet peas. The '*Chronicle* picked out twenty-three of the best ones, fourteen of which were Henry's. They particularly liked the white 'Sadie Burpee' of Henry's, 'one of the finest of all the varieties'. Turning to the coming show, 'seventy-eight exhibitors had announced their intention of being present...'. they have made 590 entries needing '3607 vases. We do not envy the judges....'. Echoing a familiar complaint that 'the distinction between many of the varieties is infinitesimal' they said that 'if the Conference would kindly eliminate half or even three-fourths of the names they would make a very great service to horticulture'.(48)

Moans about 'too many, too similar' weren't confined to sweet peas but no-one ever did anything about it. It was fine in theory but seemingly impractical to put into effect. What did happen was that as new varieties were developed, old ones got forgotten, and not always for the best. The numbers of many fruit, flower and vegetable varieties increased in the nineteenth-century. Some have since declined while others have seen an increase. It's all a bit academic however if what is generally available to the public is always limited, a problem recently being aired. In 1900, the 'problem' of over-abundance was a nice one to have. It was like an itch you could always enjoy scratching, and the coming conference was to scratch away at it.

~~~

The Gardeners' Chronicle had always tried to cover the whole of horticulture, and some botany, to which other magazines, aiming more at the amateur market, responded by writing more about less, and less critically. *The Gardeners' Magazine*, under the editorship of the sweet pea buff George Gordon, devoted nine whole pages of their July 28th edition to the sweet pea bicentenary event. *The Journal of Horticulture* ran them close with eight. *The Garden's* coverage, though less, included portraits of Henry and Atlee Burpee, with a literate front-page editorial overview of the sweet pea that could only have been by its then co-editor Gertrude Jekyll. The '*Chronicle's* total coverage was only a page and a half, but generous for them.

FIG. 19.—THE GERM OF THE BICENTENARY.
Mr. W. Cuthbertson. Mr. R. Dean.

Wider afield, there was less attention. Richard Dean had managed to get news of the upcoming event into the July 16th edition of *The Times*, [49] but despite 'all the prominent daily papers being [supposedly] represented'[50] they didn't subsequently report on it. Neither did the *London Illustrated News*. Events in China, South Africa and elsewhere were what was newsworthy and gardening, for the old traditional papers, was not. For the new *Daily Mail*, it was. Theirs was a punchy style of news mixed with sport, gossip, fashions, human interest stories, cartoons and crossword puzzles.[51] To come would be big prize competitions and full-page illustrated advertisements like one for the Eckfords' nursery in 1911.

The *Mail* was cheaper than *The Times*, with eleven times its circulation. It pandered, as it does now, to our love of a good moan, though not to the general public's desire for centre-ground politics, where a right-wing stance
~~~

would later lead it to support Hitler and Mussolini. The phenomenon of the high circulation daily, characterised by a culture of 'inflated rumours and glossed-over facts'[52] swept across Europe. It originated, as much popular culture was to do in the new century, in the United States. Unlike the gardening press, for whose reports you had to wait for a week or sometimes months to read, *The Daily Mail's* could be read by show-goers during the event, and was a nice advertisement for it. The paper took the view that flowers needed flowery language to describe them. This is what those travelling to the Show on Saturday 21st would have read:

> 'Two hundred years ago, more or less, the Sweet Pea came to England from Sicily. Yesterday it had a show and a luncheon and an international conference all to itself, in honour of the event. All this was managed at that useful spot, the Crystal Palace.
>
> The blossoms were picked at the first streak of dawn or overnight, and tenderly transported miles and miles from a score of counties before the full heat of the day – no weather for plucked flowers this! By mid-day yesterday some were beginning to faint. But they were sweet indeed, those thousands and thousands of pea blooms.
>
> Lured down the long lines from vase to vase, and stand to stand, you might smell with might and main and never have enough, never have more than an olfactory promise of delight. The sweet-pea might have studied Tony Weller on the principals of love-letter writing: it leaves off just when you are beginning to want some more.
>
> Heavenly perfume, then – and heavenly colour. All the tones of the rich summer sky, from dawn through sunrise and sunset to dawn again, were there: lakes of crimson and scarlet and purple and blue, shaded and streaked and striped and edged, and set off with filigree and saxifrage and totter-grass and asparagus.
>
> Novelties were Eckford's Jeanie Gordon, cream and rose, and Miss Willmott, salmon and pink. With Eckford's seed, also, Peter Blair, gardener at Trentham, won the Sutton bicentenary prize, worth £25, for 100 bunches of blossoms. As worthy, too were the 182 new and old varieties brought down by Dobbies from far Rothesay'.[53]

Or, more likely, from their not so far seed firm at Marks Tey in Essex or, even nearer, from Orpington in Kent where Dobbies also were. 'Heaven scent, or why the sweet pea is our fragrant favourite', was the heading for a *Daily Mail* article in April, 2008,[54] showing that some things, at least, don't change. It even mentions Henry!

Together with the gardening press, it was good coverage for what was to be in effect something of a celebration of Henry's achievements. His achievements, at least, with the sweet pea. 'Seeing the enormous strides in size, colour, and number of flowers on a stem, made by Mr. Eckford and others in England during the past twenty years' said *The Garden's* (Jekyll's) editorial, 'and seeing how enthusiastically Mr. Burpee and others have carried on the good work in the United States, it is certainly not easy to prophecy in what new direction improvements are likely to take place...'.

But Jekyll had her views. 'A flower is not more beautiful for being as round as a shilling; in fact, much of the characteristic beauty of the Sweet Pea absolutely depends on its being of an irregular shape. As it is, the best forms are quite large enough, and shapely enough, and all we can reasonably want is new shades and hues of colour. We do not want them as yellow a sunflowers, or as scarlet as Geraniums, or as blue as Gentians, since the delicate shades of colour we now have vie with those of the Tea Roses in variety, and stronger toned colouring would take away from the aerial lightness so peculiar to these flowers, as it does in the case of the very dark flowers, such as [Henry Eckford's] Boreatton and others, approaching mahogany-black. The dark chocolate flowers form an excellent foil for the lighter and brighter-toned flowers of blue, white, peach, flesh and primrose or sulphur shades, but we do not need too many of them... .

For many years Sweet Peas have been overflowing from cottage and villa gardens, until now there is scarcely a garden large or small where they are not grown. Their value for decorations indoors, as cut rather long in the stem and loosely and naturally arranged in narrow-necked vases, is now universally recognised...'.[55]

Henry, and his son John who'd co-authored a paper to be read on the evolution of the flower, were at the conference, as may Emily have also been on this special occasion. There was no mention of wives in the journal reports but women and children barely existed for them anyway. John's children at home, and those that may have already begun living with Henry and Emily could be looked after by Alice and Henry's put-upon daughter Isabella. Henry was often in London, if not often this far south of the river. It would be easy enough for them all to book into a hotel but perhaps John's fellow committee member H.J. Jones (c.1856-1928) offered to put them up and show them round his Ryecroft Nursery at nearby Hither Green.

The first thing that struck *The Gardeners' Magazine's* reporter 'on entering the Central Transept of the Crystal Palace on July 20th was the novelty of the exhibition. Even the man who sweeps the floors with a big brush said it was novel, and he has seen more flower shows at the Palace than I should like to say'.(56) It was the first ever sweet pea show. 'Never in the history of floriculture has an exhibition of one group of florists' flowers been carried to so successful an issue as this nor run with so much enthusiasm... . It must be a great satisfaction to such veteran florists as Mr. Richard Dean and Mr. Henry Eckford to see a flower like the sweet pea developed during their life time, and to assist in such a celebration...'.(57)

Following introductions, the guests adjourned for lunch. It was the hottest week of the summer, and despite the Crystal Palace being atop the highest hill in South London – we're on the same level here as the cross on St. Paul's cathedral, the diners were told – within the giant greenhouse it was hotter than the nearly 90°F shade temperature outdoors. Henry and the other unfortunate guests 'were seen perspiring most freely throughout the proceedings, and making the most strenuous efforts to keep cool by fanning themselves'.(58) It was stated, rather grandly, that representatives of Germany and the United States lent an international flavour to proceedings but this actually boiled down to two men, the Rev. W.T. Hutchins and Fritz Benary. Fritz was the son of the now late Ernst Benary, whose seed house in Erfurt still traded under the father's name (as John Eckford was to do at Wem).

The Crystal Palace interior, Journal of Horticulture 26-7-1900.

Hutchins and Benary both spoke at the lunch, though the diners' attention was tried, they being now as hot as their lunches (or hotter).

Hutchins said that though about 150 tons of sweet pea seeds a year were handled in the U.S., and sweet pea fields in California were measured in square miles, he'd never seen an exhibition like this. It wasn't so much the quantity and quality of flower but the revelation 'of its decorative qualities which surprised him'. Hutchins paid 'a high tribute to the name of Mr. Eckford, whom, he said, floriculturists were prepared to canonise as a saint in America. Mr. Benary said that he had no claim... as an improver of the sweet pea. He had merely tried [like Burpee in America] to get hold of and distribute as many interesting varieties as possible. He thought that the task of the hybridisers in future would be to make the sweet pea bloom earlier and later, so as to prevent the dropping of the blooms in hot and dry seasons. 'Richard Dean paid another tribute to the work of Mr. Eckford, who he regretted, in view of his splendid work, was not selected as a Victorian medallist'.(59) (It was to come five years later).

The exhibition of sweet peas was held in the eastern half of the nave. Competitively, Henry's prize-money was for a 'Coming of Age' class, 'provided to commemorate the fact that it is twenty-one years since Mr. Eckford commenced to improve and cross-fertilise sweet peas'.(60) First prize went to Edwin Beckett (c.1853-1935), an already well-known all-round head gardener from Elstree in Hertfordshire, and it won the cup for the best exhibit. Traders like the Eckfords and H.J. Jones put up non-competitive displays for which they were awarded gold medals (if that makes sense). The Eckfords 'exhibited a fine lot of sweet peas, arranging them in tall pure glass vases with a little sweet pea foliage; some of the rows were raised, and here and there throughout the large group of about one hundred and fifty bunches there were dotted elegant cocos [coconut] palms'. Among a mix of old and new varieties, 'Mr. Eckford also showed the value of grasses for use with sweet peas'.(61)

The conference part of the show began at 4pm. The heat was still so intense that they had a job in getting an audience. During the reading of a history paper, it was said that some people considered the 'Painted Lady' to be the original 'variety', (species?) a view 'shared by Mr. Eckford', but how, when and where it occurred was uncertain. (The variety is still obtainable). A discussion followed, participated in by John, Henry and others including Edward Laxton, the son of Henry's late sweet pea rival Thomas and destined to make Laxton a household name for fruit in the new century. The term 'sports' was used and objected to by John Eckford who, strangely,

Ernst Benary

Fritz Benary

The Benary workplace, circa 1890.

regarded them instead as natural crosses. Bees were mentioned as taking the sweet peas' pollen but not being able to fertilise the flower.

Perhaps they were also wondering why it was that self-pollinating flowers do, well, flower if the purpose of a flower is to attract pollinators. The probable reason, advanced since, is that self-fertilisation can be an evolutionary adaptation to a lack of pollinators. This could occur for any number of reasons, climate change among them, and shows the possibility that the sweet pea could once have cross-pollinated naturally. It would be nice to imagine the Painted Lady to be, with the wild Italian flower, equally descended from a much older form. However, the Italian claret and blue appears as an occasional throwback, unlike the Painted Lady which is probably an early sport of it.

Back at the conference, Walter Wright wanted to include the sweet peas with the florists' flowers of old, with all those strict ideas and rules of what constituted the perfect flower. This, at a time when the idea was becoming outdated and even the term, floristry, was changing to that in current use for flower shops. Talk of defining the flower and reducing its number would soon be seen as so much dancing on pin-heads. 'Some person has suggested that in the future we shall get fringed flowers', said Wright. 'When we survey a variety like the fluted form of Primrose we can well believe it. But it is a fearful prospect'.[(62)] It was all about to change.

The conference reassembled on the following afternoon when Charles Curtis (1869-1958) of *The Gardeners' Magazine* read a paper of his and John Eckford's on 'Evolution and improvement'. Charles Curtis had been a frequent house-guest of the Eckfords, both father and son, since the early 'nineties.[(63)] Curtis's magazine naturally continued to report at length. Leading up to the time that Henry first gained notice with his Bronze Prince in 1882, 'a great change at this time came over the methods of producers and considerable progress was made, Mr. [Thomas] Laxton and Mr. [Henry] Eckford both working along lines calculated to give the best results, by crossing existing varieties... The late Dr. Sankey was referred to as a keen hybridist, and the enthusiasm with which he and Mr. Eckford took up sweet pea crossing was noted. The variety "Mrs. Sankey" caused at one time a good deal of trouble, as people could not understand how it was that a black seed could produce a white flower. Another pea mentioned as having the appearance of the seed against it was the "Countess of Radnor".

Some notable introductions brought forward in 1893 were mentioned, when the public appetite had become whetted and the craze for new varieties had become difficult to appease. 1895, however, was a record year,

The sweet pea conference portraits, The Gardeners' Magazine, 28-7-1900.

when "Blanche Burpee" was sent out. In 1898 twenty-two new varieties were introduced, and in the following year one of the principal novelties was "Sadie Burpee". The sweet pea, the authors went on to say, had now been developed to such a degree that it was impossible to see from what direction the next improvement was to come. The labours of Mr. Eckford had resulted in great substantiality of bloom. He had not yet given us a two-inch standard, but something very near to it, and he had showed some wonderful ranges of colour. Full credit for sweet pea improvement was accorded Mr. Eckford, who had obtained fourteen out of twenty-two R.H.S. awards for new and improved sweet peas. With regard to the future, the authors said that fragrance should be the first point for consideration and we might also secure a wider range of odours... In conclusion, the authors appealed to the Conference to do its best to prevent the sweet pea becoming the flower of the faddist, and not to hedge it round with arbitrary rules',(64) this last a direct rebuff to Walter Wright.

This would have been Henry's view too. He famously failed to believe in finality with regard to sweet pea development. Throughout his life, too, he'd been attracted to irregular, asymmetrical flowers, the zygomorphs, rather than actinomorphs, the regular symmetrical ones. It's surprising, isn't it, to see how credulous people were then, supposing a connection between flower and seed colour.

The Rev. W.T. Hutchins, who had, it appears, taken the time and trouble to cross the Atlantic to deliver his paper on 'Sweet Peas in America', received an enthusiastic reception when he rose to speak. '... I am not sure' he said 'but that America discovered [Henry] Eckford. I find by a letter of 1893 from a London seedsman these words: "I wish somebody would wake Eckford up, but I guess none of the children of the West can do it. In fact, he is simply incorrigible". In my visit to England in 1895 I felt that you had not discovered Eckford. And over in France they shrugged their shoulders at the mythical mention of Eckford's name. And to this day I cannot understand the tardy recognition of Eckford's superb work on the Continent. But I hail with joy the fact that to-day it cannot be said that this floral prophet is without honour in his own country. I should call him a dear old Scotch-Welsh Englishman. In the floral calendar of America no name

The Rev. W.T. Hutchins

stands higher than his; indeed, ten years ago our catalogues blazed with miraculous lithographs of the new sweet peas, and Mr. Eckford got all the glory. There was a blue sweet pea, as blue as Chinese larkspur; and a yellow sweet pea, as yellow as a dandelion…'.

I'm not sure where the Welsh comes in. Henry lived near its borders in Wem and some of his descendants live there now. But that needed some foresight! With reference to the slowing now of America's rapid post-Civil War industrialisation, Hutchins said that 'we count among the forces that are redeeming America from the sordid lust of materialism this growing interest in floriculture, and the rapid increase of uses to which we are putting flowers, as attested by the last decade, is one of the most cheering heralds of the twentieth century… . About twelve years ago I had begun to get together a collection of varieties… [I noticed] Mr. Eckford's name in Farquhar's catalogue and in a sort of timid foot note… they intimated that they could supply, but did not feel warranted in recommending, the latest sealed packets of the Eckford introductions. The sensations I felt in those days as I succeeded, one by one, in getting the named sorts into bloom, are still vivid.

The Eckford sets of 1892 and 1893 set the United States ablaze on sweet peas. California was just in time to prove that sweet peas could be grown in America, and to meet the enormous demand the seed houses jumped from a few pounds of stock to tons. A small florist would sell a hundred pounds, an ordinary seedsman two or three tons, and one jobber has had 25 tons in stock in a season'. In several cities, sweet pea shows were now an annual event. 'We are still competing for an Eckford Cup at Springfield, Massachusetts'. It was won by an entrant living three-hundred miles away. His home town of Cohocton, N.Y., had had a sweet pea craze for three years and the whole place celebrated the win. Speaking of new American varieties he said that he thought them legitimate 'and yet, of course, we owe it to Mr. Eckford that he has made it easy for us to get them. Where a grower has from two hundred to two hundred and fifty acres of this flower in bloom, and most of it has come from the Eckford stock, of course his eye is always out for any precious rogue that may appear. And he frequently finds them, and carefully saves every seed from such a plant'. It seems that U.S. varieties then were obtained more by selection than by deliberate cross-pollination.

'Mr. Hutchins, who spoke for over an hour was heartily applauded on resuming his seat'.[(65)] Whether the applause was in admiration or as relief, or a mixture of the two, *The Gardeners' Magazine's* reporter did not say. The reporter from *The Journal of Horticulture* found it 'of peculiar interest but [it] did not evoke material discussion'.[(66)] The speaker who was supposed to have read the next two papers had already left 'on account of the lateness of the hour', and the conference ended. The profession of horticulture, generally speaking, attracts those who prefer the placid, outdoor interaction with plants to the more hurly-burly company of people. The basis of the London seedsman's critique of Henry was for his lack of social ease outside the company of his peers, his promotion of his plants but not of himself. That Henry was discovered in his lifetime was due in the main to his American prophet, the Rev. W.T. Hutchins and in Britain to the more objective yet equally passionate promotion of Richard Dean.

The *'Chronicle* lagged behind its rivals in using engravings rather than photographs as illustration. There were numerous half-tone photos in *The Gardeners' Magazine* and *The Journal of Horticulture's* Conference coverage; there was one, rare, photo in the *'Chronicle's* report, a portrait of Henry[(67)]. This familiar photo of him in old age also appeared in *The Gardeners' Magazine*, as did photos of John Eckford with his other committee members. Two of Henry's flowers were used to illustrate differences; the hooded variety 'Lady Grisel Hamilton' contrasted with the erect standard of 'Prince Edward of York'.[(68)] (As Lady Hamilton had herself worn a bonnet and Prince Edward's standard bearer would have held it erect they're well-named). The Eckfords' trade display with its coconut palms can be seen clearly in the background of a full-page photograph, arranged around the sphynxs and bears or wild-cats (it's a little difficult to tell) of the Palace's permanent display.[(69)]

The smaller *Gardeners' Chronicle* report on the two-day event made the point that it was the trade rather than 'private cultivators' who felt the need to reduce the 250-plus varieties now found in seed-lists. They wanted a Sweet Pea Society established, which could authorise a recommended list of, say, just fifty varieties... .[(70)] Has commerce always despised variety? *Gardening Illustrated*, another journal, made the point that old varieties were rarely met with anyway, 'except in very large collections, where, perhaps, they are kept for comparison'. The magazine picked out nineteen sweet peas from the Show, with detailed descriptions. Sixteen of them were Henry's: 'Salopian', 'Mars', 'Prince of Wales', 'Lord Kenyon', 'Lovely', 'Prima Donna' (which, unusually, 'frequently develops four flowers on each spray'), 'Lady Grisel Hamilton', 'Black Knight', 'Stanley', 'Sadie Burpee', 'Blanche Burpee', 'Queen Victoria', 'Mrs. Eckford', 'Colonist', 'Venus' and 'Lady Mary Currie'.

They were most taken by this last mentioned one. 'There is no better example of rich and effective colouring combined with blossoms of large size. The colour [is] rich orange-pink, prettily shaded rose, and either in the garden or when cut there is no more effective sort. This was probably shown more often than any other kind. Its fragrance, too, is superb'.(71) Surprisingly, two American varieties on show, 'Gorgeous' and 'Navy Blue', were both vivid colours that bleached in the British sun, and would suffer even more in the extremes of the more varied American climate. For another visitor to the Crystal Palace 'it would not be said that there were many deserving absolutely new varieties on view. Mr. H. Eckford, as might be expected, had several which appealed to my taste…'.(72)

The *'Chronicle* reported that 'sweet pea scent is now being prepared for public sale from flowers grown in Mr. Eckford's nursery at Wem. There were samples at the Crystal Palace… and the perfume is certainly that of the Sweet Pea, and it is very agreeable and persistent'.(73) It was used to scent sachets of sweet pea petals 'for perfuming apartments, cabinets, &c.,' and coloured photos of their grounds at Wem were mounted on card impregnated with it. By 1908, under John Eckford's management, these 'Eckford Flower Odours' had increased to fourteen different flower scents,(74) the popular lines continuing into at least the 1930s.(75) First displayed at the Crystal Palace though, as so many Victorian inventions had been. In the twentieth-century the great Crystal Palace suffered a gradual decline. In December 1936 it was destroyed in a fire that due to its site and size could be seen across London. I worked there once as a gardener and tractor-driver and remember the now derelict area being covered for much of its length by an enormous manure heap, to which I added trailer-loads from the muckings-out of the park's children's zoo. Sic transit gloria mundi… Far in the future, of course, and all unconsidered in 1900.

Silas Cole, who we first met two years ago, had probably been to the Show. Having crossed his ('Lovely' x 'Triumph') hybrids with 'Prima Donna', he'd got one plant with much bigger flowers and a noticeably waved standard. Named 'Countess Spencer' (the Spencers were his employers) it was a breakthrough, from which he only managed to save five seeds. Perhaps he mentioned his wonderful flower at the Show. But unless you see it – well, it's easy to exaggerate a little. If all went well, he'd have been told, there'd be a national sweet pea society next year, with its own Show. You'll have more flowers next year. Bring them along to that!

It was time to go now. It had been a fine do, the sweet-pea bicentennial, a bit on the warm side but as one visitor said, 'never have I seen a flower-show crowd more cheerful under such conditions of perspiration. And,' he went on, 'at eventide, when the air was cooler and the tired peaists had gone this way and that, I saw another little episode in connection with the bi-centenary. As I strolled down the grounds towards the station a crowd of holiday children, yelling [Leslie Stuart's music-hall hit] "Soldiers of the Queen" at the top of their voices, passed me on their return journey. Their lungs, at any rate, were quite unaffected by the heat... I saw them again later, and as they marched out of London Bridge Station and along the crowded street on the south side of the river I noticed that most of them carried bunches of flowers. There was nothing extraordinary about that, but one little chap hugged a bunch of sweet peas. They might have been the surplus of some kind-hearted exhibitor, but under the flare of a gas lamp I noticed the lavender hue of Lady Grisel Hamilton. Some people have asked what good the bi-centenary celebration would do. Among other things it brought a little of the pleasure, that flowers can bring, into a South London home'.(76)

~~~

A week after the Show, the seed company Messrs. Carter invited people down to see their garden pea and sweet pea trials at Mortlake in Surrey. *Gardening Illustrated's* reporter considered that there were too many varieties of garden pea and that most of them should be destroyed. (Now that they have been, the opposite view prevails). None of the popular varieties there were Henry's. Either he lacked success or wasn't bothering to hybridise them anymore. Garden peas were a well-tilled field when he joined them, and he made his mark instead by tilling the almost empty sweet pea one. Here at Carters' 'there were no two opinions as to the exceeding beauty of most of the kinds, and the shades of colour, running from pure white up to vivid crimson, justified the favour with which the Sweet Pea is now regarded in English gardens'. Of the ones that caught the reporter's fancy, about half were Henry's, viz. 'Fairy Queen', 'Mrs. Eckford', 'Venus', 'Lovely', 'Mrs. Gladstone', 'Prima Donna' and 'Firefly'. 'Mrs. Gladstone' was an old variety, not often noted now and was here, perhaps, out of affectionate respect for the widowed Catherine Gladstone, who had recently died.

The most significant of the few sweet peas that had been around when Henry started had been sent out by Carters, and today in a short address their speaker Donald McDonald made the most of it. 'Violet Queen', for example, he said 'considered the foremost of the types now so popular...
~~~

was illustrated in colours in Carter's catalogue for 1878 – four years before Mr. Eckford got a certificate for his Sweet Pea. The Queen was one of Mr. Eckford's introductions, and it evidently had in it the blood of Violet Queen. Mr. Eckford had done a lot for the Sweet Pea, and it was to be hoped he had obtained the due reward of his labours' (77) said Mr. McDonald, speaking (one can hear him) through gritted teeth. No clan fraternity here!

Brotherly love was also notable by its absence in another of Alexander Dean's letters in the *'Chronicle*. 'If the Sweet Pea is a flower of such immense importance in our gardens now, and a separate society for its supervision is suggested, certainly it needs no special encouragement. Cannot this supervision be satisfactorily furnished... by a sub-committee from the [R.H.S.] Floral Committee.... To attempt the formation of another society, or to maintain a permanent committee outside of the Royal Horticultural Society, would be to court failure...'.(78) Under the care of the R.H.S. Floral Committee, of which Alexander was a member, this Sweet Pea Society idea could be quietly drowned. Now well over a century old, the National Sweet Pea Society has done everything *but* fail.

In a long *'Chronicle* article in August, Edwin Molyneux was indulging in the popular practice of selecting favourite sweet pea varieties, grouping 'them in their respective colours, following the example set by Mr. H. Eckford in his catalogue'. His selections are notable in showing that the small but significant numbers of American varieties were now getting their due. He put 'Emily Henderson' top of the list of 'the most valuable of all' sections, the whites. Other American varieties he liked included 'America', a bright scarlet and white striped variety and, contentiously, 'Gorgeous'. Of Henry's new varieties, 'Coccinea' 'has the peculiarity of showing at times a double set of wing-petals, and may be the forerunner of a double-flowered variety – not that this would be an acquisition'.(79)

Alexander Dean was a frequent contributor to the letter pages of the horticultural press, sometimes with two or three letters on a page. One supposes that the editors were thankful for all the free copy. A pity, as his too frequent superciliousness must have been wearing. On the 18th August he raised his sights from targeting sweet peas, his brother Richard and horticulture in general to take on, bizarrely, athletics pursuits and the Bishop of London for advocating digging for boys. And why not young men, too, thought Alex, 'who would find in such recreation something profitable as well as healthful, whilst they waste enormous energies in athletic sports and pastimes, practically making of such sports a fetish, that do little physical good, and not a little moral harm. To all these an hour's trenching or digging before breakfast each day would in time represent to the nation millions of pounds profit...'.[80] You wonder if Richard ever had a quiet concerned chat with Henry and Emily about his younger brother, so unlike him and his late brother William.

Another reader wrote in to say that vegetable numbers were in more need of reduction than sweet peas. With some two-hundred culinary pea varieties, how many of them were 'Telephone, or Duke of Albany, Stratagem, or Ne Plus Ultra, under different names?... . It would not be difficult to make up a list of fifty varieties of Broad Beans, but they cannot be distinct; and when we come to Cucumbers, the list is appalling in its length, while it seems impossible to enumerate more than two or three distinct types'. There were sixty melon varieties, yet there was 'little more scope for variation in Melons than there is in the case of Cucumbers...'. There were eighty different tomatoes, yet 'can there possibly be a dozen distinct types?' The same was true for many other vegetables too, which 'vary so little. Are there not other flowers of older standing than the Sweet Pea whose varieties are much more numerous? – the Begonia, Carnation, Dahlia, Fuchsia, Pelargonium, and others. But I hear no proposal to reduce the numbers. Why the sweet pea alone should be singled out for the thinning process – the most easily grown, and the most popular of all of them, I can only conjecture...'.[81]

We now rightly bemoan the loss of fruit varieties but with vegetables it may be that there never was the choice that the old seed-lists would have us believe. Sweet peas were being got at because we always have a go at anything which rises above the norm and gets 'too big for its boots'. As the great Dr. Johnson had noted over a hundred years earlier, 'Whenever there is exaggerated praise, every body is set against... They are provoked to attack it'.[82] Before the sweet pea took its bow on the year and went off in search of some bigger boots there was the Shropshire Show to perform in. 'It is regrettable', wrote the *'Chronicle's* reporter, 'that it should be impossible to attract the immense number of people one meets at Shrewsbury by a display of horticulture alone, but it may be hoped that on each occasion many of those who have come merely to see the sports, are encouraged to make some start, and to take an interest in gardening, by the splendid exhibits they see in the tents'.[83]

This isn't to say that flower-shows, in themselves, lacked regard. Before the days of radio, cinema, television and the rest, entertainment was live, indoors or out. 'Visiting a flower show', writes the garden historian Anne Wilkinson, 'was both a social and an educational occasion as well as a pleasure. One of the biggest impacts on entering the marquee or the hall was the overpowering smell from the fresh flowers on display. In the nineteenth-century there would have been the added impact, for many people, of the blaze of light from gas or even the new electricity, for those more used to the dim glow of firelight and candles'.[84]

The Shropshire Flower Show at Shrewsbury has undergone changes since its inception in 1875. Gone are the days when late-summer crowds twice the size of the town's population would descend on the Quarry grounds. Having now to compete for the public's attention it has dropped the associated 'sports and high class music' [85] of earlier, pre-war days in favour of today's somewhat militaristic festival. But its primary function remains. In 1900, sweet peas were 'a leading feature, and a large number of bunches were staged in the various classes'. The 'Eckford Silver Challenge Cup was offered to thirty-six distinct varieties', while another prize was given for 'the best eighteen varieties of Eckford's Sweet Peas', among which the winner included 'Gorgeous', an American variety. Henry himself 'showed a fine lot of sweet peas in glasses; there were something like forty varieties, in capital condition'.[86] Henry won a gold medal for an exhibit of them with cactus Dahlias.

So; 1900. What a year! And the next one would be something special too.

THE GARDENERS MAGAZINE. December 23, 1899.

ECKFORD'S SPECIALITIES.

ONLY GENUINE DIRECT FROM WEM.

CATALOGUE for 1900 now ready, gratis and post free.

ECKFORD'S GIANT SWEET PEAS.

The purity and stamina of these home-grown stocks are so conserved by special methods of culture that they are **unequalled for strength, vitality, vigour of growth, and grand flowers.**

ECKFORD'S GIANT SWEET PEAS.

A beautiful Coloured Plate of the charming Novelties for 1900 **gratis with each Catalogue.**

ECKFORD'S PRIMULA SINENSIS.

Five unique Novelties for 1900, as follows :—

"EFFULGENT," "ELEGANCE," "SALOPIA."
NEW "PYRAMIDAL" IN MIXED COLOURS.
NEW "CRIMP-LEAVED" IN MIXED COLOURS.

The two last-named are an entirely new race of Primulas. Each, per packet, 2s. 6d., or the set of five for 10s. 6d., post free.

For description of these splendid New Primulas, see Catalogue.

ECKFORD'S VERBENAS.

Renowned strain of exceptional merit.

The Journal of Horticulture says of this strain : "Like a piece of tapestry from the singularly happy blend of colour."

Per packet, 1s., 1s. 6d., and 2s. 6d., post free.

CHOICE FLOWER SEEDS of all the leading and most popular kinds in collections at 42s., 21s., 10s. 6d., and 5s., post free.

For Specifications of above Collections and General List of Flower Seeds, See Catalogue, Gratis and Post Free.

ECKFORD'S CULINARY PEAS

Are positively unequalled for splendid flavour and enormous cropping capacity.

The following Collections are arranged to **give a continuous supply** in the open **from June to October.**

COLLECTION A—6 pints Eckford's Culinary Peas, 6s., carriage paid.
COLLECTION B—9 pints Eckford's Culinary Peas, 8s. 6d., carriage paid.
COLLECTION C—12 pints Eckford's Culinary Peas, 11s., carriage paid.
COLLECTION D—9 quarts Eckford's Culinary Peas, 16s., carriage paid.
COLLECTION E—12 quarts Eckford's Culinary Peas, 21s., carriage paid.
COLLECTION F—18 quarts Eckford's Culinary Peas, 31s., carriage paid.
COLLECTION G—24 quarts Eckford's Culinary Peas, 40s., carriage paid.

In comparing Prices it would be well to **remember that these Culinary Peas are** all practically new varieties of the very highest quality in flavour and **enormous croppers.**

Send for Catalogue, gratis and post free.

ECKFORD'S PURE VEGETABLE SEEDS.

Results from all quarters prove that Eckford's Seeds for purity, vigour, and yield are amongst the best ever sold to the public. **Liberal and well-arranged Collections for 12 months' continuous supply**—

12s. 6d., 21s., 42s., 63s., 105s., carriage paid.

Other Collections, 2s. 9d., 5s., 7s. 6d., post free.

Send for Illustrated Catalogue, gratis and post free.

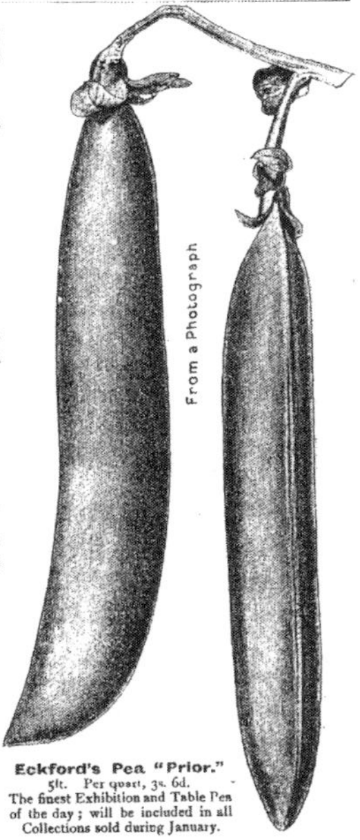

Eckford's Pea "Prior."
5ft. Per quart, 3s. 6d.
The finest Exhibition and Table Pea of the day ; will be included in all Collections sold during January.

Henry Eckford,
SEED GROWER,
WEM, SHROPSHIRE.

Eckford advertisement, Gardeners' Magazine, 23-12-1899.

References to Chapter Eleven

(1) 'Sweet Peas', *The Journal of Horticulture*, (26-1-1899), p.69.
(2) 'Sweet Peas', *The Gardeners' Magazine*, (2-2-1895), p.67.
(3) 'Early Sweet Peas', *The Gardeners' Chronicle*, (4-2-1899), p.132.
(4) Henry Eckford's seed catalogue for 1899, preface.
(5) 'The Hybridisation Conference', *The Gardeners' Chronicle*, (1-7-1899), p.1.
(6) 'Hybrids and their Raisers', *The Gardeners' Chronicle*, (8-7-1899), p.30.
(7) 'Henry Eckford', *The Gardeners' Chronicle*, (8-7-1899), p.25.
(8) 'Hybrids and their Raisers', op.cit.
(9) 'Eminent Hybridists and Raisers', *The Gardeners' Magazine*, (15-7-1899), p.429. Henry's portrait appeared in an adjoining supplement, between pp.428-429.
(10) 'The Hybridisation Conference', *The Gardeners' Chronicle*, (15-7-1899), p.50.
(11) Ibid, p.51.
(12) W.T. Stearn, in R. Desmond, *Dictionary of British and Irish Botanists and Horticulturists*, (1977), p.ix.
(13) Miles Hadfield, *A History of British Gardening*, (1969edn.), pp.394-395. An imaginative re-creation of the train ride is in R.M. Henig's *A Monk and Two Peas*, (2000), pp.1-3.
(14) 'The Pansy and Viola', *The Gardeners' Chronicle*, (15-7-1899), p.44.
(15) 'Hybrid between the Field Pea and the Sweet Pea', *The Gardeners' Chronicle*, (22-7-1899), p.72.
(16) E.R. Janes, *Sweet Peas*, 1953, (1961), p.164.
(17) 'Home Correspondence', *The Gardeners' Chronicle*, (12-8-1899), pp.134-135.
(18) 'Shropshire Horticultural', *The Gardeners' Chronicle*, (26-8-1899), p.176.
(19) Ibid, pp. 178,179.
(20) 'Bi-centenary of the Sweet Pea', *The Gardeners' Magazine*, (16-12-1899), p.790.
(21) A.N. Wilson, *The Victorians*, (2002), p.600.
(22) 'Bi-centenary of the Sweet Pea', *The Gardeners' Magazine*, (16-12-1899), p.790.
(23) Ibid, p.802.

(24) 'Bi-centenary of the Sweet Pea', *The Journal of Horticulture*, (21-12-1899), p.541.

(25) 'The Christmas Season', *The Gardeners' Magazine*, (23-12-1899), p.805.

(26) The Last Will and Testament of Henry Eckford, (7-12-1899), pp.1-2.

(27) 'Calendar for 1900', *The Gardeners' Magazine*, (23-12-1899), no paging. In the library of Edinburgh's Royal Botanic Garden it is catalogued in *The Gardeners' Magazine*, 6-1-1900, between pp. viii & p.1.

(28) 'Sweet Peas', *The Gardeners' Magazine*, (23-12-1899), p.805.

(29) 'Eckford's Specialities', advertisement, *The Gardeners' Magazine*, (23-12-1899), no paging.

(30) The Reader's Digest, *Yesterday's Britain*, (1998), p.14.

(31) 'Sweet Pea Bi-Centenary Celebration', *The Gardeners' Magazine*, (3-3-1900), p.128.

(32) 'Sweet Pea Bicentenary Celebration', *The Gardeners' Chronicle*, (27-1-1900), pp.62-63.

(33) 'Eckford's Specialities', (advertisement), *The Gardeners' Chronicle*, (3-2-1900), no paging.

(34) 'Sweet Peas', *The Gardeners' Chronicle*, (3-2-1900), pp.69-70.

(35) 'Sweet Peas', *The Gardeners' Chronicle*, (17-2-1900), p.109.

(36) 'Wem', *The Shrewsbury Chronicle*, (27-4-1900), p.8. ('H' for Henry misprinted 'J').

(37) DA 14/100/1, Minute Book, Wem Urban District Council, (1900-1903), p.22, Shropshire Record Office.

(38) Ibid., p.20.

(39) Ibid., p.54.

(40) Ibid., p.55.

(41) 'Edible Peas', *The Gardeners' Chronicle*, (21-4-1900), p.253.

(42) Brent Elliott, *The Royal Horticultural Society; A History; 1804-2004*, (2004), p.328.

(43) 'Nurserymen and the Royal Horticultural Society', *The Gardeners' Chronicle*, (26-1-1901), p.63.

(44) 'Sweet Peas', *The Gardeners' Chronicle*, (2-6-1900), p.338.

(45) 'Sweet Peas', *The Gardeners' Chronicle*, (14-7-1900), p.30.

(46) 'Sweet Peas', *The Gardeners' Chronicle*, (21-6-1900), p.50.

(47) John Betjeman, 'English Cities and Small Towns', in W.J. Turner, ed., *The Englishman's Country*, (1945), p.121.

(48) 'Sweet Peas', *The Gardeners' Chronicle*, (21-6-1900), p.51.

(49) 'Bicentenary of the Sweet Pea', *The Times*, (16-9-1900), p.9.

(50) 'Commemorative Festival of the Sweet Pea', *The Journal of Horticulture*, (26-7-1900), p.74.

(51) The Reader's Digest, *Forging the Modern Age: 1900-14*, (1999), p.29.

(52) Geert Mak, *In Europe: Travels through the Twentieth Century*, (2007), p.11.

(53) 'Honouring the sweet-pea', *The Daily Mail*, (21-7-1900), p.3.

(54) 'Heaven scent', *The Daily Mail*, (4-8-2008), p.27.

(55) 'The Sweet Pea', *The Garden*, (28-7-1900), p.57.

(56) 'The Great Sweet Pea Show', *The Gardeners' Magazine*, (4-8-1900), p.489.

(57) 'The Great Sweet Pea Bi-Centenary Celebration', *The Gardeners' Magazine*, (28-7-1900), p.469.

(58) Ibid.

(59) Ibid., p.470.

(60) 'Some Reminiscences', *The Sweet Pea Annual*, (1921), p.11.

(61) 'The Great Sweet Pea Bi-Centenary Celebration', *The Gardeners' Magazine*, (28-7-1900), p.470.

(62) Ibid, p.472.

(63) Ibid, p.473.

(64) Ibid, p.476.

(65) Ibid, p.477.

(66) 'Commemorative Festival of the Sweet Pea', *The Journal of Horticulture*, (26-7-1900), p.76.

(67) 'Mr. Henry Eckford', portrait, *The Gardeners' Chronicle*, (28-7-1900), p.79.

(68) (Sweet pea illustrations), *The Gardeners' Magazine*, (28-7-1900), p.476.

(69) 'View of Part of Great Sweet Pea Exhibition at the Crystal Palace', photo, *The Gardeners' Magazine*, (28-7-1900), p.474.

(70) 'The Sweet Pea Bi-Centenary Celebration', *The Gardeners' Chronicle*, (28-7-1900), p.70.

(71) 'Sweet Peas', *Gardening Illustrated*, (4-8-1900), pp.296-297.

(72) 'The Great Sweet Pea Show', *The Gardeners' Magazine*, (4-8-1900), p.490.

(73) 'Sweet Pea Scent', *The Gardeners' Chronicle*, (28-7-1900), p.79.

(74) Leaflet in Henry Eckford's seed catalogue for 1908.

(75) 'Eckford's Essence of Sweet Pea Blossoms', 'Eckford's Sweet Pea Petals' and Eckford's 1932 catalogue are all items in the author's possession.

(76) 'The Great Sweet Pea Show', *The Gardeners' Magazine*, (4-8-1900), p.489.

(77) 'The Sweet Pea and the Garden Pea', *Gardening Illustrated*, (4-8-1900), pp.297-298.

(78) 'Sweet Pea Varieties', *The Gardeners' Chronicle*, (4-8-1900), p.92.

(79) 'Select Sweet Peas', *The Gardeners' Chronicle*, (11-8-1900), pp.115-116.

(80) 'Digging as a Recreation', *The Gardeners' Chronicle*, (18-8-1900), p.135.

(81) 'Why Sweet Peas Alone?' *The Gardeners' Chronicle*, (18-8-1900), pp.135-136.

(82) James Boswell, *The Life of Samuel Johnson*, 1791, (1906), Vol. 2, p.374.

(83) 'Shropshire Horticultural', *The Gardeners' Chronicle*, (25-8-1900), p.156.

(84) Anne Wilkinson, *The Victorian Gardener*, (2006), p.89.

(85) 'Shropshire Horticultural', *The Gardeners' Chronicle*, (20-8-1904), p.135.

(86) 'Shropshire Horticultural', *The Gardeners' Chronicle*, (25-8-1900), pp.158-159.

Chapter Twelve

Nineteen hundred and one. The bells struck midnight on New Year's Eve to welcome in what most people saw as the start of the new century. Of the revellers around St. Paul's Cathedral in London 'the Scotch element was, as usual, well to the fore' reported *The Daily Mail*, 'and the singing of "Auld Lang Syne" could be heard all over the city'.(1) Transcending Christmas, the old traditions were kept alive by the older and less boisterous 'Scotch elements' too, with Henry celebrating its arrival in his own way at Wem.

For readers of the gardening press, the first few months of the new century were quiet ones for sweet peas. We had been served up a surfeit last year, and we needed time to digest. More importantly, this was the moment to look back over the last century. The working lives of those such as Richard Dean's and Henry's had begun in the eighteen thirties and 'forties and it was to Richard Dean that the *'Chronicle* entrusted the first page of the 5th January edition, with a look at the various plant developments. Henry had lived at the last time a gardener could hope to be reasonably proficient in horticulture in all its aspects. It's much too specialised now. It had been a golden age, *the* golden age for the professional, male gardener, an era already drawing to a close.

The century had had the decency, horticulturally speaking, to begin when Henry was just five years old. That at any rate was the view of the *'Chronicle's* editorial (which must have been Maxwell Masters'). 1828 marked 'the general introduction of hot-water heating... which rendered the protection of plants in winter easier, more certain, and less cumbrous...'. More important than 'some such event as the abolition of the window tax in 1845' or 'the subsequent removal of the imposts on timber and bricks'(2). (They must have been leaders at Penicuik because they were possibly *renewing* the hot-water system during Henry's tenure there in the 1840s).(3)

In a masterly overview of the nineteenth-century, the achievements of the Victorian era could be seen as considerable. Horticulturally it was famous for the R.H.S.'s Chiswick gardens, established in 1822, and of Kew, opened to the public in 1841, the year too of the first *Gardeners' Chronicle*.

Of the great plant collectors, it mentions Joseph Hooker, Robert Fortune and J.G. Veitch. Hybridisation, however, enriches 'our gardens at the present epoch to a greater extent than does the importation of new species... . The bedding-out system, introduced early in the century, and culminating soon after its middle in extravagant absurdity, has been very greatly modified for the better. Market-gardening has developed enormously, the quantities... of flowers grown for house decoration, would astonish our predecessors could they witness them. While agriculture has been depressed, gardening has flourished; but the great advance has been in the last quarter of a century'. (Henry's nursery started in 1888). Among botanists it highlights Herbert Spencer and Alfred Wallace, while 'if the question be raised who has had the greatest influence on the progress of knowledge throughout the world during the century, there can be but one answer – Charles Darwin'.[(4)]

Henry had lived through the reign of three monarchs and when the Queen died on 23rd January he would live into a fourth. For most people, however, Queen Victoria was the only sovereign they had known, for with her death came the end of a sixty-four year reign. It was a younger populace then than now and in its energy and optimism there was plenty of confidence in the new century. Understandable, too, in view of the recent advances in science and technology. It might even help those, like the quarter of the population still living in poverty, identified by Seebohm Rowntree in his monumentally thorough study of York. In another reflection on recent horticultural advances, the '*Chronicle* considered the demand for cut flowers, with which the sweet pea had long been associated. 'It is, however, only in quite recent years that the demand for flowers on almost every occasion, either of rejoicing or of mourning, has assumed extraordinary proportions; and gardeners and others interested have been obliged to keep themselves in touch with the ever-changing fashions... . Flower-exhibiting has also undergone many changes of fashion...'. The R.H.S. first offered prizes in open competition in about 1840, and nothing 'has done more to popularise the flowers for the time in fashion as this system of exhibiting'.[(5)]

One or two items on sweet peas began to creep back into the journals. *The Gardeners' Magazine* reported on a new 6d. brochure *All About Sweet Peas* by Robert Sydenham of Tenby Street, Birmingham.[(6)] Advertisements

'New and Beautiful Sweet Peas', Gardeners' Magazine, 9-3-1901

for new sweet pea books would soon be second only to those on roses. A full-page coloured drawing of 'new and beautiful sweet peas' appeared in their edition of March 9th.[7] Of nine varieties depicted, six were Henry's: 'Salopian', 'Countess of Lathom', 'Duchess of Westminster', 'Alice Eckford', 'Sadie Burpee' and 'Emily Eckford'. Only 'Alice Eckford', named for John's wife, was new, the others being mostly 'nineties introductions. Newness was obviously more loosely defined then.

The 1901 census,[8] that April, confirmed that Henry and Emily were still at 31 Market Street and had their daughter Isabella, now thirty-eight, living with them. The family had acquired, as befitted their status, a probably living-in domestic servant by the name of Mary Williams, a local, single woman of twenty-five. (There are many Welsh names along the border here). John, and Alice who would have been eight months pregnant with Agnes, resided with all their children still at 8 Market Street. Harry (one) and Lottie (six) hadn't yet, it appeared, gone to live with their grandparents and aunt at the corner house.

1901 continued to be a busy one for Henry Eckford, councillor, with lots to do. Connection to the Wem water-supply, seventeen years on, was still on-going. It's likely that the annual water-rates were deterring many from applying for what had always been available for free. This year of 1901 saw some of the public houses apply to be supplied, as they had applied previously to the parish council. Then, Henry had been almost a lone voice in their support. His sympathies came from the fact that both his mother and a brother had been in the trade, and that a pub may have been his birthplace. Since voicing his support, the Baptists must have reminded him of their temperance ideals, for he now opposed all the applicants![9] Fortunately, the other councillors had come round to supporting the publicans, for continual refusal would have endangered health town-wide.

In August it was agreed that a 'Mr. Bygott be allowed to use the town water in his field in Soulton Road at the annual charge of ten shillings and that he be allowed to make the necessary connection with the water main'[10]. That wouldn't possibly be councillor Edward Bygott by any chance? The minutes are rather coy! If Bygott's field was the one just beyond Henry's Soulton Road nursery, where the contemporary map shows a pump,[11] then any mains supply would have passed by Henry's nursery, to both men's benefit. A happy coincidence. The wheels of local government move slowly, however, and in November Mr. Bygott was again complaining of town sewage polluting his ordinary supply of water for cattle drinking in his field.[12]

Councillors were also hoping, in these days before telephone-exchanges, that letter deliveries might be increased from two to four times a day.(13) It's likely that they failed in this respect, although a third delivery could be collected from the Post Office.(14) As they had tried on the parish council, and with no more effect, councillors wanted the Railway Company, the L.N.W.R. , to provide a footbridge so that pedestrians could cross the line safely. They're still waiting! The week before Christmas, councillors drew 'the attention of tradesmen and others to the nuisance caused by their sweeping rubbish into the streets and leaving it there.'(15) Rubbish disposal, water provision, pedestrians' rights, letter deliveries – all issues that have come round again in our own time.

~~~

It was eleven months since the last sweet pea article had appeared in the gardening press, but there was now the first Show from the recently formed National Sweet Pea Society(16) to report on. Silas Cole's five seeds from his 'Countess Spencer' had been reduced to two by mice but these two had flowered successfully and out of interest he'd brought some down to London and the Show. In his *The Sweet Pea Book* Graham Rice writes: 'The day before the show opened rumours began to spread around the hall about a revolutionary new sweet pea. Silas Cole did little but tease the other exhibitors and it was not until a few minutes before the deadline the following morning that he staged his table decoration of his new variety – which stunned everyone.

His 'Countess Spencer' was deep rose pink in colour and a contemporary report stated that "it seemed as if nature has been so lavish that the material in its standard had to be closely pleated to hold it in position. It was beflowered and befrilled"'. (17) It wasn't known yet if it was fixed or not, and he didn't compete with it. Instead he won a prize with three other varieties, two of them new ones of Henry's; 'Miss Willmott' and 'Hon. Mrs. E. Kenyon'. Surprisingly, whoever sent in the detailed report to the press didn't think to mention Cole's fabulous new flower and no mention of it appeared in the '*Chronicle's* columns. Henry himself won a gold medal for a 'non-competitive' exhibit. A sign of the times was that many of the classes 'were designed to illustrate the decorative uses of the [sweet pea] flower for the dinner-table, for furnishing epergnes, baskets, vases, &c., and their suitability for making into wreaths and bouquets of various descriptions'.(18) This led to women, young women it seems, being
~~~

among the prize-winners in these 'floral decoration' classes. Another sign of changing times.

Silas Cole hadn't been alone in finding a large new 'beflowered and befrilled' sweet pea. That summer of 1901 saw sports of 'Prima Donna' appear in the grounds of a cut-flower grower, William J. Unwin, in Histon, Cambridgeshire and in the garden of an amateur grower, Mr. E. Viner, in Frome, Somerset. It probably occurred elsewhere too. 'I have been told by others', wrote Dobbie & Co's William Cuthbertson in 1912 'that they saw the waved form the same season, but could not perpetuate it'.[(19)] Over half a century later, in March 1969, William Unwin's son Charles was to recall his father's event in a letter to Bernard Jones. 'My memory of my father finding the mutation is very vivid... . At the time he was probably the largest grower of cut sweet pea flowers in the country, mainly for Covent Garden market. I was a small boy in Impington Church Choir and my father too was in the Choir. In those days evening service was held in the afternoon at our church and, when the Sweet Peas were in flower, it was father's habit to have a look round them after church, usually accompanied by one or more other members of the Choir. I went round with them and on the afternoon in question he had two other choir-men with him. In the rows of [Henry Eckford's variety] Prima Donna he suddenly stopped and said "Here's something different". I remember I could recognise it was a little larger flowered, but I did not notice the slight wave in the standards until he pointed it out. He traced the haulm down to the root to make sure all the blooms on the plant were the same and when he found they were, I was sent to the shed to get some bass to tie on it.

He recognised immediately that the mutation was quite new and a decided improvement and the rows of Prima Donna were quickly scanned to find if there was more than the one plant. He found two more and the next day a further two. The seeds from these 5 mutations were saved separately and the following year he was delighted when he found all 5 bred true to colour and to the new improved waved form.

Many interested people, mainly horticultural journalists, and Traders, came to see the new "break", which he named Gladys Unwin (after my eldest sister) and afterwards the whole stock was sold to the wholesale seedsmen, Watkin & Simpson, Ltd.'[(20)] 'Gladys Unwin' proved to be intermediate between the current 'grandifloras'[(21)] and 'Countess Spencer'. Once fixed, the colour unfortunately faded as cut flower, an increasingly common complaint of the new forms. Mr. Viner sold his stock to Henry Eckford, as he explained in a letter to William Cuthbertson in March, 1912.

14, Somerset Road, Frome.

Dear Sir,

....I grew Prima Donna from seed I saved myself, and quite late in the season I noticed a spray of two blooms (on Prima Donna) at the extremity of a shoot, with a peculiar crimpled character. I marked them and allowed them to seed... and I obtained seven good seeds... . The following year... five retained the wavy character... Later in the season [of 1901] I sent blooms to Mr. Eckford, to whom I eventually sold... . Of course I know Mr. Cole's variety was put on the market first, Mr. Eckford being obliged to grow another season, as the stock was small. Mr. Eckford was also informed that the variety on the market was something like his; he therefore dropped the idea of another name, and sent it out as his superior stock of Countess Spencer. I think had it not been for me, the finer form of this charming flower would never have been known.

I am, most respectfully yours,

E. Viner.[22]

Unlike Cole's 'Countess Spencer', Viner's seed was almost wholly fixed – over ninety per-cent of it came true.[23] Growers and others looking to develop new Spencer varieties would, you would think, have gone for the fixed forms in order to increase their chances of developing new fixed kinds. Although Cole's 'Countess Spencer' threw up exciting new forms, these too were unfixed. With 'Gladys Unwin' a fixed but intermediate form, it seems likely that Viner's form, perfected by Henry at Wem, was of much greater importance in the development of the modern sweet pea than has so far been recognised.

'Prior to the advent of "Countess Spencer" Sweet Pea', recalled Cuthberton in 1909, 'it was universally recognised that the stocks of new Sweet Peas put on the market by Mr. H. Eckford and others were fixed absolutely. How was this achieved? Many will be prepared to say at once it was on account of the new Eckford type being much easier to fix than the new Waved or Spencer type. This is only part of the truth. Mr. H. Eckford took infinite pains and grew his novelties many years before offering them. The same care would have purified and fixed many of the recent Spencer introductions. Mr. J. S. Eckford once informed me that "Dorothy

Tennant" required nine years' work before he and his father considered it fit to offer'[(24)].

With Silas Cole's blooms unfixed and Henry needing to bulk up his "Spencer" strain the era continued to be that of the grandiflora, principally Henry's. This was slightly over-emphasised in the summer of 1901 when the *'Chronicle* was illustrating 'Sweet Peas of the century': 'Miss Willmott', 'Mrs. Joseph Chamberlain', 'Black Knight', 'Sadie Burpee', 'Coccinea', 'Duchess of Westminster', 'Hon. Mrs E. Kenyon' and 'Lady Grisel Hamilton'.[(25)] These are all of Henry's raising, for the *'Chronicle* had asked him to select several typical forms 'as illustrations of his unrivalled skill and judgment, and as examples of what the Sweet Pea is like in the early years of the twentieth century'. Being technologically adrift of the newer journals, the *'Chronicle* illustration was in black and white and the colours had to be described and imagined.

While praising Henry, the *'Chronicle's* leader writer had an oblique critique to make of him and the other hybridists: 'Whilst we all applaud the efforts to "improve" the Sweet Pea in certain directions, such as general vigour, variety of colour, fragrance, substance of petal, number of flowers on the stalk, and the like, it is open to grave doubt whether the attempts to regularise the flower according to the florists' standard are in good taste. The flower of the Sweet Pea is naturally highly irregular in form... . Every wave and curve of the petals, whilst it contributes essentially to the beauty of the flower, has a purpose of its own, or is the result of adaptation to circumstances or hereditary endowment. If we want to improve such a flower we should follow Nature's own indications...'. Referring less perceptively to Henry's supposed bi-generic hybrid the writer stated that 'crosses between the Sweet Pea and culinary Peas have been effected, but they have not as yet established themselves in the regard of cultivators'.[(26)] So some growers, at least, had their feet on the ground!

In response to this editorial criticism, Richard Dean wrote in to say that there was no defining exactly how a sweet pea should look but that the new Society favoured certain standards.[(27)] Quite why they had these ideals, Dean neglected to explain.

An attractive display of Henry's new rose-coloured variety 'Miss Willmott' appeared in *The Gardeners' Magazine* for 10th August. When it came to naming his sweet peas, Ellen Willmott (1858-1934) was one of the few people, other than employers and family, that John and Henry would have known (Lady Hamilton was another). She was an heiress who devoted herself body and soul to horticulture. She had been on the sweet

FIG. 30.—SWEET PEAS OF THE CENTURY. (SEE TEXT, P. 94.)

Sweet Peas of the Century, Gardeners' Chronicle, 3-8-1901.

pea bi-centenary committee but her interests were wider-ranging. She is deservedly commemorated, as the garden historian Jane Brown has written, 'in all the *willmottiae* and *warleyensis* hybrids' and species 'that still fill our gardens'.[28] A commemoration that includes 'Miss Willmott's Ghost' for *Eryngium giganteum*, a spiny 1½ metre tall relative of the Sea Holly, whose seeds Miss Willmott would drop surreptitiously in people's gardens! Seed of the white 'Miss Willmott' *Verbascum* were to be offered for sale in the 1932 Eckford catalogue.

How she also came to be named for one of Henry's sweet peas was recalled years later by John Eckford's son Harry: 'Miss Willmott was a wonderful person. She did a lot for horticulture. My father had staged an exhibit at a show. Miss Willmott, who was looking at the exhibition, noticed that one of the sweet peas was not named. (This happened to be a new variety yet unnamed). She asked my father the name. He picked up a card and wrote "Miss Willmott". She was delighted'.[29]

1901 was both the year of the Spencer sweet pea, barely yet reported on, and the rediscovery of Gregor Mendel's paper on hybridisation, which most definitely *did* get noticed, this time round. The new 'Countess Spencer' sweet pea got a mention in *The Garden* magazine in August. It was one of two new varieties that had won first-class certificates from the new National Sweet Pea Society. 'It was', said the report, 'quite a distinct break, the wings are unusually large, and the general colour of the flower is a rich pink'.[30] The other award went to a grandiflora of Henry's, 'Jeannie Gordon', which had a bright rose standard with cream wings.

If the new National Sweet Pea Society hadn't been so keen to establish its credentials it would never have given an award to the 'fickle Countess' that Cole's 'Countess Spencer' proved itself to be. If the flower had stunned everyone at the London Show, well, this wasn't the mood of the journals. The new flower and its grower had to prove themselves, yet mindful of the fact that sports retaining the new waved form would also produce new shades of colours. It is easy to imagine that turning-points in history are fully realised at their inception but distance clouds the reality. Excitement there may have been, but the old grandiflora types were to hold sway for some years yet.

Attention and excitement did attend Mendel's great work, thirty-six years after its completion in 1865. *The R.H.S. Journal* won the accolade of being the first British journal to translate and publish it. [31] In his introduction, the eminent scientist William Bateson wondered, as we had in an earlier chapter, how it came to be ignored for so long. 'It is true that the journal in which it appeared is scarce, but this circumstance has seldom delayed general recognition. The cause is unquestionably to be found in the neglect of the experimental study of the problem of Species which supervened on the general acceptance of the Darwinian doctrines… . The question, it was imagined, had been answered and the debate ended. No one felt any interest in the matter…'.[32]

Maxwell Masters was quick to comprehend its significance too, précising the paper for his '*Chronicle* readership. 'Like other important discoveries', he wrote, 'it appeared before the world was ripe for it, and that it was not followed up sooner by further trials on like lines, may, we think, be accounted for by the great rarity of men who, however well qualified for, or inclined to, such a task, have the leisure and opportunity for concentrated culture and observation…'.[33]

For Henry, one of a handful of one-eyed men in the land of the blind, it was all a bit late. Not that everything had suddenly been made clear, of course. On an occasion when two pure-breeding white sweet peas were crossed together, they were found to have produced all-purple flowers.[34] This was due to a genetic interaction called epistasis, yet to be discovered. The name genetics, too, had yet to be coined. But the breakthrough had arrived.

A report on the bicentenary conference was also published this year. Richard Dean had got one out early, just as a record, and it was now republished commercially as *All About Sweet Peas* and advertised in *The Journal of Horticulture* at 1/-.[35] It gives a fuller account than the necessarily edited press version. From Charles Curtis' and John Eckford's paper, we learn that to Henry 'belongs the credit of large substantial blooms with big rounded standards in which the top notch has been filled up; the wings too, have, under his continued care, developed amazingly. A standard two inches across may yet remain to be found, but 1 $^{3}/_{5}$ inches in breadth can be secured under good cultivation in the newer Eckford forms… .

Elongation of the flower stems, with a corresponding strengthening and erection, are due largely to Mr. Eckford, some of whose varieties will, under favourable conditions… produce spikes fifteen inches long' with four blooms. (Unusually, the spikes of Henry's 'Mrs Fitzgerald' variety could be longer, with seven flowers and buds). 'With the increase of size and vigour there has arisen the hooded varieties' like 'Lady Grisel Hamilton', as a sort of protest by the flower against its development, which 'finds favour with the ladies… . In addition to all these improvements Mr. Eckford had given us a wonderful range of colour…'.[36] There is no doubt here that all the changes, including all those to the flower shape, were improvements. Others were not so sure.

The co-author of this article, Charles Curtis, went on to become editor of *The Gardeners' Magazine* and subsequently the long-standing editor of *The Gardeners' Chronicle*. He was remembered, years later, by John's son Harry, brought up in Henry's household. 'You mention', he wrote in a letter to Bernard Jones, 'that Mr Curtis' letters always had the aroma of Cigars. Certainly an unsolicited advert for Cigar makers. We always kept our business writing paper in a large airtight tin containing several Perfume Sachets. The customers when they opened our letters were sure they could detect the Sweet Peas'. (37)

Henry had continued the practice, from his Estate days, of keeping bees. Formerly, this would have been necessary to the formation of fruit. Here, the Eckford and Curtis paper informs us, 'the hives, placed almost in the centre of his home nursery', the one at Wem, proved an enjoyable side-line. Henry was safe too, in the knowledge that bees couldn't cross-pollinate his peas.

Despite continued exhortations to sow in the autumn, amateur gardeners in particular continued to sow sweet peas between mid-winter and early spring. No seed firms advertised sweet peas in the autumn. Last year, because of the July bicentenary, some of them had advertised in the summer. Henry didn't. His advertisements in *The Gardeners' Magazine* appeared in most of the weekly editions between mid January and late April; in the bi-centenary year they ran on into May. Commercial growers provided customers with what they wanted and tried to anticipate trends. To change customers' habits was someone else's job. Another form of advertising for Henry was the summer horticultural Show. The biggest of the provincial ones, just down the road from Wem at Shrewsbury, saw Henry as one of a number of 'small gold-medal' winners 'for a very fine collection of Sweet Peas, including new varieties of promise'. (38) It marked the end of another successful season for father and son. There were hints of a change, changes that would see the old grandiflora sweet peas eclipsed for most of the new century by the Spencer types. But that wouldn't have bothered Henry, who even at seventy-eight, had he stumbled upon a finishing-line would have just jogged back to the start.

~~~

Although, 'rural life remained quite violent' as Colin Matthew has written, particularly rural life in Scotland, Wales and Ireland, urban crime in
~~~

Britain had fallen remarkably between the 1850s and 1890.(39) It shifted the spotlight to such anti-social behaviour as remained, perhaps even increased as a proportion of overall crime. At all events Henry, wearing his councillor's hat in 1902, found himself struggling with these and other problems strangely familiar in our own day. 15th July: 'The question of adopting a more expeditious method in dealing with the removal of house refuse', recorded the council's minutes secretary, 'was discussed & finally left over for a time'.(40) 17th December: 'Resolved that the attention of the Police Sergeant be called to the obstruction of the footpaths in the District caused by men and youths loitering on the footpaths & by tradesmen placing goods on or near the footpaths + washing block in the streets. That he be asked to use his discretion in compelling the abandonment of the obstructions'.(41)

Wearing his other more familiar hat, Henry was quick to announce his new for 1902 sweet pea varieties in the first new year's edition of *The Gardeners' Chronicle*.(42) They were more lyrically described in March by Edwin Molyneux, who must have had them to try the previous year. 'Jeannie Gordon is free-flowering and vigorous in growth; the standards and wings are bright rose, shaded with cream. Gracie Greenwood is almost a self colour; soft cream, delicately shaded with pink, with a deeper edging. Lord Roseberry belongs to the self-coloured section, and is rosy magenta. In habit of growth and flower formation it is quite of the best'.(43)

On 4th February, a day after his seventy-second birthday, Richard Dean attended a dinner in London's Westminster where he was presented with an 'illuminated address' on velum and a cheque for £300, a tidy sum then, in recognition of 'the immense work Mr. Dean had done for horticulture during so many years'.(44) Gardening was Richard Dean's life. Along with sweet peas, his sweep was as wide as it was possible for a man of organisational flair and a practical horticultural background to have. In the 1850s, Dean was at Charles Turner's Royal Nursery at Slough. The nursery supplied Henry then, when he was at Coleshill, and it's likely that the two first made contact then, well before he began promoting Henry's first successes with Verbenas. Henry was probably at this London dinner. He liked to hide away among his fellow enthusiasts, and there would have been plenty to reminisce about.

'Year by year sweet peas grow in popularity' wrote Edwin Molyneux in *The Gardeners' Magazine*. 'The difficulty appears to be to confine oneself to a certain number of varieties, so numerous are the kinds now offered by specialists. For instance, I note that Mr. Eckford catalogues no fewer than

115 varieties, a number that renders it most difficult for a beginner to avoid having sorts that are too much like others to be necessary'. Molyneux went on to pick 'a selection of three dozen varieties for the benefit of beginners'[45]. Four were from the U.S., the remaining thirty-two were all Henry's. It seemed that those in Britain who might have competed with Henry were either unable or unwilling to. It puts one in mind of the plant hormone in the tips of leading shoots that suppresses the growth of side-shoots.

When the sweet pea was looked at less holistically, with the focus purely on flower colour, then others could creep into the reckoning. The opinions of twenty-seven growers, listing the best ones under various headings, were sought by the National Sweet Pea Society, which published its findings in the *'Chronicle* in September. Here again Henry's varieties topped the list of ninety with sixty-six. The growers were unanimous only with Henry's 'Blanche Burpee' as best white and his 'Salopian' as best crimson (just edging out his 'Mars'). The whites, as usual, were the most popular. Henry had just brought out a new one, 'Dorothy Eckford', which was to rise above the others. (Named for John's daughter, it is still available). The Society commented, astutely, that 'several varieties, like Dorothy Eckford and Countess Spencer, would have received many more marks had they been more widely known'.

A Bournemouth gardener had previously written to *The Gardeners' Magazine* to say that, horses for courses, 'Salopian', though a brighter colour initially, faded in the sun there whereas 'Mars', grown on 'hot dry soil, almost on the edge of cliffs... is all that can be desired'.[46] In this spirit, the *'Chronicle's* editorial comment, which must have been Maxwell Masters', pointed out the list's obvious flaw. 'It will be seen... that one character only, that furnished by colour, is taken into consideration. That may be permissible in a first attempt, but it is to be hoped that the Society will justify its existence by availing itself of the opportunity it has of advancing our knowledge of the phenomena of variation by correlating the diversities in colour with those exhibited in the form of the several parts of the flower, the habit, degree of hardihood, the foliage, the powers of adaption to varying conditions and circumstances, and so forth. If these details be objected to as too scientific, a moment's consideration will show that they are pre-eminently practical. If we want to promote the advance of horticulture, even in the comparatively trivial matter of Sweet Pea growing, it must be by the adoption of scientific methods'.[47]

Henry, incidentally, nearly lost his 'Dorothy Eckford', or rather the name. His nursery was said to have accidently mixed some of its seed with those of 'Miss Willmott', and the commercial recipients, thinking they'd discovered a new white, gave it new names. For a while, 'Dorothy Eckford' was also known as 'White Queen', 'White Wings', 'Lily' and 'Purity'. [(48)] Interestingly, Charles Darwin had long before noted that 'flowers which are normally white rarely vary into another colour', especially among sweet peas. [(49)] If the white 'Dorothy Eckford' had originated as a sport from 'Miss Willmott', then rather than the seed being mixed, some of the 'Miss Willmott' seed may have reverted to the dominant white. And this, it was later discovered, is what had occurred. Confusingly, some 'Dorothy Eckford' seed was also reported as producing a poor-looking strain of 'Miss Willmott', indicating that neither had been thoroughly fixed by selection. It drew some discussion in the *'Chronicle* [(50)] and remained for a while a mystery. Neither John nor Henry wrote in to explain things, being perhaps as uncertain as everyone else.

The upcoming Coronation of the new King, the future Edward VII, began to be talked about in the press. There was interest too from a horticultural perspective. 'Speaking generally' said a Norfolk nurseryman, 'the demand for flowers will be enormous, and in particular, Lilies of the Valley and Roses. Although the Rose will be "the" coronation flower, Lilies of the Valley will be in great demand, judging by the numerous extra orders for retarded [flowering held back] crowns I have recently executed'[(51)]. There were similar views from Covent Garden Market, where it was thought that roses, lily-of-the-valley and carnations would predominate[(52)]. Sweet peas were to become the Edwardian flower *par excellence* but this wasn't apparent as the era began. Scent was looked for in flowers but the sweet pea wasn't yet foremost in the public mind.

This wasn't through lack of effort by the Eckfords, who with other nurserymen continued to exhibit in the provincial Shows. Sweet peas from Eckfords' nursery were on display at the big Scottish Show at Edinburgh's Waverley Market in September. [(53)] Emily took the strain from her now seventy-nine year old husband's shoulders by giving the prizes 'for the best exhibits of eighteen varieties of Sweet Peas' at the nearer to home Shropshire Show at Shrewsbury. [(54)] One interested observer of sweet peas at Shrewsbury was Owen Thomas, one of the era's most eminent head gardeners. This was an achievement he'd somehow managed without being Scottish, hailing instead from Anglesey / Ynys Môn.

'There was', he wrote to the *'Chronicle*, 'considering the lateness of the season, an excellent and varied show of these beautiful flowers, and the collection exhibited by Mr. Robert Bolton, of Carnforth, was specially good. To those who do not wish to cultivate a great number of varieties, the task of selecting varieties is often a very perplexing matter, more especially to the amateur, and for his benefit I have pleasure in giving him a list of the finest varieties distinct in point of colour shown at Shrewsbury'[(55)]. Like Molyneux's list earlier in the season, Thomas's choice of thirty-seven varieties were almost wholly Henry's, with thirty-three. He did, however, consider that there were as yet no true blues or scarlets.

The efforts of the exhibitors to promote the sweet pea was beginning to have its effect. 'Horticultural societies and seedsmen', wrote a *'Chronicle* correspondent, 'by offering liberal prizes, have assisted greatly in popularising this pretty annual plant, the exhibition of which has greatly increased at meetings of horticultural societies...' .[(56)] The new National Sweet Pea Society however, by giving Henry's 'Dorothy Eckford' a 'First Class Certificate' at their December meeting, may have added confusion[(57)]. Such awards aped those of the Royal Horticultural Society, and you feel that Alex Dean may just have had a point. His brother Richard, seventy-two now and with sadly not long to live, resigned as the Society's Exhibition Secretary.

Not so long before, it had been rare to see sweet peas mentioned much before July and the summer shows. Now, interest in the flower was becoming year-round. 1903 had barely begun before Robert Sydenham's *All About Sweet Peas* pamphlet gained a favourable review in the *'Chronicle's* columns. Presumably updated from 1901, Sydenham was indulging in the now fashionable practise of naming the best available varieties, gained here from the opinions of thirty-three growers. Of thirty-one varieties, twenty-four were Henry's, with seven in the top-ten: 'Countess of Powis', 'Fascination', 'Queen Victoria', 'Stanley', 'Duchess of Westminster', 'George Gordon' and 'Royal Rose'. Based purely on flower colour, Henry's crimson 'Salopian' and lavender 'Lady Grisel Hamilton' were easily the most popular. [(58)]

Silas Cole's 'Countess Spencer' wasn't being ignored but the 1902 crop had been bad. It was probably Robert Sydenham who was the 'midland seed merchant' reassuring *'Chronicle* readers in February that 'a fair quantity of seeds... have long since been sent to [Cole's] seed-raiser on the continent', (implying Europe but surely America: California) and 'hopes to have a sufficient quantity to meet all reasonable demands in 1904'. [(59)]

People knew that something was astir but there was nothing to be seen. 'Countess Spencer' 'should be made a note of to obtain when offered in all probability next year' wrote a hopeful grower to *The Garden.*[60] A year later, and still hoping for the best, Robert Sydenham was assuring *The Garden's* readers in March 1904 that he should 'be able to distribute it amongst your readers or horticulturists generally'[61] that year.

An Eckford advertisement in the new *Garden Life* magazine informed readers that they would get booklets on 'How to grow and when to sow culinary peas' and 'How to sow and show Sweet Peas' free with their orders.[62] Its rather poetic!

~~~

Henry was coming up to eighty years of age. A conscientious councillor, he was still attending council committee meetings. John and Emily must have finally convinced him not to stand for re-election, for the *Shrewsbury Chronicle* of March 27th reported that, though nominated, he had decided to stand down from the forthcoming Wem Urban District Council elections. John would stand in his place.

At the last meeting of the old council, with Henry in attendance, there was 'discussed the question of the need for Higher Education in the district. Proposed... that Council provide no funds for Higher Education as it is the opinion of the Council that sufficient provision exists in Wem'.[63] It is probable that none of the councillors present had received any formal higher education beyond that of apprenticeships. With the example of Shrewsbury born Charles Darwin and others before them, they could argue that all that highly developed intellects had done was to frighten the horses. And what happens when you frighten the horses? You upset the apple-cart. They themselves had succeeded well enough without it, if you please. Why spend ratepayers money on something so intangible and of no discernable benefit to Wem? The proposal of no funding was carried unanimously.

In fairness to them, they could point to the existence of a grammar school. Also, the small amount of educational funding, by central government and other agencies, was 'a minefield of complexities'.[64] If local government can get out of providing and paying for something, it usually will. The ethos of the 'petty-minded, cost-cutting tradesmen, with a clear self-interest in holding down the cost of government' identified from earlier in the nineteenth century[65] may have hung on in places like Wem;
~~~

towns little more than villages where pride in providing a civic culture was difficult to develop.

In contrast to the first elections three years before, the *Shrewsbury Chronicle* could report that 'excitement over this election is running very high' with seventeen candidates for the eleven seats. (66) Polling took place on Monday 16th April at the town's 'National Schoolroom'.

'Feeling ran very high throughout the day', reported the *Shrewsbury Chronicle,* 'and the poll was very heavy, some 400 persons voting out of about 450 on the list. During the counting of the votes a large number assembled outside the schoolroom'. The result, declared 'amid much cheering and excitement' saw the eleven home led by Joseph Fowles, coal merchant on 264 votes, followed by John Eckford on 245 and Henry Kynaston on 243. It's instructive to see the preferences still for those big in trade, with those in the professions way down the list. 'Mr. Fowles was afterwards carried home by his admirers'(67) wrote the paper's reporter, no doubt glad for some fun in contrast to the dull event he'd had to report on three years before. Henry, Emily and Alice must have been pleased with John's result, nearly topping the poll, and Henry could take comfort in this and the fact that he had, himself 'done his bit' for six years. So, just the family, the nursery and chapel from now on!

The new council continued along the lines of the old in at least one respect. 20th May: 'Proposed... that no Education Committee be appointed for this district. Carried unanimously'. (68) Councillors proclaimed no party affiliation. The lack of any coherent political beliefs led them from reaction to radicalism in the space of a few months. 'Proposed... that this Council support the proposal of the Glasgow Corporation in favour of the taxation of land values. Carried unanimously'. (69) Society was becoming a little less deferential to the lords and masters, the landed gentry. In Wem at least, in England, thanks to Glasgow!

Returning to flowers, Henry's sweet pea varieties took most of the honours in the National Sweet Pea Society's annual summer Show. A number of women were among the prize-winners here, reflecting their gradual appearance as a force in horticulture, coupled with a particular love of the sweet pea. Henry's bright new 'Scarlet Gem' and his deeper red 'Edward VII' were both given the Society's First Class Certificate. 'Scarlet Gem' also won a silver medal 'for the best novelty [new variety] of the year'. A silver-gilt medal was also won by a display from his nursery. (70)

Over in the amateur-oriented *Gardeners' Magazine*, they were showing you how to cross-pollinate sweet peas, complete with diagrams. *The*

Gardeners' Chronicle of the professionals rarely if ever described practical techniques. Neither did the gardening books, while they remained aimed at the professional gardener. Skills were learnt in apprenticeships, or on the job. 'There is, after all', began the article, 'nothing so very mysterious in the methods by which cross-fertilisation can be carried out among sweet peas. Patience, a light touch, and a pair of sharp, finely-pointed scissors are the chief requirements of the would-be raiser of new varieties...'. (71) Women gardeners, trade secrets to the populace – movement all the time. When would it all stop? Where would it all end, some of the older gardeners must have wondered. So many changes in one lifetime! North America was on the move too, with one Californian seed firm now harvesting one-hundred tons of sweet pea seed from their three-hundred acres. The U.S. crop was now five-hundred tons annually. (72)

One unaltered British tradition was a liking for the garden pea. Henry won a rare mention, a qualified endorsement of an early variety. 'One if not the largest, of the broad-podded Peas is Eckford's Ideal, a Pea that reaches a height of 5 feet, which is a good second early variety. I have grown it for the last four years, and it has always been good, or if it has a fault it is in being tall for a second early Pea...'. (73) Another British tradition was the summer Shows, if for the one at Bishop's Stortford in Hertfordshire, at The Grange on 12th August, it was of a mere thirty-four years. 'Cut flowers occupied a large space', reported the *'Chronicle*, 'and were exceptionally fine in most of the classes. For Sweet Peas there were three classes, and those of Messrs. Eckford were very fine. The best collection of Sweet Peas for the Eckford prizes made a grand display, Mr. Hare, Miss Newman, and Mr. Beech being the fortunate winners...'.(74) Miss Newman was up among professional gardeners, although, it being cut-flowers, she may have been one of the new breed of florists, a trade almost wholly female.

Shrewsbury remained the premier Show, attracting all the top nurseries and entries from across the British Isles. Sweet peas were the principal cut-flowers. 'One tent contained Sweet Peas exclusively', enthused the *'Chronicle's* correspondent, 'and they were staged in other tents as well...'. The Eckfords' nursery made their traditional appearance, 'a glorious group of Sweet Peas, flowers of their latest novelties, including the remarkable Scarlet Gem and others'. (75) Joseph Chamberlain, from Birmingham, was a regular orchid exhibitor. Expensive and flamboyant, orchids reflected the now aged statesman's parvenu persona. Henry, we know, had already named an eye-catching scarlet and white striped sweet pea 'Mrs. Joseph Chamberlain' (which is still available).

In *Garden Life*, a new and upmarket amateur gardening magazine, is an account of a visit to Henry at his Wem nursery. It is the only article in which we get a good sense of Henry the man. He's still full of beans, although his memory is, unsurprisingly, unreliable in dating past events. It seems that the Eckfords haven't given up on Essex after all. Henry's mention of an earlier, forgotten sweet pea named after himself was probably not developed commercially. Note, incidentally, the style of this Edwardian piece. It has lost the wordiness of Victorian writing, while remaining attractively expansive and relaxed.

FAMOUS GARDENERS AT HOME
MR. HENRY ECKFORD, F.R.H.S.,
AT WEM, SHROPSHIRE.

At eighty years of age, Mr. Henry Eckford is as full of enthusiasm as ever for his favourite flower, which, as all the horticultural world knows, is the Sweet Pea. Wandering with him about his trial grounds at Wem, one afternoon in July, I speedily became aware of this fact. "Now, isn't that a beautiful thing?" "Have you ever seen anything more lovely than this?" "Just look here at that – isn't it exquisite?" It was with such remarks as these that Mr. Eckford called a halt at almost every other step, fearing lest I should miss anything that ought to be noticed. Our enthusiasm – for, as an old lover and grower of Sweet Peas, I shared my companion's feelings – was aroused to the full when we reached the couple of acres devoted to the seedlings. They are at present unnamed. It is from these that the new varieties of the next few years are being selected. I saw, as I need hardly remark, many distinct and beautiful novelties. The one that Mr. Eckford thinks the best of his seedlings is of a whole shrimp colour. The delicacy of this colour, combined with the perfect form of flower, mark it out as one of the great successes of the future. Another lovely new thing is a very soft pale lavender, a "self" colour, and a grand size; another is a new indigo, about as big again as any existing type. There are also a new mauve, a new maroon, a new crimson-lake, and a new blue, the last-named having much more substance than Emily Eckford.

In this part of the trial grounds there was a row of Scarlet Gem, the brilliant variety which had taken the highest possible honours at the National Sweet Pea Society's Show only a few days before. King Edward VII., one of the four novelties of 1903 (the other three were Dorothy Eckford, Agnes Johnston, and Mrs. Walter Wright), was also in evidence here.

Mr. Eckford is sometimes called the Grand Old Man of Sweet Peas, and the name is not unappropriate. Everybody in the horticultural world acknowledges how great a debt is due to him for his unrivalled work in connection with this most popular flower. During my stay at Wem I was able to see a good deal of Mr. Eckford, both at his nurseries and in his own home, and I was much struck by the modest estimate he puts on his own labours. Perhaps it is this pleasing trait that makes him such a general favourite. Several times he told me how he had always shunned publicity, and how his chief desire had been to hide in some quiet, out-of-the-way place, where he could devote himself to his Sweet Peas without let or hindrance. Mr. Eckford is a fine figure of a man, despite his great age. He stands well over six feet, and is broad in proportion. Whether the cultivation of Sweet Peas in conducive to cheerfulness, I do not know, but the impression Mr. Eckford left on me was that I had seldom met such a happy old man before. He was full of anecdote and jest, and now and again gave utterance to sentiments which showed him to be full of a serene philosophy. Altogether, my visit was a very pleasant experience.

As his right-hand man Mr. Eckford is fortunate in having a son whose love of the Sweet Pea is as great as that of his father. Mr. J. S. Eckford has now been associated with his father for some years. Under their joint management the business is rapidly extending... .[Since Bronze Prince] there has come a succession of good things from Mr. Eckford's hands, and horticulturists are eager to get his new catalogue every year to see what Sweet Pea novelties he has to offer them.

"I need hardly ask you, Mr. Eckford," I said, "if your love of gardening began at a very early age?"

"It began in my cradle," he replied, laughingly. "My grandfather always used to say to my parents, 'If you make that boy anything but a gardener you'll spoil him.'"

"... What made you select Wem?"

"I heard of this ground through the postman at my old home. I had told him that I wanted a place quite out of the way, where people would leave me alone, and where I could grow my Peas. He said, 'I think there's some land to be let at Wem, which would just suit you.' So I came to look at it, and I took it. What a funny thing it is how we blunder about in this world," said my companion, looking at me with a smile. "It would be impossible for a man to chalk out his course, and say 'I'll keep to it'. Yes; I came to Wem through that postman."

"How much land have you here?"

"Nine acres. Here we merely raise the stocks. The seed is grown for us from these stocks in various parts of the country which we consider particularly suited for the purpose. A great deal of it is grown in Essex."

"Sweet Peas must always have had a great attraction for you?"

"I always liked them – I don't know who wouldn't. I began with the four or five kinds that were then obtainable, but I don't know that I had very high hopes of doing much with them. Just then I was on the Floral Committee of the Royal Horticultural Society, and I was very busy raising plants of all kinds. I never said much about my Sweet Peas to anybody. I used to take them up the Royal Horticultural Society now and then and get a certificate for them, but the news soon got abroad. Some of my friends, however, were more interested in my culinary Peas, and that reminds me of a little story."

"What is it?"

"One morning, before I was dressed, I heard a tremendous knocking at my door. Looking out of the window, I saw old Mr. Cypher, the well-known nurseryman at Cheltenham. 'You here, Mr. Cypher', I said; 'what's the matter?' He said, 'I heard so much about those Peas of yours that I could not rest any longer without coming to see them.' 'Which Peas?' I asked. 'Oh, your culinary Peas, of course.' Poor old Mr. Cypher has been dead many years now. He was a great florist. It is wonderful what may be done with Sweet Peas, or with anything else, if you go on working at a thing."

"From what you have just said, I take it that culinary Peas are also an important part of your business?"

"We do more with culinary Peas than with Sweet Peas. We want a fine lot of culinary Peas to supply all our orders, I can assure you. We have never grown too many yet."

"While we are on the subject, you might give me the names of some of the best culinary Peas of your own raising?"

"Prior, a five-foot Pea; Royalty, a four-foot Pea; and Dwarf Monarch, two and a-half feet, are some of the most prolific of the late sorts. Heroine has a long, rather light-coloured pod, but the Peas are very sweet. Some of the lightest coloured Peas are the sweetest flavoured, and yet they cook sufficiently green to have a good appearance. Shropshire Hero is another heavily-podded dwarf, two and a-half feet. The Echo is a dwarf that ripens very quickly. Dawn, a three-foot Pea, of William the First type, is very prolific."

"Now will you give me your own selection of the best six Sweet Peas?"

"I never like to say that the ones I think the best are the best, because, after all, it is a matter for individual taste. But my ideas on the subject are that Emily Eckford is the best blue, Scarlet Gem the best red, Dorothy Eckford the best white, and Lovely the best pink, Hon. Mrs. E. Kenyon the best cream, and Black Knight the best dark variety."

"Do you find that your customers run after certain kinds?"

"Yes; and the public must be considered to be the best judges. That is why I feel it is a presumption on my part to tell them which are the best. Some people, in my opinion, show the most curious tastes in the varieties they order, but why shouldn't they have the satisfaction of following their own bent in the matter?"

"There is a tendency in some quarters to depreciate the hooded form of the flower?"

"I think we ought to be very careful what is done in that matter. If a flower has beauty and novelty, I don't think it ought to be discarded because of a slight inclination to hood. It may add to the beauty of the flower."

"Are Sweet Peas, in your opinion, often crossed naturally?"

"Not to any great extent. Bees do not seem to work on them much. There is a small black insect that runs in and out among them. I don't know whether that has anything to do with the cross-fertilisation of them."

"It must be a big work to go through these two acres of seedlings, and label them. When is it done?"

"We label them just before they go out of flower."

As we went through the grounds, Mr. Eckford drew my attention to the leading characteristics of some of the most striking varieties. "King Edward VII.," he remarked, "is a very rich colour, and the size of it is so grand. Colonist is a prettily-hooded flower, but it does not show itself so well as a broader-winged one. Triumph has a fully-expanded standard. We really cannot do without the Countess of Powis, although we have Gorgeous. Queen Victoria is more of a cream than a white. I think the colour in Lady Margaret Ormesby-Gore is just delicious. Lord Kenyon is a very rich pink. The Hon. Mrs. E. Kenyon is a fine flower, of a good cream colour. Royal Rose has a very pretty streak down the middle, which lights up the whole flower. Gorgeous is small, but it looks very bright when you get a bunch of it. Lovely is a good size, and has plenty of substance, with strong stems. Alice Eckford is one of the smaller-flowered varieties, but it is exquisitely pretty in a collection. Duchess of York is one of the finest of the stripes. It grows to an enormous size, is well expanded, and looks very rich bunched. Agnes Johnston is a lovely thing, with a well-opened standard."

"What are some of the greatest favourites with the public?"

"Apple Blossom and Lady Grisel Hamilton are exceedingly popular. America, one of the best of the striped kinds, is also much liked. The Hon. Frank Bouverie, which looks very well bunched; Mars, an old favourite, and a most vigorous grower;

Duke of York, a very fine one; Duchess of Westminster, which, though not one of the largest, is one of the sweetest; and Royal Rose, may all be included in that category. Lady Nina Balfour, which is inclined to grey, is a very popular variety. I have seen it with stems eighteen inches long. Be generous to it, and it is astonishing what it will do."

"The new white, Dorothy Eckford, is a great improvement, I suppose, on all the older whites?"

"It is half as big again as any of the others, and it is a pure white, without any tendency to cream or pink."

"Does it take long to fix a new type of Sweet Pea?"

"Never less than six years. Fascination, which we sent out some years ago, had been grown by us for about ten years before we could fix it."

"What new varieties shall you distribute next season?"

"Out chief one will be Scarlet Gem, which was awarded the medal at the National Sweet Pea Society's Show last week. I expect we shall also distribute Mrs. Knight Smith, a new pink. It has just sufficient curl to make it interesting. The colour is deeper on the back of the standard than in front; the effect of this is to light it up. Whether we shall bring out any others depends very much on the crop we get."

"How many varieties do you grow here for stock seed?"

"Considerably over a hundred."

"What is your system of growing?"

"We never sow any in the autumn. One reason is that we cannot get the ground ready in time. Our chief month for sowing is February. I don't recommend the amateur to sow in the autumn, as it means having to fight slugs and birds, and one thing and another all the winter, and very often it proves a failure after the greatest effort to make it a success. Out land is, as you see, full of all sorts of rubbish, but we leave some of the weeds at the bottom of the rows to protect the Sweet Peas from the sun. This year we have suffered very much from the drought, and the haulms are much shorter than usual. Our Sweet Peas are sown fairly thickly, but an amateur should leave at least six inches between the seeds."

"As you leave all the blossoms on the plants your Sweet Peas must be soon over?"

"They last about six weeks, from the middle of June to the end of July."

"You grown all your Sweet Peas with sticks?"

"Yes; they are generally from six to eight feet high. We keep the haulm in position by winding one or two rows of yarn round the sticks."

"What about manure?"

"There is nothing better than cow manure. We give the land a good dressing of farmyard manure, with a bit of superphosphate."

"When do you carry out your cross-fertilising operations?"

"Directly the Sweet Peas come into flower. It is best to choose a bright day, about noon."

"I have noticed, Mr. Eckford, that there are Sweet Peas named after your daughters and your grand-daughters, but where is Henry Eckford to be found? Surely that is an omission."

"There was a Henry Eckford once," replied my companion; "it came out many years ago, but it is now forgotten." Mr. Eckford's son interposed here to say that he thought the new shrimp-coloured variety, which I referred to at the beginning of this article, ought to come out as the new Henry Eckford. I quite agreed, and enquired when that would be ready.

"We might be able to send it out in 1905," was the reply.

Mr. Eckford's heart is wrapped up in the new seedlings. "I spend the happiest hours of my life out here," he exclaimed. "Talk about a bunch of Sweet Peas," he went on, "almost every one here is dissimilar. You cannot take a step without seeing a grand thing. I have no favourite colour, but I like the look of a beautifully expanded flower best. You mustn't run away from the shape."

Sweet Pea 'Henry Eckford', The Garden, 3-2-1906

Some of the older horticulturists who read this account of my visit to Mr. Eckford will wonder why I have made no mention of his work with the Verbena, a flower which he took in hand many years ago, and made almost as great a reputation with as he has since with the Sweet Pea. It is not that we did not talk about Verbenas – we did; but there has been so much to tell about the Sweet Pea that I have been forced to omit almost everything else.

W.C. SAMBROOK. [76]

Another good year for Henry was rounded off with the National Sweet Pea Society's silver medal for 'Scarlet Gem'. 'The finest novelty of the year' they thought, 'so remarkable in colour' and elected him as their president for the coming year. [77] Henry may have been drawn now by fading eyesight to bright colours like 'Scarlet Gem' and 'Edward VII' and later to the bright scarlet 'Queen Alexandra' and luminous orange 'Henry Eckford'.

Like the year before, 1904 had barely begun before sweet peas were being reviewed in the '*Chronicle*. An appreciation by David Williamson was the leading article on 23rd January. Williamson seems to have been the first to name them the 'queen of annuals', a title they were to hold for the next fifty-odd years.

'The Sweet Pea is the queen of annuals; no other is comparable to this flower, whether in fragrance, in purity or in grace. If all other annuals disappeared from my garden and this one remained, I would be quite satisfied with my floral possessions; for along the half-shaded borders, with an environment of Roses and Oriental Lilies, it blossoms everywhere, and "nothing can stale its infinite variety." Fair to the eye, it is still fairer to the memory. There is no other flower, with the exception of the Rose, that so tenderly links the pensive present to the fadeless past. Ever since I can remember, it has been one of my favourite flowers, and I used to envy those cottagers in this picturesque parish whose limited enclosures it adorned so gracefully, long before I began to grow it myself.'

Emerging from comparative obscurity 'by the hybridising genius of that floral magician, Mr. Henry Eckford... it has developed capabilities undreamed of by our ancestors, who were for the most part quite satisfied with flowers hardly more than half the size of its modern representatives. Its colouring, one of its finest attributes, has also been intensified in a marvellous degree. It ranges from the purest, showiest white to that latest and grandest Eckfordian introduction, the incomparable Scarlet Gem.

Grown in "hedges", which is the olden conventional custom, whether with mingled varieties or in separate colours, Sweet Peas have an invariably impressive effect; but they never have such an artistic appearance as when they are seen climbing upwards, and flowering profusely through the branches of venerable fruit-trees – as they do in my own garden – to a most commanding height. Alike in colour and in fragrance they are attractive. Even at our grandest floral exhibitions the Eckford Sweet Peas, by virtue of their lustrous beauty, arrest attention, notwithstanding counter attractions... . This I can testify from personal observation... . Its first blooms often appear before the middle of June, and we find it making heroic efforts to flower on the confines of December. The only flower that blooms for a longer period is the richly odorous Viola... .

Of the many superb varieties of the Sweet Pea raised or introduced by Mr. Henry Eckford, the following are supreme favourites with the majority of cultivators...' and Williamson went on to list them, from pure whites like 'Dorothy Eckford' and 'Sadie Burpee' to the 'predominantly pink-coloured introductions of the Man of Wem' like 'Prima Donna' and 'Lovely'. 'Scarlet Gem is probably Mr. Eckford's most brilliant acquisition.' Of thirty-one varieties listed, only two, 'Navy Blue' and 'Gorgeous', weren't of Henry's raising but were his American introductions. Williamson makes the point that the sweet pea, unlike the orchid, is not 'so aristocratic or prohibitive in price; nor does it require a heated conservatory for its adequate cultivation' but is 'within the power of the humblest cultivator to acquire and grow exquisitely...' .[(78)]

In his lifetime Henry had grown many different plants, many of them expensive new introductions but his breeding programmes had always been with plants for the multitude. Plants like Dahlias and Verbenas and, even easier, hardy plants like pansies, garden peas and sweet peas. With the rise of books and magazines aimed at the amateur gardener, Henry can be seen to have been in touch with the times. People have always gardened but a rising population meant that there were more of us, we were becoming increasingly knowledgeable and had money to spend, even on orchids! Yes, David Williamson had been right to say that orchids were expensive to obtain and to cultivate, but as cut-flower their wholesale price, at from 2/- to 8/- a dozen was now, this summer, below that of Asters (from France) and Lily of the Valley. Sweet peas were not only cheap and easy to grown but as cut-flowers they wholesaled at between 1/- and 1/6 for a dozen bunches at the bottom end of the market (partly due to a summer glut). 'Considering the hot weather,' went the Covent Garden report for 20th July, 'Sweet Peas are very good, and continue over plentiful.' (79)

In April, there was a letter to the '*Chronicle* about crosses being made between English Sweet Peas and some early-flowering ones from Algiers. These had come into flower in Aston Rowants, Oxfordshire, three to four weeks earlier than normal English varieties sown at the same time. (80) Nothing more was heard about them until, three years later, a gardener in Saffron Walden sent the '*Chronicle* some examples of his 'winter-flowering strain' of sweet peas. (81) The editor was able to recall their likely provenance which in turn prompted an explanation in the N.S.P.S.'s *Sweet Pea Annual*.

The person responsible in Algeria was The Rev. Edwin Arkwright, a missionary presumably. His flowers weren't native but early-flowering sports from 'Blanche Ferry', an old, early-flowering, Ferry, Morse variety from the United States. He had crossed them with Henry's 'Mars' (he thinks) and from them he had derived a range of colours 'shall I say imitations – of Hon. Mrs. E. Kenyon, Jeannie Gordon, Lady Grisel Hamilton, Mars, Black Knight, etc. which begin to flower about Christmas time and last for five months.

That they form a distinct group is evident from the fact that Eckford's Sweet Peas, which I sow at the same time, *i.e.*, the end of September, do not flower till May. Moreover, the leaf is considerably narrower than in Eckford's varieties and more pointed, and the stem appears to have a more woody fibre... I do not find English growers much interested in this new group; perhaps they believe that any Sweet Pea would do the same in the climate of Algiers, and that [my peas] would be of no use in England, but this is not the case, as has been proved' by gardeners in England and Wales... . (82) What happened to them? Did they eventually adopt British ways in Britain and get forgotten? Apparently not. In the 1920s, Charles Unwin reported that in some parts of America and Australia conditions favoured the early or winter-flowering class but they were of little use in the British Isles. (83)

Right. Here's another question. Has any flower bested the rose for popularity? In these Edwardian summers, sweet peas were certainly having a go. In July they were pushing their way into the Manchester Rose Show. 'It is quite evident,' wrote the '*Chronicle* reporter, 'that Manchester will shortly have to hold a show of Sweet Peas as year by year the number of exhibits from trade and private growers increase. Five exhibits came from Messrs. Henry Eckford & Sons, Wem, Salop...'.(84) Henry had started calling the business '& Sons' but there was only one son, John. Henry's other living son Peter still lived near Manchester and perhaps visited the Show. Was Henry's visit also a way of saying 'come home,' the way parents sometimes do, and join the family business? Which for Peter and his family's sake might have been a good idea.

The Mancunian rosarians were quite benignant towards their sweet pea interlopers and refrained from invading the national sweet pea show at Crystal Palace a few days later. Of new varieties on display, wrote the '*Chronicle*, one called 'Henry Eckford, which gained the Society's Silver Medal (offered for the best new variety) is of a peculiarly pleasing shade of clear orange colour, shading off to salmon.'(85) So, last year's shrimp-coloured seedling now bore Henry's name, as John had suggested to the *Garden Life* reporter. Over the years, Henry may have felt that there was never quite the right flower to bear his name, but at eighty-two time was running out, so here it was.

'When Mr. Morse, the great Californian Sweet Pea seed grower, accompanied by Mr. S. B. Dicks, inquired where it was,' said The *Journal of Horticulture* reporter, 'from its very glare and brilliancy alone I was able to say, "See that mass of fiery orange-scarlet showing up in the distance? That is the new variety".' (It is available again today and he's right, the colour's almost luminous). 'But the firm of Eckford have others to show. Achievement is one of these, and it may be called a glorified Scarlet Gem. The latter Sweet Pea, alas! suffered badly under the scorching suns that have poured upon us during the greater part of the past summer, and so belied the splendid reports of the previous year...'.(86) The new 'Henry Eckford' variety was to prove similarly susceptible to fading. Henry's 'Scarlet Gem' was soon superseded by his scarlet 'Queen Alexandra', probably partly for fading in sunlight but also because 'Queen Alexandra' proved to be an outstanding variety. Despite putting up 'a large collection in their usual good form,' the Eckfords' firm couldn't quite reach the heights in the 'non-competitive exhibits,' receiving a silver medal behind a Mr. Baker of

Wolverhampton and a display from Henry's confrère from Lewisham, H.J. Jones. (87)

The Shropshire Show in August was marred by wet weather. 'On the opening day rain began to fall as early as 5 o'clock in the morning,' reported the 'Chronicle, 'and it rained continuously until 1.30 p.m., increasing in degree. When the judges commenced their duties at 10 o'clock there was a perfect deluge... As the plants and grass became saturated, there arose a mist in the tents that partly obscured the atmosphere. To attempt to move from one tent to another was almost the same as wading through the Thames would have been.' A hundred or so professional gardeners on an excursion from Scotland 'who had never previously seen a Shrewsbury Show' would have felt quite at home, especially any from the western highlands and islands. The night before the Show 'the principal tents were lighted with electricity to enable exhibitors to continue their work,'

An elderly Henry Eckford with his workers in the seed packaging room of Eckford's Seeds, Wem.

a novelty one hopes they didn't try and repeat in the wet. Henry offered a prize for eighteen varieties of sweet peas and picked up 'a silver-gilt medal for exhibiting a grand lot of Sweet Peas in glasses. All the best varieties were included – about fifty – among which was Scarlet Gem, which was shown for the first time last year'.[88]

Reporters sometimes went on to Wem after visiting Shrewsbury and one such this year was recalled that October by *The Journal of Horticulture*.

> 'The management of the business now lies with Mr. Eckford's son, John, an active gentleman of middle life, though the veteran father visits the nursery on all favourable days. May he long remain among us as the Sweet Pea king!
>
> We hardly realise what quantities of Pea seeds five tons mean, but that is the extent of the firm's harvesting this year. And new potatoes also have attracted their attention, for a crop of 1½ ton of Northern Star had been lifted in good condition. A bedding Aster, named Dwarf Triumph, with bright crimson flowers, and growing only 4 in. high, appeared to me most excellent: a pretty and a showy subject. Pansies and Violas and other similar useful bedding plants are cultivated in small quantities, and the selection of beautiful grasses were such as one seldom sees. Particularly pretty were Panicum violaceum [*P.virgatum?*] and Tricholina rosea [*Tricholaena rosea*, now *Rhynchelytrum repens*] each of which will not fail to please...
>
> The culinary Peas and the Broad Beans are nearly as numerous as the Sweet Peas, and good varieties of Mr. Eckford's raising have long been popular. Record Pea, which grows 5 ft. high, is an admirable grower and heavy, late cropper, and is therefore likely to be as keenly sought for as Centenary was when first sent out. Eckford's Ideal and Eckford's Prior are also very excellent culinary Peas, now happily well known.' [89])Another reporter coming up to Wem after the Shrewsbury Show – and there must have been a queue of them at the nursery gates – was ecstatic about 'Queen Alexandra'. 'Until I went to Wem at Shrewsbury show time,' he wrote in *The Gardener*, 'I was under the fond impression that the novelty of the season was Henry Eckford. I doubt that now, for I have seen Queen Alexandra. Oh! For a pen adequate to describe the beauties of this lovely Pea... Hail to thee, most beautiful, most constant of red Peas!... A good many will order [it] now – nay, have already ordered. Directly a new thing in Sweet Peas is heard of, Mr. Eckford tells me, the public begin clamouring for

it. In their eagerness they will hardly allow him time to get a sufficient stock of seed... ' .[90]

Another visitor that summer was Walter Wright, for whose wife Henry had recently named a mauve sweet pea. Wright submitted a report to *The Gardener* but as they already had the foregoing, Wright's article had to wait for the slacker news-time of winter. It appeared the following February:

AN HOUR WITH HENRY ECKFORD.
A Summer Memory.

The sun shone cheerfully in an almost cloudless sky as the veteran and I sauntered though the narrow byways of old-world Wem, and away beyond the station to the nursery.

A most lovable old man, with sunny, benign face, a smile for everything and everybody, and a ramble of happy talk, chiefly about the flowers and the people of past days.

The world has surely dealt kindly with an octogenarian for whom memory has naught but sweet recollections. In his ripe and mellow old age Henry Eckford lives largely in the past, and it is a past of treasured friendships, fragrant thoughts, and flowers which, though dead, live for every fresh in his heart.

Not Sweet Peas wholly – nay, I could not but think that the flower with which his name will remain imperishably associated only holds the second place. It was of the Verbena that he talked most – the fragrant, perfumed Verbena. Of how he brought out set after set of novelties, all sent out by John Keynes of Salisbury. Of how John Keynes gave him warning of a declining demand. Of how the time came when John Keynes said "No, Henry, I can't take a set this year; there are no buyers for new Verbenas now". And of how, after all, when the eyes of John Keynes fell on that year's set, temptation overcame him, and he was fain to say, "Well, well, Henry, just one year more."

But the Sweet Pea does live in the heart of the old hero still. He talks about his early loves, his early triumphs, not in any spirit of egotism, but as he might talk of the companionship of dear friends. They were his intimates, those Sweet Peas of his hybridising youth, and he loves them as he loved them when they first came into being, not because they brought him fame and fortune, but because, out of their sweet and gracious beauty, they made his life happy.

That is the keynote of this stalwart, stately old man, as straight as a gun barrel in spite of his eighty odd years, with his kind eyes and smiling face. His flowers are his dear, his well-loved friends. He has tended them and nurtured them, and they have given him a reward in the joy they brought into his life.

As we turned into the nursery gate he was telling of how he first took flowers of some new Sweet Peas up to a meeting of the Royal Horticultural Society, and a famous horticulturist, now dead, after surveying them critically, asked: "Well, what is to be done with these, Mr. Eckford?" "That," was the reply, "is just what I have come here to find out!" The veteran tells of this incident with refreshing glee. It is a happy memory still. And there is no shadow on him when he recalls that nothing particular came of his journey, nor for long afterwards. His Sweet Peas did not stand in his mind as goods in the market place. They did not represent merely pounds, shillings, and pence to him. Far other, indeed! But recognition came in time.

We wandered in the nursery, that sunny August afternoon, in pleasant commune. For myself, I merely stood for my sire, one of those friends of his youth whose memory lives, like his flowers, ever green and fresh in Henry Eckford's heart. And I was content so to do. For with those stories of old days in my ear, and with those acres of beautiful flowers around me, with the scent of Sweet Peas in my nostrils, and the warmth of the genial summer sun on my cheeks; above all, with that gentle and serene soul murmuring the warm thoughts of a still warm heart, I could not but ask myself the question: Has ambition, with all its passionate strife and fierce discords, ever come so near to solving the problem of the highest life as the man who joins forces with Nature to create beautiful things, and, bathing his soul in their tender and refining influences, lays up sunshine and perfume in his nature that only grow the richer as the burden of life grows heavier?

It may not be a "practical" matter, this, but what a piercing, poignant one for every man who ever admits into his mind a reflection on the ultimate object of human life! If

We are none other than a moving row
Of magic shadow shapes that come and go
Round with the sun-illumined lantern held
In midnight by the Master of The Show,

is the simple philosophy that to multiply beautiful things is to multiply the sources of human happiness rendered futile? I would fain hope not, for it is as strong upon me as the recollection of that summer ramble of which I speak, that a life cannot be wasted which robs a sordid earth of some of its squalor.

And, perhaps, my question may be worth putting, and my reflection worth sharing, even in a paper that lives for "practical" things.

W.P.W. [(91)]

Walter Wright went on to become a prolific author and journalist; as founder and editor of *Popular Gardening* he was a natural successor to Richard Dean.

The sweet pea year ended with a brief line or two in the '*Chronicle*. 'In a recent number of the *American Florist* is a coloured figure of a new Sweet Pea... "Gladys Unwin". The flowers are borne three on a stalk, are large with a flat, erect, entire or undulate standard, the wings are incurved. The colour is pale rose-pink or flesh-coloured'. [(92)] 'Gladys Unwin' was, as we've seen, slightly smaller than Silas Cole's variety but unlike his it bred true. From this, Cole's one (when it was fixed), Mr. Viner's one that Henry had and possibly three or four other sports of Henry's 'Prima Donna' come the majority of our modern varieties. But the most popular one in 1905 – the variety most exhibited and the winner of most awards – was to be the white 'Dorothy Eckford', Henry's grandiflora variety from 1902. [(93)] Grandifloras were to rule the roost for quite a while yet.

References to Chapter Twelve

(1) The Reader's Digest, *Yesterday's Britain*, (1998), p.14, quote.

(2) Editorial, *The Gardeners' Chronicle*, (5-1-1901), p.8.

(3) GD18/1719/28, The Clerk of Penicuik Papers, National Records of Scotland.

(4) Editorial, *The Gardeners' Chronicle*, (5-1-1901), p.9.

(5) 'Some Gardening Fashions of the last Century', *The Gardeners' Chronicle*, (23-2-1901), pp. 117-118.

(6) 'New Books', *The Gardeners' Magazine*, (2-2-1901), p.72.

(7) 'New and Beautiful Sweet Peas', (illustration), *The Gardeners' Magazine*, (9-3-1901), between pp. 146-147.

(8) Census return, Parish of Saints Peter and Paul, Wem, 1901.

(9) DA14/100/1, Minute Book, Wem Urban District Council, (1900-1903), p.80, Shropshire Record Office.

(10) Ibid, p.131.

(11) Ordnance Survey Map of Wem, 2nd. edn., 1902.

(12) DA14/100/1, Minute Book, op.cit., p.148.

(13) Ibid, p.84.

(14) *Kelley's Directory of Shropshire*, (1905), p.277.

(15) DA14/100/1, Minute Book, op.cit., p.154.

(16) 'National Sweet Pea', *The Gardeners' Chronicle*, (27-7-1901), p.84. The National Sweet Pea Society had been founded at a meeting held at 2.30 p.m. in The Hotel Windsor, Victoria Street, Westminster on Tuesday 26th March, 1901.

(17) G. Rice, *The Sweet Pea Book*, (2002), p.9.

(18) 'National Sweet Pea', *The Gardeners' Chronicle*, (27-7-1901), p.83.

(19) 'Origin of Countess Spencer Sweet Pea,' *The Gardeners' Magazine*, (3-8-1912), p.597.

(20) Letter from Charles W. J. Unwin to Bernard Jones, 7-3-1969.

(21) The term 'grandiflora' was used later for the pre-Spencer varieties to distinguish them from the smaller, earlier ones.

(22) 'Origin of Countess Spencer Sweet Pea', *The Gardeners' Magazine* (3-8-1912), p.597.

(23) 'Mendelism as applied to Sweet Peas,' *The Sweet Pea Annual*, (1909), p.9.

(24) Ibid.

(25) 'Sweet Peas of the Century', (illustration), *The Gardeners' Chronicle*, (3-8-1901), p.87.

(26) 'The Sweet Pea', *The Gardeners' Chronicle*, (3-8-1901), p.94.

(27) 'The Properties of the Sweet Pea', *The Gardeners' Chronicle*, (17-8-1901), p.138.

(28) Jane Brown, *The Pursuit of Paradise*, (1999), pp.128-129.

(29) Letter from Henry (Harry) Eckford to Bernard Jones, c.1982.

(30) 'Two new Sweet Peas', *The Garden*, (3-8-1901), p.75.

(31) 'Experiments in Plant Hybridisation', *Journal of the Royal Horticultural Society*, (9-1901), pp.1-32.

(32) Ibid, p.2.

(33) 'Experiments in Hybridisation by Gregor Mendel', *The Gardeners' Chronicle*, (21-9-1901), p.226.

(34) P. Raven, R. Evert, H. Curtis, *Biology of Plants*, (1976 edn.), p.139.

(35) 'All About Sweet Peas', *The Journal of Horticulture*, (11-7-1901), p.33.

(36) R. Dean, (ed.), *The Sweet Pea Bicentenary Celebration Report*, (1900), p.29.

(37) Letter from Henry (Harry) Eckford to Bernard Jones, (1-8-1980).

(38) 'Shropshire Horticultural', *The Gardeners' Chronicle*, (24-8-1901), p.160.

(39) Colin Matthew, *The Nineteenth Century*, (2005 edn.), pp. 132-133.

(40) DA14/100/1, op.cit., p.160.

(41) Ibid, p.251.

(42) 'Eckford's New Sweet Peas', (advertisement), *The Gardeners' Chronicle*, (4-1-1902), p.vii.

(43) 'Select Sweet Peas', *The Gardeners' Magazine*, (8-3-1902), p.141.

(44) 'Presentation to Mr. Richard Dean, V.M.H.', *The Gardeners' Chronicle*, (8-2-1902), p.98.

(45) 'Select Sweet Peas', *The Gardeners' Magazine*, (8-3-1902), p.141.

(46) 'Sweet Peas', *The Gardeners' Magazine*, (5-4-1902), p.217.

(47) 'Sweet Peas', *The Gardeners' Chronicle*, (13-9-1902), p.198.

(48) 'Sweet Pea Dorothy Eckford', *The Gardeners' Chronicle*, (2-8-1902), p.85.

(49) Charles Darwin, *The Variation of Animals and Plants under Domestication*, Vol.2, (1868), p.20.

(50) 'Sweet Peas', *The Gardeners' Chronicle*,(4-7-1903), p.3.

(51) 'Coronation Items', *The Gardeners' Chronicle*, (14-6-1902), p.395.

(52) Ibid, p.396.

(53) 'Royal Caledonian Horticultural', *The Gardeners' Chronicle*, (13-9-

1902), p.205.

(54) 'Shropshire Horticultural', *The Gardeners' Chronicle,* (23-8-1902), p.147.

(55) 'Shades of Colour in Sweet Peas', *The Gardeners' Chronicle,* (18-10-1902), p.284.

(56) 'Showing Sweet Peas', *The Gardeners' Chronicle,* (13-9-1902), p.202.

(57) 'National Sweet Pea', *The Gardeners' Chronicle,* (3-1-1903), p.16.

(58) 'All About Sweet Peas', *The Gardeners' Chronicle,* (17-1-1903), pp.41-42.

(59) 'The Countess of Spencer Sweet Pea', *The Gardeners' Chronicle,* (28-2-1903), p.137.

(60) 'Sweet Peas', *The Garden*, (7-3-1903), p.157.

(61) 'The Best Sweet Peas', *The Garden,* (28-3-1904), p.206.

(62) Eckford advertisement, *Garden Life*, (3-1-1903), p.v.

(63) DA14/100/2, Minute Book, Wem Urban District Council, (1903-1907), p.13, Shrop. R.O.

(64) Colin Matthew, *The Nineteenth Century*, (2005), p.130.

(65) Martin Daunton in C. Matthew, ed., *The Nineteenth Century*, (2005), p.63.

(66) 'Urban Council Election', *The Shrewsbury Chronicle,* (27-3-1903), p.8.

(67) 'Urban Council Election', *The Shrewsbury Chronicle,* (10-4-1903), p.8.

(68) Minute Book, (1903-1907), op.cit., p.41.

(69) Ibid, p.109.

(70) 'National Sweet Pea', *The Gardeners' Chronicle,* (18-7-1903), pp.44-45.

(71) 'Crossing Sweet Peas', *The Gardeners' Magazine*, (18-7-1903), p.486.

(72) 'Sweet Peas', *The Gardeners' Chronicle,* (4-7-1903), p.6.

(73) 'Peas in 1902', *The Gardeners' Chronicle,* (28-2-1903), p.131.

(74) 'Bishop's Stortford Horticultural', *The Gardeners' Chronicle,* (22-8-1903), p.146.

(75) 'Honorary Exhibits', *The Gardeners' Chronicle,* (22-8-1903), p.146.

(76) 'Famous Gardeners at Home', *Garden Life*, (26-9-1903), pp.463-465.

(77) 'National Sweet Pea', *The Gardeners' Chronicle,* (19-12-1903), p.430.

(78) 'The Sweet Pea: An Appreciation', *The Gardeners' Chronicle,* (23-1-1904), p.49.

(79) 'Markets', *The Gardeners' Chronicle,* (23-7-1904), p.67.

(80) 'Algerian Sweet Peas', *The Gardeners' Chronicle,* (30-4-1904), p.284.

(81) 'Sweet Peas', *The Gardeners' Chronicle,* (12-1-1907), p.28.

(82) 'The Télemly Sweet Peas', *The Sweet Pea Annual*, (1907), p.23.

(83) Charles W.J. Unwin, *Sweet Peas,* (1926), p.36.

(84) 'Manchester Rose Show', *The Gardeners' Chronicle,* (23-7-1904), p.66.

(85) 'National Sweet Pea Society', *The Gardeners' Chronicle,* (23-7-1904), p.66.

(86) 'Gadding and Gathering', *The Journal of Horticulture,* (27-10-1904), p.377.

(87) 'National Sweet Pea Society', *The Gardeners' Chronicle,* (23-7-1904), pp.66-67.

(88) 'Shropshire Horticultural', *The Gardeners' Chronicle,* (20-8-1904), pp.135-139.

(89) 'Gadding and Gathering', *The Journal of Horticulture,* (27-10-1904), p.377.

(90) 'Eckford's New Sweet Peas', *The Gardener,* (10-9-1904).

(91) 'An Hour With Henry Eckford', *The Gardener*, (4-2-1905), p.780.

(92) 'Sweet Pea', The *Gardeners' Chronicle,* (26-11-1904), p.371.

(93) 'The Best Sweet Peas', *The Gardeners' Magazine*, (3-2-1906), p.87.

Chapter Thirteen

Saturday, 14th January, nineteen-hundred and five. It was the start of the sweet pea sowing season and the Eckfords' nursery were informing the press of their new varieties.

'Only three new sweet peas are to be distributed by Mr. H. Eckford, this season', wrote *The Gardeners' Magazine*, 'and though many lovers of these fragrant annuals will be somewhat disappointed that Henry Eckford and Queen Alexandra, not to mention other seedlings, are not yet obtainable, they will agree that the new-comers are excellent. The three are Romolo Piazzani, a brilliant blue, and an improvement long waited for... Whether Black Michael will supersede Black Knight remains to be seen, but there should be room for both; the third variety is David R. Williamson, named after the well-known horticultural cleric of Kirkmaiden, and it is another beautiful blue... We are pleased to note that Mr. Eckford is in his new list promoting strongly the aims of the National Sweet Pea Society [of which he'd just been president] as he writes "in the following pages a good many Sweet Peas are marked 'discarded.' I still grow and supply seed of these, but it is my intention to eventually exclude them from the list, as other and better varieties in similar colours have taken their place".' This may well have been Henry's view but John was in charge now and intended to stay up with the latest and newest, whoever their raiser, for he had no great flair as a hybridist.(1)

Richard Dean was, among his attributes, an indefatigable writer both to and for the gardening press. In the columns of the less well-known *Agricultural Economist* in February his 'best sweet peas' had nine of Henry's in the top twelve. The list, however, was led by the new large, wavy-flowered (and unfixed) 'Countess Spencer' of Silas Cole. 'Unfortunately', wrote Dean, 'Countess Spencer is apt to sport, and as a substitute for it I name Gladys Unwin', raised by William Unwin. Well, you might think, these Spencer types had arrived a bit late in Henry's life and he'd been caught on the hop. That wouldn't be fair. Throughout his life Henry had always been cautious and meticulous. His one big calculated gamble had been with sweet peas. He had, or was, to buy up all of that Somerset 'Spencer' seed, which bred true. For now, he would have the satisfaction of seeing Dean's top Eckford choice, 'Dorothy Eckford', become the most popular variety of the year.

'That the sweet pea continues to increase in favour no one will deny... Amateur and cottager who love their gardens for what they produce are ardent cultivators of this popular flower'. (2) So wrote *The Gardeners' Magazine* in February, in the somewhat sentimental style of the amateur's magazines that the '*Chronicle*, under Masters, was never to copy. If 'a cottage in the country' was a dream for some, it was a place from which they had once been only too glad to have escaped. The sweet pea was now a life-style choice of the urban dweller; just the thing for a small garden. The Victorian potted plant was making way for gay, simple throw-away flowers like carnations (3) and sweet peas. The gardenless could purchase them from street-corner traders or the new florists' shops. All destined to fill the vases of the new Edwardian home or, newer still, the Edwardian buttonhole. White, ('those white sweet peas you are wearing, and that become you so admirably...' in the words of a contemporary gardening book,(4)) was a popular colour, worn by choice or imposition by respectable Edwardian women-folk.

The vogue for simplicity, of cottage-garden plants in the town, was now also a reaction to the complexities of urban living, where women in particular could express themselves within the home in ways they found difficult or impossible to do outside it. Floral arrangements too, while common in the old mansions of the aristocracy, were now being taken up by the newly affluent or those hoping to become so. To all this was added a touch of frivolity, an influence from the new fun-loving King. This too favoured flowers like carnations and sweet peas. Whereas, however, carnations had been in Queen Mary's bouquets for King Edward's Coronation, and would be again in Princess Alexandra's for George V's, the sweet pea was not. It never actually ever seems to have featured in any royal occasion. The sweet pea lacks the gravitas for celebration or sympathy. It is altogether too frivolous, too lacking in substance, too un-grand. In Edwardian Britain, there was the added disadvantage of its popularity. Have you ever seen a garland with privet in it?

While the sweet pea was becoming something of a cult on the floral side, so was the potato among vegetables. As the sweet pea was never to eclipse the rose, so neither could the potato best the garden pea. Alex Dean had lists of the best garden peas in the *'Chronicle* in March. Drawn from seven experts including himself, none of them featured any of Henry's varieties. Not even that of Mr. Pope's who'd won, you may recall, Henry's prizes for garden peas when exhibiting Henry's peas. In fairness to Henry, their selections leant a little towards the exhibition varieties. Alex didn't name any of the firms associated with the varieties because, as he said, he didn't know them all and was perhaps disinclined, with his anti-commercial bias, to find out.(5)

John Eckford's new policy was now apparent in an advertisement in the same issue of the *'Chronicle*. Where the Eckfords had formerly, almost without exception, advertised only their own varieties, they now promoted a rival's. The new, fixed, Spencer-style 'Gladys Unwin' was listed alongside Henry's new red-maroon-flowered 'Black Michael' and bright indigo-flowered 'David R. Williamson'.(6) David Williamson sought to preen the new feather that Henry had put in his cap with a letter to the *'Chronicle*. 'As in the case of thousands of cultivators like myself, I cannot doubt the Sweet Pea was the favourite flower of their boyhood; in the light of its development and intensified attractiveness, it can hardly fail, under such exquisite conditions, to remain the chief abiding floral fascination of their maturer years', he began, sending the odd floral flourishes of Richard Dean into oblivion.

'Some of our grandest Roses do not open with facility, are wanting in fragrance or artistic formation; the Sweet Pea is almost infallible in those respects. Nor is this queen of all annuals so susceptible as many other flowers of greater splendour and impressiveness, to atmospheric influences...'. Meaning, city soot and smogs. A 'Sweet Pea of recent origination... is the already far-famed King Edward VII, the brightest and most effective of all the finest existing crimson-coloured varieties'.(7) Henry didn't think so. According to Harry, Henry's grandson, 'we had two new varieties ready. Both scarlet and better than any on the market. But one of these was better than the other. We decided to introduce the second best calling it Queen Alexandra and the following year the better pea, calling that King Edward VII... . It was never known how it happened but the better variety went out the first year as Queen Alexandra and although the following year King Edward VII went out, it had a short life'.(8)

One with a longer life was Henry's namesake 'Henry Eckford', which was given a boost for its market appearance next year. The R.H.S.'s floral committee, meeting on 20th June, unanimously favoured it with the Award of Merit, describing its flower as 'very large, of perfect form, and a rich orange-scarlet colour'.(9) Anticipating success, John seems to have persuaded Henry to put it out early, when stocks were still low but the price would be high. 'An excellent variety seen in general collections this year for the first time', said *The Gardeners' Chronicle*, beneath an illustration of the flower in their edition of 15th July.(10) Having been the previous year's National Sweet Pea Society's 'Novelty of the year', with 'Scarlet Gem' gaining the title in 1903, this year would be different. The Eckfords' lost their hold on the best new variety to 'Helen Lewis', one of the new Spencer types, as best new variety at the N.S.P.S.'s annual Show at the R.H.S.'s Hall at Vincent Square. They did, however, pick up a gold medal for their trade exhibit, 'the majority of which have been raised by the exhibitor'.(11)

At the Show, Henry was honoured with the greatest award it was possible to receive in horticulture. Amazingly, no reference was made to it by whoever sent in the long detailed report to the '*Chronicle*. It was left to *The Gardeners' Magazine* to break the news, on 8th July. 'We are delighted to learn that our old friend Mr. H. Eckford of Wem, Salop, has been honoured by the Royal Horticultural Society with the Victoria Medal of Honour in Horticulture. The Society honours itself by making this award to the man who has created sweet peas as we know them...'(12) The award must have been a well-kept secret, for there was no hint of it in a pre-show *Gardening World* interview with Henry, published on the same day. Their report emphasised the point that the sweet pea had its craze in the United States before we had ours.

'...Naturally [Mr. Eckford] has received more recognition in the British Isles than on the continent, but it is questionable whether our cousins in America have not sung his praises to a greater extent than we have in this country. The Sweet Pea was first grown extensively in America, where it is usual to grow anything worthy of special attention on a large scale. The area under Sweet Peas has certainly been immensely increased in this country since our cousins set us the example... . Mr. Eckford has always maintained a calm and unassuming demeanour with regard to his accomplishments and achievements. He has simply been allowing his creations to speak for themselves. Like many others of Scotch nationality, he maintains a large measure of reserve concerning his own work...'.

Henry's answers in his *Gardening World* interview, though given in quotation marks, seem too stilted for natural speech and probably got changed in the telling. However, his answers often begin 'yes, that is so', and 'that is so' which sounds like an authentic mannerism of Henry's. "Many of the varieties which I raised and put into commerce," Henry noticed, "have been grown to the present day by all classes of growers, who can with difficulty be persuaded to replace them by varieties with blooms of larger size. The beautiful flowers having once become popular, growers continue to ask for them year after year...".

> "Do you think that the Sweet Pea has reached its limits of perfection, or can it still further be improved?"
>
> "Since [the 1890s] there has been no evidence that the Sweet Pea has reached its limits of improvement. For instance, colour, form, size of bloom, and constitution are yet in the making, and it would be impossible to foresee what improvements can yet be made...".
>
> 'Mr Eckford himself has no fairy tales to unfold, either concerning the origin of the several varieties or his part in connection therewith. From what I have been able to glean, he has simply a keen eye for observation, and is ever ready to seize upon anything of advantage to the Sweet Pea, and desirable in the matter of size, colour, form, constancy, etc.
>
> He commenced by singling out the individual varieties worthy of attention, and has devoted himself chiefly to the selection of the most promising individuals, keeping all the points in his mind's eye which go to the make-up of a first-class Sweet Pea. He has simply seconded and guided Nature, as it were, in the evolution of the Sweet Pea, which Nature has moulded in such feeble form as to require support, and painted in such a variety of delicate colours... . By shaking up the kaleidoscope, as it were, a fresh set of varieties is the result, exhibiting subtle and delicate colours of inimitable hues which painters can but imitate. After many generations, extending over more than 200 years, the Sweet Pea remains the same frail and fragile plant, rejoicing in a profusion of flowers emblematical of perpetual youth and beauty yet as fragrant as it was 205 years ago when first introduced to Britain...'.(13)

Although the interview had taken place before the announcement of Henry's V.M.H. award, *The Gardening World's* editor was lax in not mentioning it. Alex Dean had also recently been so honoured and they both joined the fifty-eight other recipients which included Richard Dean, one of the original holders. One of Richard Dean's very last articles, written in July, appeared in *The Agricultural Economist* in August. It was, appropriately you may feel, on sweet peas, and showed Richard's passionate and undiminished Victorian belief in progress. 'It is high cultivation which has developed forward five flowers on a stem, instead of only two, or at most three, as were seen a few years ago...'.[14]

~~~

W. Atlee Burpee, born in Nova Scotia, Canada, had settled in Philadelphia, in the eastern United States, to found the seed firm that still bears his name. He and Henry had known of each other for some time, Atlee having visited Henry a few years earlier, in 1901. In 1893 Henry had named a white variety 'Blanche Burpee' after Atlee's wife and in 1898 he named another white 'Sadie Burpee', for a daughter presumably. By way of return, Atlee had undertaken, on his earlier visit, to be godfather to three of Alice and John's daughters who were christened then, an event witnessed by another daughter, Charlotte (Lottie). 'I remember being dressed up', she recalled, 'and all standing in the drawing room waiting for him. When I was six (and I was the eldest). There were six of us. Terrible! Six girls and one boy...'.[15] (Five girls actually; the sixth came along in 1907).

W. Atlee Burpee and Henry Eckford

Atlee visited again in that summer of 1905. The one boy, Harry, was now five. 'I remember as a boy seeing Mr. Burpee when he stayed at Wem' he recalled. 'Mr grandfather [Henry] was 6 foot
~~~

A photograph from 'The Florists' Exchange', shows Henry and Atlee Burpee with Henry's son and daughter, John and Isabella Eckford, in the background.

tall but was a small man compared with Mr. Burpee'.(16) Tall people and tall plants and climbers go naturally together. You don't have to bend down all the time! Do alpines attract short people? Perhaps they do. Burpee's report of the visit appeared in *The Florists' Exchange*, an American paper not normally available in Britain. Atlee Burpee must have sent Henry a copy, for an old cutting appears among the Eckford artefacts collected by Bernard Jones. Two photos of Henry and Atlee illustrate the article. In one, Henry and Atlee stand together; in the other they are in among rows of sweet peas with Henry's children Isa (Isabella) and John Eckford, both now in their forties.

> 'On July 13 I spent a day with Henry Eckford, the sweet pea specialist, on his grounds, at Wem, Shropshire.... The boutonniere I am wearing in one of the photos herewith is of the unique new Henry Eckford. I remarked to his son, John Stainer Eckford, that he was particularly fortunate in having such a remarkable variety to name Henry Eckford, just the year when his father was honoured with the Victoria medal in Horticulture. I remarked further that I did not think there would ever be a more wonderful

creation in sweet peas. He said, however, that he thought with this new blood he was now really on the way toward getting a true yellow.

The many friends of Mr. Henry Eckford in America will be pleased, I am sure, to see how well he is looking in these photos, and to know that his interest is unabated. He walked through the grounds with me with his old-time enthusiasm, and yet, of course, takes life considerably easier than when I visited him last. This he is enabled to do because his son has inherited the enthusiasm of his father, and also his ability.

Mr. Eckford explained to me their reasons for discontinuing the sale of their sweet peas to the trade, but of his own accord said that he would be pleased to make an arrangement with our house to be his exclusive selling agents in America, we, of course, agreeing not to fill any orders from Great Britain. It is needless to say that we gladly made this arrangement; and, in fact, I had in mind when I visited Wem to ask if he could not make concession, so far as to let us this coming year (which would be our thirtieth anniversary) handle his two most remarkable novelties – Henry Eckford and Queen Alexandra. He feels quite confident that he will have sufficient stock of both of these to put them out for 1906 at the price of two shillings and six pence per packet of twenty seeds. I told him that this was, of course, a practically unknown price in America, but that these two varieties were so really exceptional that I had no doubt there would be a large sale for them. The other variety, which will complete his set for 1906, is not yet positively determined, but will probably be either Miss Eckford or Lady Hatherton... .

The young lady seen in one of the photographs sent is the original Miss Eckford. Once before her father named a sweet pea after her, "Isa Eckford", and she laughingly told me that if I would look in her father's catalogue now I would see that she was marked "discarded." She will, however, not be "discarded" as Miss Eckford for many a long year. [The 'other variety' eventually came out as a new 'Isa Eckford'].

> My day at Wem was altogether too short, and I regretted very much that I had not the time to accept their cordial invitation to stay another day. I had to leave, however, to make Queenstown…'.

Atlee Burpee's description of the 'Henry Eckford' variety included some novel (novel for Great Britain) recommendations. 'It is an ideal buttonhole flower, and unequalled for night decoration, as it lights up so well either by gas or electricity…'.[17] Flowers as male adornment weren't noted much by the gardening press. Acceptable in America perhaps but a little ostentatious for the man in the street over here. More the province of the well-to-do; the 1905 Harrods store catalogue stated that button-holes and sprays were on sale at the counter or could be made to order.[18]

A parent's recent death may have been the reason that Emily stayed out of the family photograph, for it was probably a legacy that now enabled Emily to purchase the joint properties of 9 Market Street and 31 Noble Street. Henry had acknowledged the fact in a codicil to his Will, made a few days before Atlee's visit, adding that 'it is my wish that my grand-daughter Lottie Eckford shall be my wife's companion after my death as much as possible…'.[19] Bricks and mortar investments, the mores of the British middle-classes in the twentieth-century, peculiarly juxtaposed with the Victorian ideal of a dutifully celibate, unmarried (grand) daughter.

> Back in March, the following notice had appeared in the columns of *The Gardeners' Chronicle*. 'All who have sympathy with the proposal to present a testimonial to Mr. Henry Eckford, in recognition of his services to floriculture are asked to communicate with Mr. H. J. Wright, 32, Dault Road, Wandsworth. Our readers are well aware that the evolution we have witnessed in the Sweet Pea has been mainly due to the work done by Mr. Eckford at Wem.'[20]

This was followed by regular updates as the fund increased. It was funny really, because although this was a good (and cheap) way to address Henry's friends and admirers it was one accessible to Henry too, who could follow its progress. He might, if he'd wanted to, see if it would beat the £300 raised in a similar fashion for Richard Dean's testimonial in 1902. Henry, being who he was, would have pretended not to notice. By 3rd June the fund had grown to £38-15s[21] and a fortnight later to 812 shillings (£40-12s).[22] As we'll see, the fund had to close at £58-7s-9d; not record-breaking but a tidy sum, nevertheless. Richard Dean hadn't a business behind him and the money, which he had received as a cheque, would provide for his dependants after his death. In Henry's case some silver momento was deemed more appropriate.

The crooked sixpence that his mother had given him all those years ago, so he'd never be short of money, had worked its magic for Henry, now comfortably off. John's son Harry, who lived with Henry, remembered his grandfather in his own old age: '... he would take the sixpence out of his waistcoat pocket for me to play with. Many times I thought how easy it would have been for me to flatten it, so that I could buy something...'.(23) With the testimonial, it was now just a matter of finding an appropriate occasion. 'Mr. Eckford being a native of Edinburgh, it was generally thought that no more fitting occasion than the period of the great International Show could be chosen for presenting him with some silver and an illuminated address which have been subscribed to by upwards of 200 admirers...'. and premises at Edinburgh's St. Andrew Square were chosen as the venue.(24)

From the centre of one coastal capital to the environs of another, 'Sweet Peas were a very fine feature at the autumn show of The Royal Horticultural Society of Ireland, held at St. Helens, Booterstown... . Mr. H. Eckford, Wem, Salop, showed a big trade group of sweet peas, and Mr. J. Eckford acted as judge of sweet peas. Ireland bids fair to do as well with sweet peas as with roses' wrote *The Gardeners' Magazine* in August(25). On the ball as ever, the Eckfords were capitalising on their closeness to Ireland, via Liverpool. Reporting on the event in 1906, the N.S.P.S.'s *Annual* noted that John Eckford 'expressed himself as greatly pleased indeed at the excellence of the blooms shown; indeed, he considered them quite as good as any he had seen elsewhere, even at Shrewsbury the previous year...'(26) Early August does seem an odd time to have an autumn Show!

The late summer's Shrewsbury Show, 'the event of the season' in the view of the gardening press, was held this year on Wednesday 23rd and Thursday 24th August. 'History repeated itself this year,' wrote the *Journal of Horticulture*, 'in the first day being wet and the second day beautifully fine. From daybreak until dusk perfect weather prevailed on the Thursday, and as a result all records were easily beaten. At an early hour visitors began to pour into the town, and the influx of heavily loaded trains continued until well on to mid-day. Even at this time the Quarry appeared to be full, and a little later the people were literally wedged together. From an elevation the grounds resembled nothing so much as the sight one sees on dislodging the top of an anthill. Great as Shrewsbury Show is, and numerous as are its attractions, undoubtedly its most wonderful feature is the spectacle afforded by so huge a congregation in such a small area'.[27] The feeling from within the Quarry itself may have been different!

The 'heavily loaded trains' were seventy special trains running into Shrewsbury 'and fortunately there was no accident',[28] wrote the '*Chronicle*, indicating that, at that volume, there usually was, or at least that there used to be. A respect for its dangers had been learnt by passengers, compared to the railways' early days. Takings of over £3,000 broke all Show records. Many of the exhibitors had come from Scotland. Henry's prize for eighteen sweet pea varieties and his Champion Cup was won by one J. Gibson from Duns, in the Scottish borders. 'The Scotch growers, now famous in the sweet pea world, [took] the leading honours... . A fine display of Sweet Peas came from Mr. Henry Eckford, of Wem, whose flowers were strikingly deep and rich in colour, as well as good in form. They were, too, admirably staged...'.[29] Henry was one of seven winners of the large gold-medal, the top medal.

The *Journal's* reporter was one of the safe Shropshire travellers who went on to Wem by train. 'The nurseries lie close to the station, so that anyone visiting the famous Salopian town during July or August ought to try and go so far also as Wem to see the Sweet Peas.

This year, however, we saw next to nothing in the Sweet Pea line at these world famous quarters, for the simple reason that with an extra early harvest all the Peas were past, and even the seed harvest was nearly gathered in. This was regarded with much favour by Mr. Eckford, though he also regretted the bareness of that scene, which a few weeks earlier had been one of truly regal gorgeousness, and of honeyed perfume. The seeds were safe: they were dry, ripe, firm, clean, and under these conditions their germinating power is at the highest, and the seedlings are generally stout and vigorous'. Atlee Burpee, visiting in July, had seen the flowers in bloom and the *Journal's* reporter 'with the editor's permission', included an edited extract of Burpee's *Florists' Journal* report. Although agreed to by the editor, he wasn't perhaps over-pleased and the report wasn't published until November.[30]

~~~

'It is not only a sad thing', wrote *The Gardeners' Chronicle*, 'but one difficult to realise, when one who has been specially active for something like fifty years is taken from us'. Richard Dean had died on the eve of the Shrewsbury Flower Show, at the age of seventy-five. While his brother Alex had been the strangest person to enter our narrative, Richard's presence
~~~

has, by contrast, been typified by kindness, bonhomie and enthusiasm. A trained and experienced gardener, a born secretary, a competent writer and speaker, his association with horticulture across the British Isles was immense. 'There has indeed', continued the '*Chronicle*, 'been scarcely a horticultural movement of any kind during the last forty or fifty years in which Richard Dean has not taken an active and generally a prominent part...'.[31] He even found time outside horticulture to become secretary of the Postal Reform Committee[32] that helped promote the establishment of the Parcels Post in 1883.

His death was widely reported by the horticultural journals; he had written for most of them. 'Richard Dean', wrote Edward Owen Greening, editor of *The Agricultural Economist* and a contemporary of Richard's, 'the foremost writer on florists' flowers', Chrysanthemums in particular, 'our generation has known... was... full of enthusiasm for causes. Like many other good and generous men, he was not a man of business in the sense of money-making for himself. Therefore he died poor...'.[33] His funeral on the afternoon of Saturday 26th August attracted around a hundred and fifty people. 'The day was bright, peaceful and beautiful', noted one reporter, 'with just a faint touch of autumnal mellowness in the air...'.[34]

It would have been unwise for Henry, considered quite elderly then at eighty-two, to have travelled the one hundred and seventy miles from Wem to Ealing. He may have felt duty bound, however, for it was only a few days after the funeral that Henry appears to have suffered a stroke. It necessitated a quick change to the plans for his testimonial. The following notice appeared in the gardening journals on 16th September: 'The serious illness of Mr. Henry Eckford, V.M.H., made it imperative that the presentation arrangements should be altered. It was decided that the illuminated address and the most handsome tea and coffee service, on a salver, should be taken to Wem. This was done on Saturday, and the presentation made quietly on Saturday morning, the recipient being in bed, and exceedingly weak. Mr. Eckford was most deeply affected, and could scarcely thank his friends who had subscribed to the gift. He made his gratitude clear, and also his sorrow that he could not meet his friends in Edinburgh according to arrangement. Everyone will desire to sympathise with this great man in his illness. It is good to know that he suffers no pain, except during periods of coughing, and that he knows the members of the family as they tend to him at his bedside...'.[35]

Well, you might think, if only their good intentions had led them to organise things a bit earlier! A few days back in *Auld Reekie*, a wander round the Show, chin-wags at a get-together with old friends – just the ticket. What had been important in Victorian Britain (and Edwardian Britain wasn't about to change overnight) lay, as Oscar Wilde observed, in being earnest. Earnest, energetic, practical, substantial. These were the era's bywords. Add to them a belief in progress. Those among the aristocracy, working classes or intellectuals and outsiders like Wilde who disowned them (other than practicality – Emerson's 'muffins and not the promise of muffins' was ubiquitous) were obliged to acknowledge their presence.

Earnestness and energy should have hurried out testimonial committee along (if anything can hurry committees along). But this is to partly misunderstand the times, which were slower than ours. Transport was slower. Communications were slower (if speeding up). People may have walked and conversed more slowly. Dialogue in plays was slower. Individual movements in long classical music concerts could be encored. An ethos that stressed substantiality of process, from which results duly emerged, suited science and technology but not the arts it seems, which appeared to suffer. Think of ploddish and laboured, pre-Jekyll, garden design. Think of the (admittedly usual) dearth of eminent British composers, and of artists in this atmosphere, typified by John Ruskin's objection to the American James McNeil Whistler's *Nocturnes,* painted in only a day. It was an age in which gifted writers produced structurally weak door-stop novels.

The intention of the testimonial committee had been to get Henry to Edinburgh. And things had to be done properly. The fact that he never got there, which some imaginative forethought would have prevented, cannot therefore be blamed on the said Henry Eckford Testimonial Committee's chairman, secretary, treasurer or members! This is a text of their 'Illuminated Address':

> Mr. Henry Eckford, V.M.H.,
> 19-September-05
>
> Sir,
>
> We, the undersigned, representing several hundred admirers of your work and character, who have associated themselves with the presentation of the accompanying Testimonial, beg to offer you, in their name and our own, a small acknowledgement of your great services to horticulture.

The 'Illuminated Address'

Your remarkable success in the development of the Sweet Pea has made your name famous the world over.

Taking in hand a favourite old flower of British gardens, you have secured a large number of new and beautiful varieties, giving delight and gratification to thousands of people. In addition, you have made great improvements in other popular plants. In the course of your beneficent life-work you have come into contact with innumerable flower lovers, to all of whom you have endeared yourself by the sweetness of your disposition, the dignity of your demeanour and the uprightness of your character.

It is our earnest hope that you may continue to share for many years to come in the joy which your productions gives to others. And we trust that the evening of your days may be brightened by the knowledge that you, equally with the lovely offspring of your skill and devotion which adorn our gardens, enjoy the boundless affection and esteem of mankind.

We are, Sir,
Your Obedient Servants,
Percy Waterer
Chairman.

Committee:

Miss H. C. Philbrick.	Mrs. S. B. Dicks.
Mr. C. W. Breadmore.	Mr. H. J. R. Digges.
Mr. R. P. Brotherston.	Mr. T. Duncan.
Mr. E. T. Cook.	Mr. C. W. Greenwood.
Mr. C. H. Curtis.	Mr. E. F. Hawes.
Mr. W. Cuthbertson.	Mr. H. J. Jones.
Mr. R. Dean, V.M.H.	Mr. G. H. Mackereth.
Mr. J. Harrison Dick.	Mr. E. Molyneux, V.M.H.
Walter P. Wright, Treasurer.	Horace J. Wright, Secretary.[36]

Journals like the *'Chronicle* would have made preparations for obituaries. Looking at Henry's *Curriculum Vitae* would have reminded them that some gardens hadn't appeared in their columns for years – decades actually, in the case of the once famous Trentham and not once yet for Oxenfoord. This was now rectified by short articles on Edinburgh's Botanic Gardens and one with a drawing of Oxenfoord Castle and gardens (both in the 9th September edition). An article and illustration of Trentham's front flower garden, their Italian garden, looking rather new and manicured and probably an old picture, appeared in December.

The Duke of Sutherland was reported to have given the Trentham mansion, on which his family had formerly spent so much money, to the local council as a resource for higher education.[37] This was more a reaction to a change in the economic climate than of some benevolent rush of blood to the head. Land and property was becoming something of a liability for the older landed-classes. In Wem, for example, the now nominal Lord of the Manor there, Lord Barnard, had offered Henry's local council the Town Hall and his rights connected with markets, fairs and tolls, in 1900.[38] (On the Sutherlands leaving Trentham for good in 1911, most of the rather solid and dull mansion of Trentham Hall was destroyed by Staffordshire County Council, without anything much better being put in its place).

The Garden was the last of the journals that year to run a major article on sweet peas. They had a coloured plate of the scarlet 'Evelyn Byatt' too (not one of Henry's), an area in which the *'Chronicle*, with its rather soulless monochrome etchings, was losing ground. A November review of 'The Sweet Peas of 1905' was written by Alexander Malcolm, a new name from Duns. The Scottish border counties were becoming something of a mecca for the flower. 'One striking fact regarding Sweet Pea novelties is the wonderful development of what is known as the Spencer type. There is a subtle charm in this form which seems to draw the attention away from the other sorts with the exception of a few distinct colours among the ordinary and hooded types'. This is the first article in which references to Spencer types outnumber those of grandifloras. Only four grandifloras were chosen by Malcolm; 'Queen Alexandra', 'Henry Eckford', 'Evelyn Byatt' and 'Helen Pierce'. Of these, only the first two were Henry's.

'Queen Alexandra is too near Scarlet Gem to need further description. The raiser says it is not inclined to burn and go purple so quickly with age as the latter.' Of Henry's namesake 'there can be no doubt that this variety is a most distinct novelty, not only in colour – which is a lovely orange-salmon , with a distinct thin line of carmine down the centre of the standard – but the standard is of a beautiful round form. Mr. Eckford is sending it out next year, and readers will do well to place their orders early'. [39] Of the seven Spencer types, none were Henry's. 'Nora Unwin', listed as a Spencer was really one intermediate between the two types. 'Mr. Charles Foster' (soon becoming Mrs.; all misters became either Mrs. or Miss!) was then unfixed and two more were to disappear; unfixed too perhaps, as so many of the early Spencers were, or too similar to others already in existence.

Another article that November, with Henry's photo in the centre, appeared in *The Agricultural Economist*. 'Notable Personalities. Mr. Henry Eckford, V.M.H. The men who are greatest in any particular direction are commonly known by their labours alone, and when these live beyond the span of life of the workers, no more fitting monument could be desired. Of their personalities, the generation among which they live know practically nothing: they seek not notoriety, and resent it when it is thrust upon them. Their work is their life, they desire nothing more than to be left in the peace of comparative obscurity. Of such as these is Henry Eckford, whose labours in the improvement of the sweet pea will endure when many future generations have come and gone.

Those whose familiarity with sweet peas lies within the range of modern varieties can have no conception of the remarkable work that has been achieved by this eminent man. They can admire the perfect form of the blooms, the grace and elegance with which they are poised upon the stems, the wonderful vigour and floriferousness of the plants, the wide range of colouration, and the delicious fragrance of the blooms, but all these things either as units or as a whole cannot of themselves tell us of the evolution which the last quarter of a century has seen effected at the old world town of Wem in Shropshire, by Henry Eckford, one of the last of the grand old men of British gardening, who have lived for flowers alone…'.

Although there were several in advance of him in the development of the sweet pea, 'the work they did was presumably largely if not entirely in selection… . None of Mr. Eckford's work could be termed in the slightest degree haphazard; he worked hard at cross-fertilisation, recorded all his crosses and tabulated his results, which were remarkable for their consistent excellence when once he had got on the proper track. As regularly as year succeeded year he sent forth to a delighted world new varieties, mainly differing in colour, but also steadily advancing in size and the form of the flower.

A conspicuous fact in relation to the evolution of the sweet pea is that in the hands of Mr. Eckford every good point has been accentuated and none lost. If we observe the result of the labours of the florist among many flowers (roses and carnations may be taken as examples) we see that there has been gain in size, form and colouration, but loss in fragrance. With the sweet pea the perfume has been fully retained, while size, form and colour have improved. Certainly some varieties have a more attractive scent than others, but nevertheless the sweet pea of the present day is as sweet in perfume as the one introduced to us from Sicily, upwards of 200 years ago. The colour of this was purple, and from it, with the slight aid of one or two other species, our modern varieties have descended.

For a period of thirty years, Mr. Eckford has been among the sweet peas, but it will readily be understood that a man of his ability and enthusiasm found time to turn his energies into other directions also. He has raised some excellent bedding pansies, beautiful verbenas, and several culinary peas that combine heavy cropping properties with splendid flavour in an exceptionally high degree. But it will be for his labours among sweet peas that Henry Eckford's name will live forever in the history of British horticulture... .

It is to him that we owe the institution of the National Sweet Pea Society. By this I do not mean that he was a prime mover in its promotion, but that his grand work in the evolution of the flower has made such a society not only desirable but necessary. The actual starters were the late Mr. Richard Dean, V.M.H., and Mr. William Cuthbertson, others coming in immediately to their assistance. The workers did great things, but they had a splendid foundation upon which to erect a fabric.

Now, alas! Henry Eckford lies awaiting the call of the Master. It may come at any moment, but he is ready. His life has been one of persistent work, his intercourse with others characterised by courtesy and sweetness, his business relations marked by probity. If the love and respect of his fellows can smooth his passage across the bar it will be peaceful indeed, as peaceful, mayhap, as his future glorious home'.[(40)]

The early Edwardians shared with the Victorians an almost obsessive fascination with death. Death was all around, true, but it always had been, throughout human history. Perhaps now, for the first time, ordinary people had both the time to ponder it and the wealth to honour it. With advances in health-care and a surfeit of death in World War One and the following Spanish Flu, attention would once again be focused elsewhere. In the present case, the paper couldn't wait for Henry's demise and brought out an obituary, for such it was, in indecent anticipation of it.

Incidentally, the author's reference to inter-species hybridisation ('the slight aid of one or two other species') in sweet pea development is, as we now know, an error. It has been later claimed[(41)] that scent in sweet peas diminished in these early days of its development, but, as here, there is no contemporary evidence to justify it. The theory may have stemmed from a decline in quality, through poor management, of surviving early varieties. To the article's reference to 'thirty years... among the sweet peas' – Henry's serious hybridising work had begun twenty-six years before but he had been among sweet peas, as we've seen, all his life.

'The Autumn', the Meteorological Office observed, had been a 'dry, singularly cold and rather dull season' this year, the sun hidden frequently from view. 'Taking the season as a whole, it was the coldest autumn since 1887, or for 18 years'. With the approach of winter the weather became, perversely, warmer, wetter and more unsettled.(42) In December the midland naturalist Edith Holden was to write in her now famous diary that 'the weather up to now has been very mild and open, we have had only one slight snow-shower this winter and only one spell of severe frost', adding the motto 'In December keep yourself warm and sleep'.(43)

Henry never recovered from the stroke that had struck him down in September. He died, in his own bed in his home on the corner of Market Street and Noble Street, on the evening of Tuesday 5th December.

It was probably John who reported the death to the Registrar the following day. He had been present at the event, he said, which occurred at 10.30p.m. In the way of these things, it may have taken a time to be realised, though probably not occurring at the 10p.m. reported in the press. Dr. Leader of The Old Hall in New Street had called to certify the death. It had been a short six or seven minutes walk for him via the High Street; right by the bank into Noble Street. The Registrar was Thomas Kynaston, of the family dominant in Wem's public and commercial life. He would have been known to the Eckfords if only via Thomas's brother and father on the Council. So any condolence would have had the warmth of personal association.

The funeral took place on the afternoon of the following Saturday, 9th December. A report duly appeared in *The Shrewsbury Chronicle*. 'Full of years and honours Mr. Henry Eckford... was a man of fine physique. He was well over six feet in height, and at the age of 80 he used to ride his bicycle to and from his nursery. For the last few months he had, however, been confined to his house, and gradually sinking he passed peacefully away...'. Then followed a short history of his life. 'After Mr. Eckford settled at Wem, he carried the sweet pea to great perfection, and his work won him world-wide reputation. His grounds at Wem have annually been visited by a large number of people from all parts of the country... . A tea and coffee service... had been subscribed to by upwards of 200 of his admirers in every part of the civilised world', the civilised world consisting in the main of the isles of Great Britain, the United States of America, Germany, Algiers and, possibly, France.

'Mr. Eckford took a keen interest in the welfare of Wem...'. Henry's employees acted as bearers. The 'chief mourners were the immediate relatives, and a large number of tradesmen and other inhabitants of Wem also attended. The Rev. H. Cowling (pastor) conducted a service in the Baptist Chapel, and also officiated at the graveside. Many floral tributes were sent, the contributors including – Mrs. Strong and family, Mr. and Mrs. B. Stinchcombe, Mr. C. F. Richards and family, Dr. and Mrs. Leader, Mr. and Mrs. John Jones and family, Mr. C. H. Kynaston and family, Mrs. Franklin and family, Messrs. Lewis and Hassall, and the employees'. It was important then to record the senders of wreaths, as it is for some today.

The mourners had to travel just outside town for the burial, which took place in the south-west corner of the then new Church of England cemetery on the Whitchurch road. Henry would have and has still, in the now closed burial-ground, its second-largest memorial, featuring the draped and empty urn, a symbol of death. Such neo-classicism, outdated in Henry's youth when the poet Thomas Hood had called it 'a marble tea-urn', was once again in vogue. Lines from Psalm thirty-seven were engraved on its plinth: 'The steps of a good man are ordered by the Lord'.

'On [the] Sunday evening', *The Shropshire Journal* report concluded, 'Mr. Cowling conducted an impressive memorial service in the Baptist

Henry Eckford's gravestone

Church, and preached an appropriate sermon. There was a crowded congregation'[(44)].

Henry's fellow tradesman, John Ikin, who ran a drapers business in the High Street, took responsibility for the funeral arrangements. Another Ikin, William, a councillor, was present at the council meeting on the following Wednesday. 'Proposed by Mr. Lee and seconded by Mr. H. Kynaston that a vote of condolence with the family of the late Mr. Henry Eckford be recorded on the Minutes of this Council and that the Clerk convey the Council's sympathy to the family. Carried unanimously'.[(45)] The day before, in south-west London, the National Sweet Pea Society were holding their AGM at the Hotel Windsor. 'At the commencement of the proceedings', reported the '*Chronicle*, 'The decease of the society's late President, Mr. H. Eckford, was touchingly referred to, and it was decided to forward a letter expressing deepest regret and condolence with Mrs. Eckford'.[(46)]

The information on Henry that John must have given to *The Shropshire Journal* was sent to a number of journals. Most of the principal gardening papers ran it, modifying and adding in their differing ways. Not so *The Garden's*, under E.T. Cook, and the only obituary notice without a photo of Henry, which borrowed from the article that the former editor, William Robinson, had run back in 1897 and reprinted by Cook and Jekyll in 1900. Only the introduction was new. 'We are very sorry to receive the news of the death of Mr. Eckford, whose name has become familiar in almost every gardener's home...'.[(47)] *The Gardeners' Chronicle's* obituary was written to Maxwell Master's usual concise requirements, being mainly the supplied record of Henry's work history. A *Curriculum Vitae* was seen as obligatory by all the journals and is a boon to the historian. 'His services to horticulture', it ended, 'are shown by this bare enumeration to have been great. Personally he secured the respect of all with whom he came in contact, and the affection of those with whom he was brought into closer relations. The business will, we learn, be carried on without change'.[(48)]

The Journal of Horticulture was more expansive and eulogistic. 'It is with regret', they wrote, 'yet without surprise, that horticulturists learned through the daily newspapers a week ago of the decease of Mr. Henry Eckford, the doyen of the Sweet Pea fanciers. And who is not included in that band these days? We imagine that the man does not exist who fails to appreciate the grace of form, the immense and beautiful diversity of colours, and the sweet, ethereal fragrance of this modern flower. It is truly modern: was not the late florist the "Sweet Pea King?"

Mr. Henry Eckford was one of the most meek and loveable of men...'.[(49)]

George Gordon V.M.H., all-round gardener and sweet pea enthusiast, was, as editor of *The Gardeners' Magazine* able to elaborate on John's submission with both knowledge and enthusiasm: '...one of the most notable of horticultural personages... . A genial kindly hearted, stalwart, modest man was Mr. H. Eckford; a man who stuck to his business and carried on his life's work with a patience and perseverance that won for him the admiration not only of those who knew him intimately, but also of those who in many lands have long enjoyed the fruits of his labours. In the United Kingdom, in the great seed-producing districts of Germany, in the seed farms of California, and in the United States generally... there will be thousands upon thousands who will think of him... because of the beautiful flowers... he has given to the world. From the garden point of view Mr. Eckford made the sweet pea, and that flower, in its infinite variety of colour, will ever be his finest memorial... .

The progress of the sweet pea in popular estimation was at first slow, and it was not until about 1901 that the demand for finer forms grew keen. Then popularity came with a rush, and both America and England called loudly to Eckford for more and better sweet peas. In 1902-3 and 4 the demand grew and California took up the Eckford varieties for seed purposes. But 1905 was about the record year; it was the Blanche Burpee year. From then on to the present time there has been no question as to the popularity of sweet peas, and no question as to whom that popularity is wholly due...'.[50] Well, almost wholly due... . *The Gardening World's* obituary on the 16th was a mere twenty-five lines long but even so was longer than that of *The Gardener's* of the same date. Death was bad for business, the newer journals saw, signifying a move away from the fascination held by the Victorians and early Edwardians.

Letters of condolence came in from far and wide. Some were from relatives making their first entrance in this narrative. Others came from kindly customers and fellow seedsmen, some known to us and some new. Atlee Burpee sent a telegram as soon as John's letter arrived after its eleven-day journey to Philadelphia. He followed it up with a letter.

> 'My dear Mr. Eckford:- It came to me as a great shock upon reaching the office half an hour ago when our cashier handed me my personal mail and on top was your note of December 7th announcing the death of your father... . Not having had any news from you or other friends in England for several weeks I was in

hopes that your father's condition was no worse and that with his splendid constitution he would yet pull through.

I well remember how in 1882 life seemed black and dark to me when my father, who was only fifty-six years of age, died. I know that you have been close to your father likewise.... Sincerely your friend, W. Atlee Burpee'. (Atlee would himself die aged only fifty-eight).

Charles Curtis, from 2 Adelaide Road, Brentford, whose letters smelt of cigars was well-known to John but he began in the formal custom of the time: 'My Dear Eckford... . Your father's uprightness + his steadfastness of purpose will live long in the memory...'.[51] Another 'Seed Grower and Florist' (and author) was James Douglas, VMH, (1837-1911), now at Great Bookham in Surrey but originally from Ednam near Kelso in the Scottish borders. 'Dear Sir,' he wrote, 'I have known your father for nearly 50 years and I am truly sorry to hear we shall see him no more in this world. I do not know if Mrs Eckford is alive, if so she has my sincere sympathy. I feel for all of you...'.

Others less close wrote formally and respectfully. George Forbes, a flower-merchant from Manchester wrote 'Dear Sir, I was very sorry indeed to receive from you the intimation of your Father's death...'. The seed-firm of John Moss and Son at Kelvedon wrote 'of your distinguished father [whose] memory will remain fresh among lovers of the garden for many a long year...'. Charles C. Hurst (1870-1947), a later renowned hybridist and geneticist, wrote from Burbage in Leicestershire to order some flower-scent bottles as Christmas presents. 'Please accept my sympathy' he wrote in a P.S. 'I have pleasant recollections of meeting him at the Hybrid Conference at Chiswick in 1899'.

A letter from one William Wood, gardener at The Gardens, Kelso, arrived in January. He was one of those that the Eckfords' sent seed to prior to their appearance on the market, in the evident hope of what we'd now call feedback. They had sent him the varieties 'Queen Alexandra' and 'Henry Eckford'. 'Thank you for your generosity. I hope we may be able to repay you with blooms later... what a want you must feel without the presence of the genial, kind and grand old man.

I shall never forget the pleasure his visits to Newton Don gave us nor yet the happy expression on his face when, just before beginning to "cut" for the show, after a walk round our pen he exclaimed "We are in for some grand stuff today".'

Further letters from southern Scotland included two from the Reverend David Williamson, residing at the Manse of Kirkmaiden (and about as south as you can get in Scotland; more southerly than Durham in England).

He found time, in a busy period, to write thoughtfully if a little self importantly.

> 'December 19th… . I need not say how grieved I was to hear the sad news of your gifted Father's death, when I opened my copy of the Gardener last week.
>
> I know from your recent kind letter to myself that he had been very ill, but at that period you were not altogether hopeless of his ultimate recovery. But God's will be done.
>
> About two years ago your Father asked me to write a special article on the Sweet Pea, to be included in his Reminiscences. I sincerely trust that this work was completed before his death, and that it will yet appear. It would be an extremely fascinating book. [It would be an extremely fascinating find!] I have contributed a short poem to the Sweet Pea Annual written at the Editor's request'. (A dire four-verse 'To The Sweet Pea' appeared in the subsequent 1906 annual). In reply to a letter from John he wrote again. '27th December… . All that you say of his peculiar amiability of character, his… courtesy and splendid sympathy (a very rare endowment) is intensely true. For my part, I feel that I will never forget his kindness. All lovers of nature and of the garden will reverence his… memory… .
>
> I have an article in the current issue of The Gardeners' Chronicle which I hope may interest you'. (It appeared as a short, perhaps edited, letter on the sweet pea variety 'Henry Eckford'). 'I think you will like the verses I have contributed, at the Editor's request, to "The Sweet Pea Annual" ', Williamson continued, proudly if forgetfully repeating what was obviously uppermost in his mind.

Alexander Malcolm, the new name who'd written the November *Garden* article, was a competent amateur hybridist from the Scottish borders. Like Henry, he would eventually receive the Victoria Medal of Honour in Horticulture. He'd met Henry, too, at the Royal Caledonian Society's Show at Edinburgh, where Henry hadn't recognised his own variety 'Miss Willmott', 'as shown in fine form and colour by Mr. Malcolm…' .[52] Slight changes probably which can be due to natural variation in response to a different soil or climate – gardeners always assumed it was due to their superior abilities! He wrote, unbidden, from the 'Town House' in Duns:

> 'Dear Mr. Eckford, I see from the Garden papers that your father has passed away. I can only saw how sorry I feel at such a loss. His life's work will never die and the foundation he has laid is already built on to a wonderful extent, and no one can dream of the improvements that will be seen in time to come. It is a source of just pride to foster the memory of such a father. When the mind gets dulled as in his case, it is a merciful dispensation when providence relieves and allows rest and peace.
>
> Excuse me venturing to write a line of sympathy with you and yours at such a time. I am, Yours faithfully, A. Malcolm.'

Bert Ballinger, you'll recall, had taken over at Boreatton Park after Henry had left for Wem. He'd won prizes for sweet peas at local shows and the families had kept in touch. Alice or Emily had sent him the news. He wrote a sad and flustered reply.

> 'Drumble Lodge, Boreatton Park, Baschurch… . Dear Mrs. Eckford, Many thanks for the memorial card you so kindly sent we are very sorry to hear of Mr. Eckford's death please accept our sympathy in your loss.
>
> We received a telegram this morning to say that Mother died last night only a week's sickness. With kind regards, I remain. Yours faithfully, A. Ballinger'.

Three relatives' letters survive. One was from a cousin living in Surrey that the Eckfords had obviously lost touch with. The young mother wrote in what she considered the appropriate manner, yet which, by contrast with that of the others, is seen as dated; more mid-Victorian than mid-Edwardian. (John was called Jack in the family).

> 'Osborne House, Nutfield Road, South Merstham… . Dear Cousin Jack, You will be surprised to hear from me no doubt but Father has written to tell me of your great loss of your dear Father. We miss our loved ones oh how much we miss them but oh do not grieve he is not dead but only Iyeth sleeping in the sweet sunshine of his Master's breast and far removed from sorrow, toil and weeping. He is not dead but only taketh rest.

What a time that will be when we meet our loved ones in heaven. I do hope your dear mother keeps well also your wife and children, you have 5 I think Dad told me. I have only 2 boys, real boys too.

How is Casey, give my love to her. I wonder if she remembers me. I hope you'll succeed in the business.

With all kind love and sympathy from myself and Husband in your great sorrow. May He comfort you all and your dear Mother.

I remain your loving cousin, Alice Warr'.

Casey might be a niece. There were plenty of relatives. An uncle wrote from Essex.

'Shirley, Old Road, Frinton on Sea. 16th Dec. 1905. My Dear Nephew, ... it seems to me as though I had lost a most dear and valued friend and to whom I always look to as giving me a start in life. I know you will excuse me but I must say this to you, try my Dear John to be as near like him as you possibly can. You will accept my sympathy also [that of] Mrs. B. and Nells... please convey the same to Mrs. Eckford and your sister. Trusting that your Wife and Family also yourself are well and wishing you the compliments of the season... . Your affectionate Uncle J. W. Blackwood'.

It sounds slightly odd for an admittedly elderly uncle to be dispensing advice to a forty-one year old nephew. Perhaps John, in his uncle's view, had strayed from the straight and narrow. Another letter from a Blackwood, this one in east London, was more down-to-earth.

'55 Senrab Street, Stepney... . Dear Jack,... please convey my deepest sympathy to all, hoping you, your wife and Family are all quite well. I am pleased to say we are all quite well with the exception of colds... your affectionate cousin W.J. Blackwood'.

It was probably helpful that the Eckfords' had a business to run, keeping their minds from dwelling wholly on present events. Some of their customers would have been aware of them, however, and like Thomas Fenwick of Clematis Cottage in Nether Witton near Morpeth, added a postscript to their orders. 'I was very sorry to read of the death of Mr. Eckford. I felt as if I had lost an old friend. I think it is between 50 and 60 years since I used to grow his verbenas and what beauties they were'. It is going back a long way but Henry's first Verbenas went on the market in the spring of 1866, just under forty years ago.

Another customer was a Commander Humphrey, still commanding from his now permanent shore-leave in Oxfordshire.

> '24th Dec… . Dear Sir, I am obliged for your letter dated 18th. inst. addressed to my gardener and enclosing a postcard for my address. My gardener has handed them to me. My permanent address is Commander P.E.M. Humphrey, RN, Camden Lodge, Caversham, Oxon.
>
> I was sorry to hear of the death of Mr. Henry Eckford (Senior) who, I presume was your father. He has done more for the Sweet Pea, my favourite flower, than anyone, and I gather from your letter and card that you intend to carry on his good work, and I am very glad of it, as I preferred his seeds to any other, and hope to continue to get my supplies from you.
>
> Will you kindly let me know if you are continuing to offer the Eckford Cup for competition and if so, particulars of the same'.

An order from nearer home came from J. Flut, head-gardener at Netherleigh House, Eaton Road, Chester.

> 'I received your letter this morning. I was very sorry to hear you had lost your Father. I am sure you will miss him very much, you must accept my sympathy. I shall be very pleased to have what seed I can off you as usual. I would be pleased if you would send me the following Sweet Peas, one Sixpenny Packet of each.
>
> George Gordon.
> Stanley.
> Lady Mary Currie.
> Coccinea.
> Hon Mrs. Kenyon.
> Miss. Willmott.
> Duchess of Westminster.
> Lady Grisel Hamilton.
> Captain of the Blues.
> Duke of Clarence.
> Lovely.
> Salopian.

Prince Edward of York.
Dorothy Eckford.
King Edward VII.'

This is the only Eckford seed-order known to have survived. It is instructive in that all, bar the last two, are nineteenth-century varieties. 'Stanley' takes us back to 1892, 'Captain of the Blues' to 1889. It endorses Henry's comment, in his July's *Gardening World* interview, of growers asking for the old popular varieties rather than for the newer ones. The demand for the latest varieties came from the younger customers and those in the expanding amateur market. Older professionals like Mr. Flut stayed with the tried and tested. Not that Mr. Flut necessarily saw himself as stick-in-the-mud. As you get older, what happened sixteen years ago is what happened yesterday. And, Mr. Flut would argue, wasn't he purchasing seed of the most fashionable flower of the new century?

Our last letter, the first to make the Eckfords smile again was probably the one received mid-month from the upper-class lady of 'The Grey House' at Gravesend in Kent. Normally a gardener would have written. Perhaps she was down on her luck. It gives us a chance to witness the peculiar third-person manner they sometimes affected. 'Dec. 13th 05. Lady Georgiana Legg encloses three shillings to Messers Eckford, she is sorry she had forgotten this little account before. She writes to express her sympathy for the loss they have sustained but she is glad the Country will not suffer as the firm will continue'. Her letter was kept and treasured by the Eckfords along with the others. Three shillings (15p), incidentally, was a lot of money, when full-time work might only bring in £1 (or less) a week.

Obituaries and, especially, letters of condolence show us the more attractive side of our natures. The more agreeable traits of the lately deceased are brought forth, those less so forgotten or diminished. Genial, amiable, kind, modest and affectionate – these are qualities remembered from his later life. There was probably no-one left alive to remember his early years. They are qualities that can be concomitant with success but are not usually regarded as the forces driving it. George Gordon in *The Gardeners' Magazine* came closer when he spoke of Henry's patience, essential for a good gardener, and his perseverance. H. Digges, from Ireland, wrote of Henry's self-control, of his being 'a stranger to excitement, passion, haste'. (53)

Add tenacity too, and obsession – Wem had been somewhere he could be left alone to continue his life's work. In 1891 a certain ruthlessness, too, had roped his son John into business partnership in rural Shropshire; the son who had left home at an early age to pursue a business career among the bright lights and excitements of London; a life he left, as he later said, 'with great reluctance'. (54) All in all, Henry's was a more complicated character than his admirers allowed. But look too closely and we might see too much of our own selves. The external factors shaping Henry's life – his family background, the constraints and opportunities of the times – these things too played their part. We'll now have a last look at the Soulton Road grounds as Henry would have known them, see how his sweet peas fared and how Henry Eckford, the firm, prospered without its founder and guiding mentor.

References to Chapter Thirteen

(1) 'Eckford's New Sweet Peas', *The Gardeners' Magazine*, (14-1-1905), p.38.

(2) 'Sweet Peas', *The Gardeners' Magazine*, (11-2-1905), p.95.

(3) Jennifer Davies, *Saying it with Flowers*, (2000), pp.87, 96.

(4) Alfred Austin, *The Garden that I Love*, (1906), p.104.

(5) 'Vegetables', *The Gardeners' Chronicle*, (18-3-1905), p.166.

(6) Advertisement, *The Gardeners' Chronicle*, (18-3-1905), p.xiii.

(7) 'Home Correspondence', *The Gardeners' Chronicle*, (8-7-1905), p.32.

(8) Letter from Harry (Henry) Eckford to Bernard Jones, 4-10-1979.

(9) 'Floral Committee Meeting, 20-6-1905', *Journal of The Royal Horticultural Society*, (1906), p.Lxxxiii.

(10) 'Sweet Pea Henry Eckford', (illustration), *The Gardeners' Chronicle*, (15-7-1905), p.47.

(11) 'National Sweet Pea', *The Gardeners' Chronicle*, (8-7-1905), p.36.

(12) 'Mr. Henry Eckford, V.M.H.', *The Gardeners' Magazine*, (8-7-1905), p.424.

(13) 'Occasional Interviews', *The Gardening World*, (8-7-1905), pp.543-544.

(14) 'Sweet Peas', *The Agricultural Economist*, (8-1905), p.243.

(15) Letter from Charlotte Ryley, née Eckford, to Bernard Jones, 10-9-1979.

(16) Letter from Harry (Henry) Eckford to Bernard Jones, 15-8-1979.

(17) 'A Day With Henry Eckford, Sweet Pea Specialist', *The Florists' Exchange*, (12-8-1905), p.161.

(18) Jennifer Davies, op.cit., p.192.

(19) Codicil, 8-7-1905, The Last Will and Testament of Henry Eckford, (7-12-1899), p.3.

(20) 'Testimonial to Mr. H. Eckford', *The Gardeners' Chronicle*, (18-3-1905), p.168.

(21) 'The "Eckford" Testimonial', *The Gardeners' Chronicle*, (10-6-1905), p.361.

(22) 'The "Henry Eckford" testimonial', *The Gardeners' Chronicle*, (24-6-1905), p.395.

(23) Letter from Harry (Henry) Eckford to Bernard Jones, c.1980.

(24) 'Presentation to Mr. Henry Eckford', *The Gardeners' Chronicle*, (2-9-1905), p.185. The notice appeared simultaneously in *The Garden*, p.136.

(25) 'Sweet Peas in Ireland', *The Gardeners' Magazine,* (19-8-1905), p.527.

(26) 'Sweet Pea exhibitions and exhibitions in Ireland', *The Sweet Pea Annual,* (1906), p.30.

(27) 'All Records Beaten at Shrewsbury', *The Journal of Horticulture,* (31-8-1905), p.198.

(28) 'The Shrewsbury Show', *The Gardeners' Chronicle,* (2-9-1905), p.185.

(29) 'Shrewsbury Horticultural Show', *The Journal of Horticulture,* (24-8-1905), pp.182-183.

(30) 'New Sweet Peas', *The Journal of Horticulture,* (2-11-1905), p.406.

(31) 'Richard Dean', *The Gardeners' Chronicle,* (26-8-1905), p.168.

(32) 'The Late Mr. Richard Dean, V.M.H.', *The Garden,* (2-9-1905), p.135.

(33) 'Notable Personalities', *The Agricultural Economist,* (10-1905), p.243.

(34) 'The Late Mr. Richard Dean, V.M.H.', *The Journal of Horticulture,* (31-8-1905), p.211.

(35) 'Henry Eckford Testimonial', *The Gardeners' Magazine,* (16-9-1905), p.589. The notice appeared simultaneously in *The Gardeners' Chronicle,* p.216.

(36) A transcription is in *The Sweet Pea Annual,* (1906), p.78.

(37) 'Gift of Trentham Hall for Purposes of Higher Education', *The Gardeners' Chronicle,* (16-12-1905), p.425.

(38) DA14/100/1, Minute Book, Wem Urban District Council, (1900-1903), pp.54-55, Shropshire Record Office.

(39) 'The Sweet Peas of 1905', *The Garden,* (11-11-1905), pp.304-305.

(40) 'Notable Personalities', *The Agricultural Economist,* (11-1905), p.336.

(41) See, for example, 'The Old Sweet Peas', *Journal of the Royal Horticultural Society,* (1-1965), pp.23-29.

(42) 'The Weather', *The Gardeners' Chronicle,* (9-12-1905), p.416.

(43) Edith Holden, *The Nature Notes of an Edwardian Lady,* (1905; pub. 1989), pp.177, 181.

(44) 'Wem. The Late Mr. Henry Eckford', *The Shrewsbury Chronicle,* 15-12-1905, p.7.

(45) DA14/100/2, Minute Book, Wem Urban District Council, (1903-1907), pp.329-330, Shrops. R.O.

(46) 'National Sweet Pea', *The Gardeners' Chronicle,* (16-12-1905), p.431.

(47) 'Obituary. Henry Eckford, V.M.H.', *The Garden,* (16-12-1905), p.378.

(48) 'Obituary. Henry Eckford', *The Gardeners' Chronicle,* (16-12-1905), pp.431-432.

(49) 'The Late Mr. Henry Eckford', *The Journal of Horticulture,* (14-12-1905), p.547.

(50) 'Mr. Henry Eckford', *The Gardeners' Magazine,* (16-12-1905), p.811.

(51) The Burpee and Curtis letters are in the author's possession. The subsequent letters are from transcripts kindly lent by Barry Eckford.

(52) 'Mr. Alexander Malcolm, V.M.H.', *The Gardeners' Chronicle,* (16-12-1929), p.112.

(53) 'Henry Eckford, V.M.H.,' *The Sweet Pea Annual,* (1906), p.39.

(54) 'The Cult of the Sweet Pea', (advertisement), *The Daily Mail,* (21-2-1911), p.10.

Chapter Fourteen

THE HENRY ECKFORD MEMORIAL CHALLENGE CUP.

Value 50 Guineas.

John Stainer Eckford began his first year as head of the family firm by acting with emotional haste, something his father would never have done. 'We are informed by Mr. Horace J. Wright' said the first new year's edition of *The Gardeners' Chronicle* 'that the firm of Henry Eckford has placed at the disposal of the National Sweet Pea Society a silver cup, value 50 guineas, in memory of the founder of the firm. It will be offered for 12 bunches of Sweet Peas, distinct, to amateurs... at the Show to be held on July 5th next, and the society will also give a gold medal to the winner...'. A photo of the two-handled 'Eckford Challenge Cup', a female figure surmounting the lid, accompanied the article.[1]

The photo also appeared in the National Society's second *Sweet Pea Annual* out in January, as did a statistical analysis of their previous year's Show. Henry's last year guiding the Firm showed that of the top ten varieties exhibited, nine were Henry's, with 'Dorothy Eckford' the easy winner, being shown 105 times. The varieties winning the most number of first-prizes were also headed by 'Dorothy Eckford', here sharing with the non-Eckford variety 'Gladys Unwin'. Henry's varieties also filled the next six places. Of those chosen as the best varieties in the nineteen colour classes, Henry, again headed by 'Dorothy Eckford', had all bar three and one of those was closely run. The rather brash striped and picotee-edged classes we rather looked down our noses at, the American varieties 'Dainty' and 'America' allowed to take the picotee and red-striped honours respectively. Henry's 'Princess of Wales' from 1884 remained the best of the blue-stripes.[2] The *Annual* also showed John as one of a number of vice-presidents (Ellen Willmot was another) and a member of a rather large general committee, though not on the executive. John also had a whole-page advertisement in the *Annual*, as did a number of other firms.

William Robinson's new 1906 reprint of *The English Flower Garden* spoke of Henry as if he were still living. Understandable perhaps but lack of updating was common with such books. Walter Wright's list of grandifloras in his 1922 edition of *Beautiful Flowers and How to Grow Them* was unchanged from a decade earlier, though grandifloras were almost forgotten by the war's end. John, Alice or Isabella kept a scrapbook of gardening press-cuttings during John's first two or three years in charge, when it was almost all grandifloras. One of the cuttings from an unidentified journal is by him and is likely to be his only published article. He covers ground already trod but in his own style, with an eccentric spelling of the poet Keats.

'...as Keate in his day was charmed, are there not living to-day those who in the years long ago loved the quaint, old-fashioned flower, and remember the spot in the old garden where they grew and sent forth their fragrance? So that if the sweet pea had its admirers before its beauties were really discovered, it is small wonder that it is now so popular... . It is the flower among all others [to] give to your friends, send... into the convalescent home, the infirmary, and hospital... . Personally, it has made me many friends among many nations...'. (Atlee Burpee was God-father to three of his daughters). John went into culture – 'We sow as many areas as can be got ready every Autumn, the remainder in the Spring. The Autumn-sown, as a matter of course, flowers much the earlier', and gave a list of twenty-four 'splendid and inexpensive varieties', recommending a mix of both his and his rivals' varieties whose seeds he was now selling. Lacking his father's lifetime experience and abilities, and being a different person with a different outlook, he was, by uttering his father's taboo word 'inexpensive' showing his thinking – if you can't beat them, join 'em.

A lifelong Wem resident and Baptist member, H. John Phillips, born in 1905, had a friend, Harold Farlow, who worked for John at the nursery. In conversation with me in 2001, Mr. Phillips, bright as a button, recalled that the business made gas to heat and light the work premises in Market Street. They also employed a lot of women there, full-time, to make sweet pea scent (not derived from sweet peas but from bitter-oranges, apparently). Harold Farlow was one of twelve men who worked for John at Tilley. John Phillips said that the business had fields in France until the outbreak of war in 1939.

In July of 1906, David Williamson was just getting into his expansive stride in the '*Chronicle* when he was pulled up sharp by the editor. 'Nothing is more marvellous in the history of horticulture', Williamson began, 'than the evolution of the Sweet Pea, which has only been equalled by that of the Chrysanthemum'. Surely, interrupted Maxwell Masters, 'the development of the Begonia is, morphologically, even more remarkable than either?' Making a mental note to look that long word up, Williamson ploughed on. '...Mr. Lester Morse, the great American hybridist... recently sent to me from San Francisco, just before the conflagration, [the great destructive earthquake of 18th April] a most attractive classification of modern Sweet Peas, in which my own Eckfordian namesake, among many others, has the honour of being described...'.(3) Williamson was, as we've seen, inordinately proud of the bright magenta-blue flower that Henry had named for him.

Articles by the likes of the avuncular David Williamson were expected sights to sweet pea enthusiasts. One that must have had them blinking their eyes was the appearance among them of Alex Dean. Here he was, out with the National Society on one of their summer outings! The members could hardly believe their eyes. Either Alex had mellowed on Richard's death, or it had made antagonism to Richard's sweet pea society pointless.(4)

A lot was expected of sweet peas now. They had to be able to grow well in a range of conditions, including neglect; which they generally did. They had to look good under artificial light on the Show bench and in the home, and their attractive airy grace had to be robust enough to withstand rough-handling as a commercial cut-flower. *The Horticultural Trade Journal* represented the commercial growers, whose requirements would differ from those of both the amateur and the non-commercial professional gardener. The journal was usually quite blunt about what was likely to sell and what wasn't, but the height of the summer's growing season found them in reflective mood.

> 'When the late Mr. H. Eckford first exhibited a collection of improved varieties they did not create any great stir, but later on, when we got further varieties from Mr. Burpee, and such enthusiastic reports from America concerning their value, public interest gradually increased, but it is only since the [bi] centenary show and the establishment of the Sweet Pea Society, that real enthusiasm has been created. It was rarely that we saw them before the middle of June, no one thought of getting them in before their natural season. Now, however, we see them in the market early in February, and a regular supply is kept up all through the season. We also see them at all the principal flower shows from April onwards and there are numerous growers now trying to rival each other in the production of new varieties; but with all the rivalry the Eckford varieties still take the lead and though the pioneer of the sweet pea has passed away he has left a worthy successor in his son, Mr. John S. Eckford.
>
> In this year's novelties we have Agnes Eckford, a pretty shade of blush rose; Sybil Eckford (one of last year's, is a fine variety). No collection would be complete without Henry Eckford, it being one of the most pleasing and distinct shades of colour, but I am afraid it will hardly prove useful for ordinary market work. We have yet to get a rival to Dorothy Eckford as a white and Miss Willmott still holds first place as a pink for market. Of other new varieties Princess Maud of Wales, Maud Guest, Horace Wright, Mrs. Rothera and Lord Nelson are promising'. Only the creamy-pink 'Mrs. Rothera' (later renamed 'Sutton's Queen') and the navy-blue 'Lord Nelson' were non-Eckford varieties.
>
> '...Taking the varieties that are most appreciated, *Miss Willmott* probably holds first place; Countess Spencer and Gladys Unwin have sold fairly well. Of the mauve shade *Lady Grisel Hamilton* remains a favourite. We have no other white to equal *Dorothy*

> *Eckford*. Of the yellow or buff shades *The Hon. Mrs. Kenyon* is the best, and it is a great favourite with florists. In scarlets or reds *Salopian* is still extensively grown, but *King Edward VII* is taking its place and *Queen Alexandra* may prove even more useful. The only fault about these bright colours is that in hot weather they soon fade.
>
> The deep blues are not much wanted. Newer varieties which may prove useful are Evelyn Byatt, Helen Lewis, John Ingman, Helen Pierce a peculiar dark veined blue on a white ground, *Romolo Piazzani* a good mauve, Dora Breadmore is pretty for some work but would hardly be likely to sell in large quantities. Of deep colours *Horace Wright* is one of the best. Paradise, a variety of the Spencer type is a good thing and some of the market men had a great fancy for it. Miss H.C. Philbrick, and Lady Cooper, are pale lavender and should be worth a trial. *Scarlet Gem* and *Henry Eckford* are beautiful varieties but hardly of sufficient substance for market work. White Wonder, Bolton's Pink and Bolton's Blue may prove valuable. None of the striped or fancy varieties are of much use'.[(5)]

Italicising the Eckford varieties, among those 'most appreciated' you will see that, though well regarded, they are just outnumbered (eleven to thirteen) by the competition.

One thing that the probable writer here, James S. Brunton, editor and founder of *The Horticultural Trade Journal*, doesn't mention but which you may recall from an earlier chapter, was that he had Eckford blood in him. His father, also a James, had been a childhood friend of Henry's and had married Henry's sister Isabella. He had therefore been Henry's nephew. This doesn't seem to have clouded his judgment for he is making much the same noises as everyone else and is honest about the Eckford varieties' value as commercial cut-flower.

Having avoided the limelight, Henry hadn't had what we'd now call celebrity status. The name was, however, one the public was conscious of, as John's daughter Dorothy later recalled: 'I remember hearing someone say that a letter from abroad addressed only to Henry Eckford, England always reached him, so that shows how well he was known'.[(6)] If Henry hadn't told John to trade on it, he would have been wise to.

Among the many visitors to the Wem nursery that summer were at least five journalists, keen to catch the glory days before they faded from both view and memory. Reports of their visits appeared in *The Gardeners' Magazine, The Garden Home, The Gardener, The Journal of Horticulture* and *Gardening World*. You wonder how John managed to get any work done, having to show them all around. And show them all he did, drawing their attention to much the same things, which, not trusting their own judgment (*The Gardeners' Magazine's* correspondent, The Rev. Joseph Jacob was a daffodil enthusiast) they all duly reported. They also all seem to have come down by train, and were keen to extol the virtues of this increasingly sophisticated mode of travel.

> 'It was a fortunate thing for me' wrote Jacob, 'that on my way home from the great sweet pea show at Westminster, I chanced to sit opposite Mr. Eckford in the dining car from Euston. He told me his flowers were at their best, and gave me a very cordial invitation to go and see them the next day.
>
> So it was that Friday morning (July 6) found me walking along the short bit of road between Wem Station and the famous Sweet Pea grounds... . Arrived inside, I found Mr. Eckford, with whom was Mr. Digges, and we at once began out tour of inspection. We naturally started among the seedlings. There are nearly four acres of them...'. Reporters' estimates of the acreage of new unnamed seedlings varied from three to five, indicating that John wasn't too certain either. About a third of this part was devoted to the Spencer-style seedlings; John, like his father before him, seeing the way forward. Including the acreage devoted to named sweet pea varieties, rogued to keep them all pure, the visitors' estimates of the area varied at between nine and eleven acres.
>
> 'As we walked about the rows, every now and again Mr. Digges would stop and pick up something and bring it to Mr. Eckford. I soon saw he was looking out for disease. I fancy he must have come across some, either in his own or some friend's garden... . But the great expert was profoundly ignorant... . He and his father before him had grown sweet peas on this identical ground for the past eighteen years and had never had any!!' From Donnybrook, Dublin, Mr. Digges had been on Henry's testimonial committee and John was to name one of Henry's last varieties for him this year, the claret and maroon 'H.J.R. Digges.' In the bigger area of named sweet pea varieties, a photographer had recently erected

> a viewing platform, still standing. From this 'we had a birds-eye view... just when the flowers were at their best. Rows of pink, and red, and purple, and lavender, and white, and all beautifully blended' with the thin hazel twigs supporting the plants which 'reduced everything to a delicious harmony of colour...'.(7)

Jacob was one of only two reporters to remind their readers that garden peas were grown here. Their cultivation appeared to be mundane work compared to the labour of love with the sweet peas. Yet garden peas were the backbone of the nursery's success. The reporter from the new, upmarket *The Garden Home* was another railway enthusiast and in an expansive article sung its praises as well as those of the flowers around the Eckford nursery entrance. His report also included a photograph.

> 'It is a far cry from London to Wem, but by the well-appointed service of the London and North-Western Railway the journey is easily accomplished within five hours. On the return journey (viâ Crewe) we passed through several counties and watched the ever-changing scenes. Shropshire has always been famed for its sheep and wool, Cheshire for its milk and cheese, Stafford for its pottery ware, Warwick for its sylvan scenery, Bucks for its cattle and poultry, and Middlesex for its market gardens.
>
> On arriving at historic Wem we were met by Mr. John Stainer Eckford, who conducted us to his famous fields of Sweet Peas. Their perfume was borne on the air towards us. On the right of the entrance gate was a border of Geraniums, Lobelias, and Marguerites, backed by Cypressus Syringa; [there's no such conifer; perhaps *Cupressus sempervirens* is meant] and on the left were Poppies, Lupins, double Arabis, Golden Privet, with a background of Fir. As we passed along the pathway we observed on the right about 80ft. of glass by 20ft. wide and 20ft. high, which is utilised for various flowers such as Cinerarias, Primulas, Calceolarias, Cyclamens, &c. Then we enjoyed the scent of the Sweet Briar in the evening air...'.

Although glasshouses and a footpath are identified from the contemporary O.S. map, it's nice to have them described, the first visitor to do so.

The reporter also nosed around, a bit out of his depth, among the nursery's other flowers.

'In the trial borders were Snapdragon and Shirley Poppies, for early summer flowering. Here was the Nemesia strumosa, a Cape flower which is treated here as a half-hardy annual...' (that's because that's what it is!). Also noted were *Phacelia companularia*, 'Red Flax' (*Linum grandiflorum*), the 'Cloud Plant, Gypsophila glandiflora' (the Chalk Plant, G. *paniculata*, cloud-like in flower, is probably meant), Candytuft, *Salpiglossis*, *Erysimum perowkianum* (*E. perofskianum* actually, one of the alpine wallflowers), the Californian Poppy and 'some sweet-scented Stock... . A great feature of the present day Sweet Pea is the stem, which has increased in length from a few inches up to 18 and 20, which is a very common length today. One flower – a scarlet – was 2 ½ inches wide, and the stem measured 19 ½ inches...'.

The reporter also mentioned John's move to Tilley, the hamlet just south of Wem. With Alice pregnant again after six years, they decided on a move to larger premises. Perhaps they had thought too of having Charlotte and Harry back but in the end the two stayed with their aunt and gran. John, Alice and their remaining four children moved to Ferndale House in Tilley, a property which they rented from the Kynastons. Their last child, Joan Stainer Queenie Eckford (known as Queenie) was born on Monday 8th July, amid all the journalistic coming and going. The *Garden Home* reporter described the new home:

'It is a villa after the Tudor style, with arches and pergolas covered with Rambling Roses and other flowering climbers. The lawn in front is beautifully arranged, in which the Geranium has the pride of place. Behind the residence, but florally screened off, is a five-acre field of Oats, on which Sweet Peas will be sown next season. The kitchen garden covers a large area, and here everything is grown and tested for stock. Within the dwelling is a very hospitable "ingle-nook" fire-place, wherein Mr. Eckford and family enjoy a quiet siesta "far from the madding crowd"'.[8]

With fields at nearby Pankymoor to come, John was literally surrounding himself with work, while the madding crowd, one journalist strong, was already inside the house. Perhaps John was trying to escape up the chimney! The somewhat backward-looking mood of the 1870s had given rise to a 'vernacular revival' movement in English domestic architecture, of which Ferndale House was an example. (A similar mood in Scotland revived its harling-faced houses). It enlarged the cottage style into a middle-sized house for the middle-classes.(9) A fashion somewhat pretentious, it has to be said; pretentiousness made more so here with Ferndale House being rented rather than owned.

~~~

The white sweet pea variety 'Dorothy Eckford' led a dominance of Eckford varieties at the N.S.P.S.'s annual summer show. It appeared 121 times, with another six Eckford varieties among the top ten most exhibited varieties. The number of first-prize awards, however, were charting the future. Eckford varieties could only number 'Dorothy Eckford', 'King Edward VII', 'Black Knight' and 'Agnes Johnson' in the top ten. The others were all Spencer or Spencer types, led by the Watson and Breadmore's orange Countess Spencer sport 'Helen Lewis'.(10) It was unreasonable to expect John to have inherited his father's intuitive eye. Henry had been the leader on both sides of the Atlantic and the decline of the Eckford varieties was only to be expected. Their disappearance was rapid. In 1907 there were just three Eckford varieties in the top ten number of first prize winners.(11) By 1908, two and a half years after Henry's demise, the Eckford varieties were out of the reckoning.(12) The only qualification to the Spencer dominance is that the new waved varieties would be expected to dominate at exhibitions, with grandifloras still being grown in gardens. But grandifloras would slowly disappear from here too.

1907 began with the usual new year advertisements for Eckford seed in the columns of *The Gardeners' Chronicle*. As before, they ran weekly from mid-January to the end of March. This year, however, the 30th March was to be the Eckfords' last advertisement in the '*Chronicle*. John made the decision to switch his advertising to the new popular papers for amateur gardeners, *The Gardener* (which would become *Popular Gardening* in 1920) and *Amateur Gardening*. John would also target a larger audience with an enormous whole-page spread in *The Daily Mail*, during the *Mail's* build-up to a sweet pea competition of theirs in 1911. For John, the old established gardening press was the past and, possibly, too much bother as well. Being the son and not the father, John would settle down to the pleasant life of a reasonably prosperous country gentleman, respectable chapel-goer and councillor as he was. One hobby was shooting, an amusement less favoured by the Shropshire wildlife, slaughtered and injured over decades. He was a man about town too, open to whatever other charms Edwardian society might lay at his door.
~~~

John's professional expertise could be relied upon to both breed true and market the forms selected by his father. Recent Eckford novelties were covered in another article by the amateur enthusiast the Rev. David Williamson in the same *'Chronicle* issue as that containing their last Eckford advert. He covers much the same ground as the previous year's *Horticultural Trade Journal* while adding 'Earl Cromer' and 'Queen of Spain' to the Eckford varieties. Even these had been mentioned in at least the previous year's *Market Growers' Gazette* and *The Gardener* respectively, with fuller descriptions of all four novelties, 'Queen of Spain', 'Agnes Eckford', 'Horace Wright' and 'Earl Cromer' in *The Garden Home* in December. Perhaps John thought of relying on the journals to do his advertising for him! *The Garden Home* was particularly eulogistic.

In his article, Williamson wrongly assumed that 'Gladys Unwin' and 'Nora Unwin' were varieties derived from 'Countess Spencer'.

> 'Shortly before his death', Williamson began, 'Mr. Henry Eckford gave to the world several varieties of the highest distinction, which should be sufficient of themselves to immortalise his name. Supreme among these is Dorothy Eckford, undoubtedly the largest and loveliest pure white Sweet Pea in cultivation, which, although not hooded like its immediate predecessor, Sadie Burpee, surpasses it in purity and dimensions. Henry Eckford, while somewhat disappointing to exacting cultivators, is perhaps the nearest approach to a pure orange colour that has yet been introduced...
>
> The Scarlet Gem, when introduced by Mr. Eckford, made a veritable sensation, and though it is not, when compared with some others, a very vigorous grower,... ought to be grown in a half-shady situation. It is, in my opinion, a nearer approximation to a true scarlet variety than Queen Alexandra, which, in my own garden, where I had it in bloom for the first time last year, did not seem to me to be quite so luminous in colour... . [It] is, however, unquestionably very effective – I should call it the brightest of crimson-scarlet varieties – and... it does not burn in the sun. My own beautiful Eckfordian namesake, parma-violet in colour, with purple shading, has proved popular beyond my utmost anticipations, and Mr. John S. Eckford tells me in a recent letter

that there is hardly a foreign country to which he has not sent it, including China and Japan...'.[13]

One imagines the Far East recipients to be British missionaries. Their wives might have received, too, 'Eckford's Flower Odours', bottles of scent like the new 'Essence of Sweet Pea Blossoms' advertised in the autumn's new Eckford seed catalogue for 1908; 'seasonable and acceptable presents for Christmas and the New Year'. The front-cover featured the new Eckford sweet peas; 'May Perett', ivory, tinged buff and 'James Grieve', a pale yellow or sulphur-coloured flower. Another Eckford novelty, 'H.J.R. Digges', claret and maroon, was featured on the back cover with three new varieties of Dobbies, probably in respect of Dobbies William Cuthbertson, a great admirer of John's late father.

The new *Sweet Pea Annual* appeared early in the new year of 1908. It contained a photo of John, reminding readers that John, a founder member was also 'a Vice-President and a staunch supporter of the National Sweet Pea Society'.[14] With the sweet pea craze still to peak, the *Annual* noted the 'great advance in popularity of the waved varieties during the [past] year'.[15] The craze for sweet peas had been limited to Britain and the United States, despite efforts by the German growers to fan interest on the European mainland. This was now likely to change. 'Such a stir is being made by Sweet Peas in the British Isles', mused *The Horticultural Advertiser* in September, 'that one cannot help wondering whether the same excitement exists elsewhere. On the Continent, things are not very lively with regard to the butterfly flower, but they are waking up.

A representative of M. Vilmorin & Co. [of France] was present at the N.S.P.S. show on July 16th, and during the short chat I had with him he clearly showed that he knew more than a little about modern Sweet Peas...'.[16] It may have been this growing French interest, and expertise, that would eventually lead John to leave his Essex seed grounds for the more suitable climate in France.

The Eckfords' hadn't been in Essex for a while. Henry had moved the seed grounds to Clive but they had recently been moved back. A W.H. Adsett of the American journal *Horticulture* paid them a summer visit. 'It was my good fortune lately to visit the seed farm of Henry Eckford. Mr. Eckford has found it necessary to augment his facilities, has now a farm in Essex, a famous seed growing county, where fifty-three acres are devoted to sweet peas. The rows placed end to end would reach a distance of 180 miles. The outlook is much more satisfactory than was the case last year.

Countess Spencer, which in the early stages of its career was distinguished for its "sporting" tendencies, was seen in batches of over an acre, and every one true in type and tint [that's because it was Viner's seed and not Cole's].

Another striking variety was Triumph, a bold bloom of shapely form of an orange pink hue. Others in the same field were Mrs. Walter Wright, the popular mauve; Unique, an excellent striped variety; Prince Edward of York, a useful scarlet, and John Ingman. Lovely was justifying its name, the pretty shade of pink showing up in strong contrast to some of the darker tints in the adjacent rows. The Queen of Spain was another variety which arrested attention, not only on account of the admirable color, [a pearly pink] but also owing to the vigorous growth. A large amount of space was devoted to Queen Alexandra, whose fine array of bold scarlet blooms provided a warm breadth of coloring over a broad area. Another popular variety was seen at its best in Henry Eckford'.(17) 'Unique' and 'John Ingman' were non-Eckford varieties. Only half of the Eckford varieties mentioned are recent; 'Lovely', for example, goes back to 1895, 'Triumph' and 'Prince Edward of York' to 1897. America's craze had peaked and there seems to be affection now for the older varieties, despite America being well up with the new Spencer forms.

John was keen for journalists to visit, for a magazine called *The Throne and Country* 'chanced' to send its reporter there too. 'Chancing to be at Althorne the other day I took the opportunity of paying a visit to Mr. Henry Eckford's famous Sweet Pea farm where a most lovely sight met the eye. Acres and acres of sweet peas were growing for the production of seed. One or two fields were each over twenty acres in area, and full of sweet peas. Each variety is kept to itself, several rows of each being grown so that the fields appeared as though crossed by great bands of bloom – a most lovely sight. Mr. Eckford grows an immense quantity of seeds of standard varieties. He finds there is a greater demand for these than for the newer "waved" sorts. The former are not so easily affected by weather conditions, and they seem to last better when cut.'

Althorne is a village in south-east Essex on the north side of the Vale of the River Crouch. It is a long way from the triangle formed by Kelvedon, Marks Tey and Coggeshall, Britain's then largest area of hardy-annual seed growing fields, in north Essex. The Althorne fields were, naturally, near a railway and easy access to the London markets. 'It is really remarkable', their reporter continued, 'what a tremendous "boom" there is in sweet peas now... . New varieties are put on the market every year, and are quickly taken up by an expectant and enthusiastic public'.(18)

"Three popular Sweet Peas", The Garden, 13-2-1909

The shapes that hybridists had gone for in sweet peas had often been criticised, but usually by outside observers. 'The coarseness that is creeping into the modern waved section' offered to the public was discussed by Walter Wright and Charles Curtis on a visit to the Eckfords' Soulton Road grounds in August – 'not because there were coarse flowers there, but because the topic came up' Wright later recalled. The problem was considered 'more a question of form than of size.... There may be a thirty-inch stem and there may be four huge flowers but their association is such that there is no real beauty in them...' .(19)

Late autumn saw the arrival of the new 1909 Eckford catalogue, and showed John embracing the new technology. Alongside 'telegrams: "Eckford, Wem" 'came the notice 'Telephone: N°.4' (later N°.6!). Two Eckford novelties were displayed on the front cover; 'Dodwell F. Browne', a bright crimson Spencer type and 'Mrs. Charles Masters', a salmon-pink and cream grandiflora. A third, 'Annie B. Gilroy', deep cerise, shared the back cover with Robert Bolton's 'Mrs Henry Bell'. A photo on page twenty-five, entitled 'A summer scene on Eckford's seed farm – hoeing sweet peas', shows, before a background of working horses, workmen in long-sleeved shirts, waistcoats and long trousers, one even in a jacket (though that's probably John, there for the photo) and all wearing hats. It was explained to me years later at work, by one of the over-dressed old-school gardeners, that 'what keeps out the cold keeps out the heat'.

~~~

With the new year of 1909 came the new *Sweet Pea Annual*. Of 'the late Henry Eckford', wrote a contributor, 'time serves only to increase our appreciation of his work. Not only did Mr. Eckford raise new varieties, but the stocks he distributed were all fixed. Now that the market is flooded with so many unfixed strains, we can understand, as never before, the importance of this fact, for had Mr. Eckford, in the early stages of the development of Sweet Peas, distributed unfixed strains – and it was as easy to do so then as it is now – the popularity of our favourite flower might have been rendered impossible. It is certain that it would have been greatly hindered...'.(20)

About the only way that John could exceed his father's record was as a councillor. On 21st April the Wem Urban District Council elected him as their vice-chairman for the coming twelve months.(21) The extra responsibility meant some re-jigging of his out-of-work activities and he stood down from the NSPS's general committee.
~~~

In 1910, the sweet pea craze was coming to a peak in Great Britain. It was the last year also that the National Society tried to produce a catalogue of sweet pea names. It was all getting to be too much! 'Judging by the number of lists issued by growers of sweet peas', wrote Thomas Bolton (Robert Bolton's son Thomas presumably) of Exeter in a February edition of *The Gardeners' Chronicle*,

> '1910 bids fair to outstrip any previous year, especially with respect to the number of so-called novelties put before the market. A dozen years ago cultivators were spared the worry of having to make a selection from such a number of lists, the late Mr. Henry Eckford's catalogue being in those days the guide to all good things in the Sweet Pea world'. (Meaning, that it wasn't any more). 'Dorothy Eckford, Miss Willmott, Lady Grisel Hamilton, and Henry Eckford were probably four of the best and most distinct varieties sent out by the Wem firm. The last-mentioned variety had held its own, as far as colour is concerned, but will this season be easily surpassed by the newer variety, Nancy Perkins, of similar shade, a much larger flower, and with the addition of a waved standard, without which latter recommendation no Sweet Pea now appears to have much of a chance of recognition...'.(22)

In March, the '*Chronicle* were mistakenly naming 'Picotee' as one of four Eckford 'novelties' or new varieties. The true Eckford ones were 'Vicomte de Zanzé', a deep bright rose; 'Mrs. E. Gilman', a pale rose, and 'Mary Vipau', a rich rose Spencer type.(23) We are probably coming to the end of Henry's varieties now, they taking a while to 'fix', and these may be John's hybrids and selections. 1910 was the highlight of John's council career, for on 20th April he was elected chairman of Wem Urban District Council for the coming twelve months.

In the short time that the Eckford Memorial Challenge Cup had been in existence the Surrey Scot Thomas Stevenson of Addlestone had won it twice. In the autumn, he combined the sweet pea craze with the Christmas market to bring out his *The Modern Culture of Sweet Peas*. Even he, however, championing sweet peas, thought that things had got a little out of hand.

> 'Among amateur gardeners', he wrote, 'the Sweet Pea is nowadays the one flower with them, many discarding every other flowering plant to make room for more of their favourites. I could point to the gardens of several of my amateur friends where there is

> nothing else in the garden – vegetables and everything being displaced by Peas. I don't say that this is as it should be, but it just shows the hold, or I might say the deep root, that the "cult" of the Sweet Pea has taken on the general public…'.[24]

The National Sweet Pea Society's over-emphasis on exhibition blooms had pushed the new waved forms. Here, Stevenson points out that 'good, free-flowering varieties of quite decided colours [are] very much the best for garden decoration', naming Henry's 'Dorothy Eckford', 'King Edward VII' and 'Queen Alexandra', along with the now old 'Lady Grisel Hamilton' of 1899 and the cerise-coloured 'Coccinea' of 1901 among a recommended mix of waved and plain forms.[25] For market growers and home decoration he includes 'Dorothy Eckford', 'Queen Alexandra' and 'Coccinea' again, with the inclusion of Henry's mauve 'Mrs. Walter Wright', among his selections.[26] 'I have heard it said', Stevenson writes, 'that buyers do not care for the Spencer form of flower… . Given a suitable season, however, they will go up in popular favour, even for the market…'.[27] So, even at the crest of the sweet pea mania, outside exhibition halls it was the plain or grandiflora type that dominated, not the new waved form. Closer to the wild, original form, grandifloras had caused the craze in the United States, as they were again doing in the U.K.

Another autumn sweet pea event was the arrival of the Eckford seed catalogue. John was claiming three novelties for 1911, 'Astä Ohn Waved', 'Cerise Waved' (a cherry-red) and 'Luminosa' (deep coral), a grandiflora. There was already the American lavender 'Astä Ohn Spencer', from Morse in 1909, making John's variety a duplicate. 1909, incidentally, was the year that the Frenchman, Louis Bleriot had become the first to fly across the English Channel, so winning a £1,000 prize offered by *The Daily Mail*. The vogue for sweet peas had not escaped the *Mail's* notice either, and in February 1911 they announced a £1,000 first prize for a bunch of sweet peas, open to amateurs employing no more than one gardener.[28] A young gardener, Alec White from the village of Sprouston (near the village of Eckford!) in the Scottish borders drew it to the attention of one of his clients, the parish minister, who decided to enter for it, [29] as did some 30,000 other people nationwide.

The day after *The Daily Mail* announcement, John took out a full page advertisement in their paper.[30] This seems an extravagant expense but full-page ads were one of the *Mail's* fortes, and as it was indirectly promoting the *Mail's* own competition, John probably got a good deal. John was, after all, quite sharp and 'in the know' about the competition. The advertisement was an illustrated article by a journalist describing a visit to 'the sweet pea farm of Henry Eckford' where he was shown around by John. It's quite informative. If a *Daily Mail* reader sat down and concentrated he or she would learn that 'the aim in sweet-pea culture is not wholly to produce new varieties; it is likewise to improve existing varieties, and this in three directions – shape; colour; size... a great advance has been made in... the increase in the number of and the grouping of the blossoms on the stalk, and the lengthening of the stem. A long stem or stalk is very desirable in that it makes possible the use of much deeper vases...'.

John claimed to send out 100,000 - odd seed catalogues a year. 'Twenty years ago 500 catalogues – modest ones at that – sufficed to supply the customers of Henry Eckford. To-day two hundred times five hundred are not enough...'. The advertisement ended with a drawing of the Eckfords' 'Countess Spencer' sweet pea, showing its wing petals waved, the standard upright. 'Eckford's original True Countess Spencer. The forerunner of all the Waved Sweet Peas...'. A slight exaggeration, as you'd expect, but not far off the mark.

In July, the winner of the £1,000 prize for a bunch of sweet peas went to Alec White's parish minister, the Rev. D. Denholm Fraser. (Actually his wife Nettie won – he'd entered one bunch in her name and one in his).[(31)] Almost a novice grower, his other bunch won the third prize of £50! With the help of Alec White he'd rapidly assimilated the expertise that had been built up in the Scottish borders. Clever experiments simulating the flowers' journey to London all helped. His first-prize varieties, 'John Ingman, Mrs. Hugh Dickson, Arthur Unwin, King Edward Spencer and a dark sort, like Nubian or Tom Bolton'[(32)] were all waved types, to appeal to the judges. They all came from different companies though none were Eckfords'. The Eckfords could however claim ancestry, especially of Burpee's 'King Edward Spencer'.

John was claiming more than his fair share of 'novelties' in the Henry Eckford seed catalogue for 1912: 'Constance Oliver', 'Dorothy Tennant Waved' and 'Miss Guest'. 'Constance Oliver' was a 1908 introduction and not John's, though he could have been claiming a superior strain. The new 'Dorothy Tennant' was a waved form of Henry's old 1890 rosy-mauve self. Although the original grandiflora had taken seven years to 'fix', seven years on from Henry's death the Eckford novelties must now all be John's. Although they were to make little or no impression, sweet peas were in 'the fire of transformation'[33] as Walter Wright had put it, and varieties now appeared and disappeared almost without notice.

One item in the catalogue didn't, however, escape the notice of the gardening press. 'Eckford's £1,000 in prizes for Sweet Peas' proclaimed *The Gardener* on 13th January, with a longer article in the same day's *Gardeners' Magazine*.

> 'Those who are interested in the cult of the sweet pea, whether owners of ducal or cottage gardens, will this year have sufficient encouragement to sustain their interest in their favourite flower. We have received information from various parts of the country to the effect that sweet peas will be more liberally encouraged by the various societies than has yet been the case, and now comes the announcement that Mr. Henry Eckford, of Wem, whose great work in connection with sweet peas is known to all, offers £1,000 in cash prizes to be competed for in the month of July. Mr. Eckford has very wisely so arranged this large amount that, while the prizes will be more than sufficient to secure an enormous competition, the various sections of cultivators will compete on equal terms...'.

The paper went on to explain the various classes, one for everyone, from the richest to the poorest, including one for 'a girl or boy under 16 years of age... who must grow the flowers without assistance... [the *Mail* had had a "children's bunches" too]. The judging will take place on Thursday, July 18, and each competitor must send the bunch by parcels post to reach Wem on Wednesday, July 17. The conditions governing the competition are fully set forth in Mr. Eckford's catalogue, which has just been published'.[34] John Eckford was now taking out large three-quarter page advertisements in *Amateur Gardening*, and one headed 'Another £1000 for Sweet Pea lovers' also appeared on 13th January. A fortnight later, *Amateur Gardening* informed its readers that there would be five classes. In each of the first three classes there would be one large prize of £200 and fifty of £1. In the remaining two classes the large prize would be of £100, with twenty-five £1 prizes.[35]

The Journal of Horticulture's coverage came later, on 1st February. '... Mr Eckford is going to provide £1,000 for special prizes, but offered in a way that will meet with more general approval than was the case in the great "Daily Mail" prize last year... There is bound to be a big demand for entry forms, and for Eckford's seeds'.[36] The stunt-like nature of *The Daily Mail* event had obviously not been to the liking of the sweet pea cognoscenti and John had taken their concerns on board with his spread of prizes. There would be a run on Eckford seed because to enter you had to send flowers from varieties listed in their last two catalogues.

For the *Mail* to give away £1,000 was one thing. For the Eckford nursery it was quite another. Where would the money come from – the entry fees? Was that the hope? No-one thought to ask. The *Mail* event had been free to enter, as would John Eckford's. The *Mail's* one had also required a 300′ marquee, with two rows of tables along its entire length; the help of five hundred boy scouts, accommodated in nearby tents; the expertise of eminent judges, the facilities of the great Crystal Palace, and arrangements made with the Post Office and railway companies to avoid chaos. You could see that, even with a fraction of the *Daily Mail* entries, a Wem event was going to be an awful lot of extra work for a busy commercial nursery, with few resources other than land.

Out soon after the Eckford catalogue came the N.S.P.S.'s *Sweet Pea Annual*. John had stopped advertising now and there was no mention of his sweet pea competition here. There was no mention of last year's *Daily Mail* one either, despite the *Mail's* proprietor, the now ennobled Alfred Harmsworth, placing his wife Lady Northcliffe as their president last year. They were above that sort of thing. Entrants to the Eckford Sweet Pea Competition were given detailed instructions on how to send their flowers, with illustrated instructions on how to pack them in a light wooden box, '23 inches long, 8 ¼ inches wide, and 1 ¾ inches deep, inside measure'.(37) All serious stuff. What happened on the great day, July 18th, is, generously speaking, unclear. Alone of all the gardening press, some of the winners' names appeared in the 7th September edition of *The Gardener*, the paper most wholly associated with the fashionable sweet pea.

'There are some familiar names among the winners of prizes in the Eckford competition' they wrote. 'Edward Cowdry, Robert Hallam, Dr. J. E. Phillips and Mr. James Hall, to wit. But the winners of first prizes: Mr. E. Evans, of Ruabon; Mr. Wm. Davies of Malpas, Mr. Robert Smith, of Cunnock, N.B. [North Britain, i.e. Scotland] Mr. Samuel Woollams of Whitchurch; and Miss E. Davies, of Malpas, are not so well known'.(38) It wasn't so surprising for main prize winners to be unknown. You only had to remember last year's *Daily Mail* winners. But the big money winners may have been unknown (and I think this is what *The Garden* is hinting at) because they didn't exist. As the late John Good told me, 'the rumour was that nobody won anything'.

It's likely that all the 'prize winnners' did exist. William Davies, for example, from nearby Malpas, was a gardener to a Dr. Phillips there, and 'Miss E. Davies' was his thirteen-year-old daughter Edith. The (short-lived) *Wem Herald* of Thursday 25th July reported that William had won the first prize of £200 for professional gardeners, and that Edith had won the first prize of £100 for children under sixteen. They also reported that Samuel Woolams (or Woolam) of The Chemistry, Whitchurch, had won £100 in the amateur gardeners' section. What is odd is that, in a nationwide competition with almost certainly thousands of entrants, four of the five main prize-winners should come from within a fifteen-mile radius of Wem. This gives rise to the suspicion that the competition was fixed; that the growers were known to John and would, for a reasonable consideration, allow their names to go forward as prize-winners for prizes that were never given.

What *did* happen now was that John became unusually busy on the various sub-committees of Wem Urban District Council.(39) Was he hiding away among friends? Another thing that happens now is that John is no longer listed as a member of The National Sweet Pea Society. Perhaps he had already decided on this – a progression from his leaving their general committee in 1909. But it was odd to leave the organisation which had meant so much to him; John Eckford, a vice-president, a founder member with so many friends here, whose photograph had appeared above his description as a 'staunch supporter' in 1908. Following on so closely from his £1,000 sweet pea competition it was certainly an odd coincidence. Baring the odd passing reference, John's name wouldn't be mentioned here again. Of course, clubs and organisations have to live in the present and look to the future. Old faces do get forgotten. But John was to be around in the sweet pea world for a long time to come. There was still a 'Henry Eckford Memorial Class' at the 1913 N.S.P.S.'s Summer Show, (40) with its Cup, N.S.P.S.'s medals and money prizes. This continued to 1915, when the Shows stopped for the war.

It was as if John had died in that war.

Sweet peas were now so popular (and profitable) that there was temptation to cheat. John Eckford's claim on 'Constance Oliver' as a new Eckford variety was but one high-profile instance of deceptions apparently common, like the similar practice of giving other people's varieties another name and passing them off as yours. In reaction it led to complaints and calls for control, picked up by the '*Chronicle* in October 1913. The first was from Miss Hemus, a raiser with thirty-four varieties in the catalogues. It was for the N.S.P.S. to be run by disinterested and 'independent amateurs'[41].

The *'Chronicle* picked up on other complaints. 'Ever since the introduction of the large waved Sweet Peas' said the Duchess of Hamilton, opening the Glasgow Flower Show in September, 'the decline of perfume has been marked... So long as the demand has been for varieties of flowers to win on the exhibition table, raisers have kept the qualities of form and size continually before them to the almost entire exclusion of others such as habit and scent...'.[42] Both complainants women, interestingly. It was the time of the woman's suffrage movement and women were finding their voice. A decline in scent had already been noted. 'How can we stop the scent going in Sweet Peas?' *The Gardener* had written in November 1912. 'It certainly is going, and it really looks as if we are within measurable distance of the time when Sweet Peas will be no longer sweet: I suppose that if scent were bred for it could be got, but before our raisers think seriously of it there must be public opinion to move them'.[43]

One way of assessing popularity is to look for detractors. 1913 also saw the publication of *The Garden of Ignorance*, the first book by a South-African born writer, Marion Cran, soon to win fame as the first radio gardener. '*Sweet-Peas* are flowers I have strenuously avoided mentioning throughout the book, inasmuch as your sweet-pea lover is as narrow-minded an enthusiast as your rose lover. I have no doubt that if my soil were rich, or if I had two or three gardeners and an unlimited purse, I would be a sweet-pea enthusiast; but seeing that these lovely flowers are gross feeders and need quite as much attention as the rose garden I have left them severely alone, being unable, from some innate contrariness, to tackle them without a distant hope of doing them justice. Books have been written about sweet-peas; every garden paper has its columns devoted to the subject, and anything I may try to say is impertinence. Seeing these flowers even boast a society of their own... it is plainly obvious that they need no words of mine to urge their claims'.[44]

~~~

John Eckford attended his last Council meeting a few months on from the outbreak of the Great World War, on 17th March, 1915. There was a local election now, it seems. But with a war on *The Shrewsbury Chronicle* had more important news to print, and ignored it. By the following year, grandiflora sweet peas were being similarly ignored, an observer noting
~~~

'the almost universal adoption of the Spencer type now occupying their place'.(45)

In October 1921, the following appeared in the pages of *The Gardeners' Chronicle*, of which, incidentally, Charles Curtis was now the editor: 'To celebrate this, its coming of age year, the National Sweet Pea Society inaugurated a fund for the purpose of instituting a medal in memory of the late Mr. Henry Eckford, the pioneer raiser of Sweet Peas. It was agreed that the medal should be awarded annually to someone who had done good work on behalf of the Sweet Pea and the National Sweet Pea Society...'.(46) More elaborately, 'the award will not necessarily be made to exhibitors of Sweet Peas of superior culture, but may be awarded for achievements in hybridising, for literary or research work, for investigation of diseases or pests, or for any form of meritorious work helpful to the development and popularity of the Sweet Pea or the National Sweet Pea Society'.(47) Robert Bolton of Surrey, an old friend of Henry's, was the first recipient. Henry's nephew James Brunton would receive it in 1933. After a few years, the medal fund ran into problems and donations were sought. None, however, came from John Eckford who was not only no longer a member but who was having financial worries of his own.

The marriages of five of John and Alice's six daughters, happy events though they were, would have been a financial drain. Charlotte married George Ryley in nearby Oswestry in October, 1913. The wedding in Wem Parish Church of Emily to James Barclay of Orkney in October, 1922, maintained a Scottish identity for Emily. Three years later, in the same church, Dorothy's bouquet in her marriage to Norman Fairweather of Stretham in June, 1925 was, naturally, of her grandfather's sweet pea 'Dorothy Eckford'. A bridesmaid held a bouquet of sweet pea 'Royal Scot', a popular crimson variety that would still be listed and promoted in the Eckford's catalogue for 1932. The 'Dorothy Eckford' sweet pea, along with 'Lady Grisel Hamilton', were probably Henry's two finest. In their day they elicited no criticism, only praise, and no collection of the old grandifloras should be without them. Kathleen (Kitty) married Harold Prosser in Cardiff in January 1926 while Agnes (Nan) married Arthur Caslaw in Wem Parish Church in July 1932. Let's hope the grooms' parents helped out a bit with expenses! Like Kitty, Harry also married over the border, in Dukestown, in December 1926, leaving Joan (Queenie), the youngest, as the only unmarried one. Queenie lived on into 1991, the last of her generation.

By the 1920s, the sweet pea craze had passed. Plants can go in and out of fashion, like anything else. In the nineteenth-century, flowering plants like Camellias and Verbenas, architectural plants like the monkey-puzzle tree *Auracaria auracana* and foliage plants like ferns all had a vogue. Sweet peas merely followed on from these. Henry could, and would, charge sums for some of his latest sweet and garden pea seed that were complained of then and would just make no commercial sense if enacted today. Thought expensive, they were however mere echoes of the prices charged in Holland for hyacinths in the eighteenth-century or tulips in the seventeenth. 'As much as £200 has been known to be given for a [Hyacinth] root' then, Loudon informs us,[50] while 'tulipomania' endangered the actual Dutch economy.

In Victorian Britain, growing pineapples successfully was your benchmark as a gardener. The garden pea, along with *the* Victorian flower, the rose, seem to have been the most continually popular plants, as, with daffodils, they surely are now, although we've tended to look to commerce to grow our peas for us.

Were sweet peas out of favour now, in the 1920s; a commercial liability? 'What of the present time?' wrote W.J. Unwin's son Charles in 1926. 'Is the Sweet Pea as popular as ever? My answer is decidedly "yes". Perhaps the feverish enthusiasm of the first ten or twelve years after *Countess Spencer* was introduced has abated, but a far wider and more sober interest has taken its place. As a seedsman, I can definitely state that more Sweet Peas are being sold and grown'.[51] The flower was so high in the public eye, in fact, that Queen Mary, the wife of King George V, visited Vincent Square and the NSPS's summer Show there in 1933. She may have become a regular visitor, her presence noted again in 1936. Conscious as she was of protocol, the Show organisers may have had to run around after her with a roll of red carpet. She lived to 1953, 'the last of the Victorians'.

The continuing popularity of sweet peas kept Marion Cran for one, in opposition. In her 1933 *I Know a Garden* she contrasts them unfavourably with nasturtiums which 'will give you an infinite range of colour, far more thrilling, far more brilliant, than the sweet peas people make such a fuss about and grow on bamboo poles; they are easy, too – which sweet peas are not! You throw the seeds down where you want them to grow and leave them to do the rest...', an advantage that most hardy annuals have over the cultivated *Lathyrus*. 'Some flowers', she noted, 'keep their old-fashioned names more tenaciously than others; heartsease is passing out, I fear, but sweet pea, for instance, remains sweet pea with becoming obstinacy. No one calls these butterfly flowers of the summer-time Lathyrus odoratus',[52] an observation as true now as it was then.

Whatever problems were now besetting the Eckfords' nursery weren't due to people's loss of interest in the sweet pea. Problems started in Wem in the late 1920s, though there was no hint of any in the forty pages of a rare surviving Eckford catalogue for 1932. Here, sweet peas with names like 'Eckford's Scarlet', 'Nannie Eckford' (a large white), 'Mrs. John Eckford' (rose) and 'Margaret Eckford' (apricot pink on a cream ground) showed that the nurseries were still happily hybridising. They probably always had been, year after year, if to no great acclaim since Henry's day. (Alice and John's daughter Agnes was called Nan or Nannie; Margaret was their son Harry's wife).

Financial difficulties finally came to a head in 1934, the following notice appearing in *The Shrewsbury Chronicle*: 'Salop Bankruptcy Court. An expensive and unproductive advertising scheme was stated to be the cause of failure of John Stainer Eckford, of Ferndale House, Tilley, Wem, trading as "Henry Eckford" seedsman, who appeared for his public examination in bankruptcy before the registrar (Mr. Allan G. Hughes) at Shrewsbury on Wednesday.

His statement showed liabilities expected to rank of £386 13s 5d., and a deficiency of £203 5s. 5d.

Examined... he said he had been in the seed business since 1888.

- In 1929 a company was promoted to take over the business, with a nominal capital of £6,000. He was allotted 1,000 shares. Only £1,400 was subscribed in actual cash, and the business failed about 15 months later.
- He purchased some of the fixtures and implements and a small amount of stock from the liquidator, and carried on.
- The business paid until an unfortunate expedition into advertising.
- "It has not paid to advertise in your case", observed Mr. Hughes.
- Eckford said that the total cost of an advertising campaign was £303, and the value of the orders received in response only £42. The advertising agency obtained a judgment against him.

- In his time, Eckford said, he had spent many thousands on advertising, and it had generally been successful.

The examination was closed'.[53]

Wherever John had lost money, it wasn't through advertising in the traditional outlets. There were no Eckford advertisements in the *R.H.S. Journal* or the *Sweet Pea Annual,* nor, it seems, in *The Gardeners' Chronicle* or the more likely *Amateur Gardening* and *Popular Gardening* either. The depression years of the 1930s weren't good ones to be out of work in. Not here, far from the fortunate south of England. Perhaps the business could be revived somehow. After all, it isn't a mine or a shipyard we're dealing with...

Eight months later, under 'Bankruptcy discharge. Wem application adjourned', *The Shrewsbury Chronicle* reported:

> 'John Stainer Eckford (70) of Ferndale House, Tilley, Wem, applied to Judge H.W. Samuel, K.C., at Shrewsbury County Court on Monday, for his discharge in bankruptcy.
>
> It was stated that an order was made in May, 1934, and a first and final dividend of 6s 2 ½ d. was paid on proofs amounting to £385.
>
> Mr. P.M. Millward (Official Receiver) opposed the application, which adjourned for a month to enable Eckford to pay a sum which would bring the total dividend paid to 10s.'[54]

John's application must have been successful. But there had had to be changes. The business no longer had the Soulton Road grounds or, it seems, any commercial connection to Wem proper. The firm was now trading solely from Tilley as 'Henry Eckford, FRHS',[55] keeping their address as 'Wem, Shropshire' for customers. If a nursery continued in Soulton Road, it wasn't being run by the Eckfords. The loss of their seed farms, still presumably in France, would have turned the Eckford firm from a national and international business into just another local nursery. This doesn't seem to have happened. Not yet.

The view from south London, where Alice and John's daughter Dorothy now lived, was different. As Dorothy later wrote: 'The 2 wars killed the Business – we were not allowed to grow flowers – it had to be vegetables – but then it was necessary I suppose'.[56] The outbreak of World War Two in 1939 and the loss of the French fields must have been the last straw for John and Isabella, both now well into their seventies. Seed-growers, including the Eckfords, continued to offer a limited selection of flower seed, despite the ban. The last line in what must have been their last catalogue, however, an 'Abbreviated War Time List of Eckford's Glorious Sweet Peas' in 1943[(57)] shows the reality. 'We shall be unable to supply Sweet Pea plants this season' it read.

Marion Cran meanwhile, still writing with an enviable gaiety at the end of her life, considered her attitude towards the sweet pea misunderstood. In *Hagar's Garden* she explained herself by way of a dialogue. ' "Here I grow my sweet-peas! You do not like them, I know, but I love them".

"Oh no", I protested, "you have me wrong! Sweet-peas are lovely flowers, but I did get very bored once when I lived near a rich man in Surrey who spent money and labour enough for a racing stable... on his Show sweet-peas. He didn't care what they cost or how much of a bore he was until he had won the last gold medal possible. And then he didn't bother any more! I am sure yours are not grown in that spirit."'[(58)] Over in the United States, the reverence for sweet peas and the molly-coddling that exhibition blooms required attracted the attention of E.C. Segar and his violent, anarchic *Popeye* strip-cartoon in the 1930s.[(59)] Mess with Popeye and he'd 'lay ya among th' swee' peas'. Segar also created 'Swee' pea', Popeye's 'adoptid boy-kid', a tiresome infant requiring Popeye's constant attention.

~~~

Emily, Henry's second wife, died just short of her ninety-ninth birthday on Wednesday 25th March, 1942. Sixty-two when her husband had died, she never remarried and lived on, almost certainly with Isabella in Wem's Shrubbery Gardens 'for several years' as *The Shrewsbury Chronicle* put it.[(60)] She was buried in the same parish cemetery as Henry, [(61)] though in her own grave. Isabella died just seven months later, on Thursday 15th October. She lies beside her step-mother. Both graves lack headstones, the absence of which is usually blamed on the war. For Isabella it seems indicative. Her mother not her mother, her children not her children, her work imposed, her life other people's. In death, ignored. Not entirely ignored. The funeral drew her nieces and nephew, among others, though not her brother John, 'absent through indisposition'.[(62)]

John outlived his sister by sixteen months. He died aged seventy-nine at home in Ferndale on Tuesday 18th April, 1944 and was buried alongside the graves of Emily and Isabella. Close together in life, they remain so now, Henry's towering gravestone some distance away. *The Shrewsbury*
~~~

Chronicle report, with some reading between the lines, went as follows: '... a native of Berkshire, he had lived in the Wem district since 1888' (at his father's insistence) 'and took an interest in the business of his father... Mr. J.S. Eckford became one of the largest growers of sweet peas in Europe' (following on from his father) 'and he also had much success with violas' (as his father had). 'He was a F.R.H.S.' (a member of the R.H.S., like his father) 'a former chairman of Wem Urban District Council' (following his councillor father) 'and the first joint secretary of Wem Show...'.

At the funeral, on the following Saturday, were family members, former members of the nursery and warehouse staff 'and many others...'(63). An important figure, an important funeral, his grave marked then or later with a stone memorial. Alice lived on to 1961 and the age of ninety-two,(64) the years between John's birth and her death making an impressive ninety-seven. The span in Henry and Emily's case, 1823-1942, was a quite extraordinary one, both in terms of years (119) and of events witnessed. In 1823 Britain was still the rural society that it always had been. Those that could afford to, travelled by stage-coach. The only railways were wagons pulled by pit-ponies in mines whose cavernous mouths fed equally on men, women and children. Edinburgh's Georgian New Town was twenty-odd years away from completion. London's Buckingham House hadn't yet been transformed into Buckingham Palace (and was much criticised when it was). Culturally it was the age of Byron, Keats and Shelley and of the Germans Goethe and Beethoven, although there would be no country called Germany for another fifty years. A twelve year-old Charles Dickens was sticking labels on to jars of blacking in a Thames-side warehouse near London's Charing Cross (where Hungerford Bridge now is).

When Henry died in 1905 he had witnessed a transformation in science, industry and technology. Britain had become an urban society. Horticulture had experienced its golden age in Britain, led by the great plant collectors and by its head gardeners and plant hybridists, of whom Henry was one of the most prominent. There was as yet no radio (or 'cats' whiskers' or 'wireless' as the early crystal sets were known). And of course there was no television. Henry may never have seen a motor-vehicle. By 1942 Britain had experienced a great world war and was in the midst of yet another. It was the dawn of the nuclear age. Diversion was provided by fairs still but also by the cinema and by the wireless with shows like *ITMA* and *Happidrome*. As we used to say, 'you can't beat progress' and then stopped saying it.

Perhaps Henry's grandparents, like the one who told his parents "if you make that boy anything but a gardener you'll spoil him", perhaps when Henry was older they told him about the old days. About *the black auchty-twa*, the great freeze of 1782, when Henry's father had been a little boy of three. You just said the year auchty-twa and everyone knew what you meant. Visiting Edinburgh, they would have passed by the dwellings of the rich and famous, like that of the Boswells who knew Dr. Johnson. They would have passed too among the homeless and destitute, estimated at 5,000 of the city's 50,000 population in 1760.[(65)] It would have confirmed Henry's decision to stay on the land, like his father, who had made a good living from it. For Emily, recalling her own Berkshire childhood of the eighteen-forties and 'fifties, life in inter-war Britain was contrast enough. To compare Shropshire in the 1930s with stories of Georgian England and Scotland must have been beyond imagining, yet they were stories she could have heard at first hand, from both Henry and her own parents.

~~~

The longer hours worked in former times show that work was the main focus of people's lives, with little time, money or energy to pursue other interests. For gardeners like Henry, attention was expected to be paid to a wide variety of tasks. Too wide a variety, quite often, as these two letters to *The Gardeners' Chronicle* from 1898 and 1912 reveal:

> 'Being in want of a situation, I inserted in a local newspaper the following advertisement:- "Gardener (head), age 30, experienced in Vines, Melons, Cucumbers, Peaches, Chrysanthemums, Orchids, Stove and greenhouse plants, kitchen and flower gardening, &c."
>
> I received this reply. "October 15, 1898. In answer to your advertisement… I want a head-working gardener to take a position of trust. I shall be glad to see your references from your present and previous employers, and state your reason for leaving in both cases. In addition to the allowance I make, viz., free house, coal, vegetables, and lamp oil, I only wish to give 19s. or 20s. per week, to include wife's services in scrubbing office-floor and stairs once a month, boiling poultry food daily, washing during our visit here, attending to my wants on weekly visit, and preparing house for our annual visit. I live here only about three or four months in the year, but pay weekly visits, and keep one
~~~

> horse to meet me at the station, it being looked after by a youth who assists in the garden, and whom the gardener is expected to board at 6s. to 7s. per week. Everything will be in the gardener's charge. The total area of my property is about four acres, divided into an acre of kitchen garden, two glasshouses about 50 feet long, each containing Vines, &c.; house and other buildings, plantation, orchard, lawns, etc. What family have you? Of course, I should object to the children running about the garden at any time. Are you a total abstainer [from alcohol]? If my terms are not acceptable, state what would be".'[66]

Fourteen years later, little had changed.

> 'We both hear and read a great deal nowadays concerning the education of gardeners, and the best lines to be followed in the matter. It seems to me that in the case of a great many gardeners a very important part of their education would be to teach them how to live on the miserable pittance that so many employers pay them. Small wonder that many leave the profession for other callings, which, though less pleasant than gardening, afford higher wages and regular hours. Some employers, when engaging a gardener, consider they have also a right to the services of the gardener's wife. Furthermore, the term gardener is a most elastic one, and includes the working of electric light plant, the care of poultry, stock, and agricultural matters in general, as well as in some cases singing in the church choir, and playing cricket. That these are some of the multifarious duties expected of a gardener may be seen by reference to the advertisement columns of the horticultural journals'.[67]

As a head gardener on big estates, Henry had been in a position to delegate away many of these onerous duties, leaving him free to concentrate on his hybridising work. All the same, he must have been relieved to finally be his own master in 1888. These complaints about private-service conditions continued throughout the twentieth-century. Particularly irksome was the 'gardener's wife' aspect, assuming the gardener to be a married heterosexual male, his wife content to be someone else's unpaid skivvy.

In the late nineteenth and early twentieth-century, increasing knowledge led to divisions in horticulture and professionals have become deeper but narrower in their knowledge. A good nineteenth-century gardener would have been a competent joiner in order to maintain glasshouse and frameyard lights. He would have had a command of lawn culture (today's groundsman or woman) and of fruit-tree grafting and pruning (today's specialist fruit-grower). He could have both designed a bedding scheme (an office technician's job now) and planted one out (done by today's amenity horticulturist). These are just a few examples. Concomitantly, with rising incomes and increased leisure-time have come vast numbers of amateur enthusiasts, whose knowledge is generally wider but shallower (unless they're sweet pea aficionados!).

The plants and ideas of the Victorians dominate our parks and gardens today. Much of the attraction of the cultivated garden or park is in their contrast with a wilder hinterland. We need the uncultivated in order to better enjoy the cultivated. The vogue for natural plant associations and of wild areas in gardens, even of the return of the wild and grandiflora sweet pea is, in part, a reaction both to an over-urbanised landscape and to the loss of original, simple beauty in so many over-hybridised garden plants.

> What would the world be, once bereft
> Of wet and wildness? Let them be left,
> O let them be left, wildness and wet;
> Long live the weeds and the wilderness yet.[(68)]

That was Gerald Manley Hopkins, London born, Highland inspired. Emerson, the American, was making a similar plea in the 1850s, anticipating Hopkins and the Victorian back-to-nature movement by a decade or two. There was still then a largely unspoilt United States of America for Emerson to contrast with England: 'There, in that great sloven continent, in high Alleghany pastures, in the sea-wide, sky-skirted prairie, still sleeps and murmurs and hides the great mother [nature], long since driven away from the trim hedge-rows and over-cultivated garden of England.'[(69)]

Two cries from the nineteenth-century that re-echo for us in the twenty-first.

~~~

In the 1960s, Wem was getting round to discovering and honouring its former eminent residents. A new private housing estate, built next to Henry's former fields in the Soulton Road was named Hazlitt Place by the urban parish council, despite the rural district council's objection that it was too distant from Hazlitt's Noble Street address. (A plaque went up on Hazlitt's house in around 1981). *The Shropshire Journal* of 5th January, 1968
~~~

suggested another name, one more appropriate to Soulton Road. 'The Eckford name lives on in horticultural circles all over the world, including America, where there is an Eckford Society of sweet pea growers. Yet very real links with Wem seem to have been strangely forgotten in the town itself.... As Henry Eckford put the town on the map, so his name should be put on the town map'.(70) All to no avail, however, on this occasion.

Another chance came in the 1980s with another housing estate going up near Wem Mills, on the south side of town. The site, having no direct connection with Henry or even John was nevertheless named 'Eckford Park' in Henry's honour. Should the English smugly assume that nowhere else could events be managed with such a fine degree of incompetence let me remind them of events that occurred during the construction of Edinburgh's Georgian New Town. Remember Henry's testimonial that was supposed to have taken place in Edinburgh's St. Andrew Square? Well, a church of St. Andrew was supposed to have gone there too, a focal point to match that of another in Charlotte Square, the two squares at either end of George Street. Money and influence saw the attractive St. Andrew church demoted to a position part-way along George Street (among a present-day hotel and bank offices), a grand town house (dwarfed by later buildings) in its place.

In Charlotte Square, the town council rejected the eminent Robert Adam's design for a church, choosing one inferior and cheaper. This emerged, in the correct place, as the impressively eccentric St. George's Church (now West Register House, where I researched Henry's Oxenfoord and Kenwood employments). As was later said, 'the New Town has suffered a twin misfortune in its churches, for St Andrew's is the right church in the wrong place, and St. George's is the wrong church in the right place'.(71) Incidentally, an avenue of trees would compliment both George Street and on into Melville Street. It would break up the streets' solidity and severity.

Along with Eckford Park, the 1980s in Wem also saw the emergence of an Eckford Sweet Pea Society, led by the much missed John Good. A centenary celebration was planned to commemorate Henry's arrival in the town in 1888. With the death of Henry's grand-daughter Charlotte in 1985 there was now no longer any family member alive who would have known him. (Any living grandchildren through Peter were unlikely to have met him). A reporter from *Traditional Homes* described the centenary events of July, 1988. 'Those of you who know this small Shropshire town will be aware that, though noted for the worthiness of its citizens and the propriety of its commercial enterprises, Wem is not usually a first choice for unbridled celebration... Imagine then, my surprise to find this sober little town in carnival mood: flowers everywhere, the brass band oompahing, the church bells ringing, the shop windows overflowing with flowers and the pubs with bonhomie'. There's a Show in the town hall, he was told, an exhibition and a flower festival, with the mayor naming 1988 'Henry Eckford Year'.[72]

Every year since, on the third weekend of July, the Eckford Sweet Pea Society organises an exhibition and sweet pea Show. In the nineteen-nineties, the road name signs were changed by the town council. They all now display a bunch of sweet peas, a lovely touch. Stand close, and in a quiet moment you can catch the scents of history.

The End.

References to Chapter Fourteen

(1) 'Eckford Memorial Cup', *The Gardeners' Chronicle*, (6-1-1906), p.10.
(2) 'The Best Sweet Peas', *The Gardeners' Magazine*, (3-2-1906), p.87.
(3) 'New Sweet Peas', *The Gardeners' Chronicle*, (7-7-1906), pp.12-13.
(4) 'Twickenham and Feltham', *The Sweet Pea Annual*, (1907), p.29.
(5) *Horticultural Trade Journal*, (25-7-1906), heading and paging unknown.
(6) Letter from Dorothy Fairweather, née Eckford, to Bernard Jones, (3-11-1979).
(7) 'Sweet Peas at Wem', *The Gardeners' Magazine*, (4-8-1906), p.517.
(8) 'The Headquarters of the Sweet Pea', *The Garden Home*, (18-8-1906), pp.41-42.
(9) Andrew Saint in C. Matthew, ed., *The Nineteenth Century*, (2005 edn.), p.269.
(10) 'Audit of the great Sweet Pea Show of 1906', *The Sweet Pea Annual*, (1907), pp.36-41.
(11) 'Audit of the Sweet Pea Show, 1907', *The Sweet Pea Annual*, (1908), pp.56-60.
(12) 'Audit of the Sweet Pea Show, London, 1908', *The Sweet Pea Annual*, (1909), pp.54-59.
(13) 'Sweet Peas of Recent Introduction', *The Gardeners' Chronicle*, (30-3-1907), p.208.
(14) Photo, *The Sweet Pea Annual*, (1908), p.30.
(15) 'Audit of the Sweet Pea Show, London 1907', *The Sweet Pea Annual*, (1908), p.58.
(16) 'Random Sweet Pea Notes', *The Horticultural Advertiser*, (25-9-1907), p.13.
(17) 'British Horticulture', *Horticulture*, (29-8-1908), p.283.
(18) 'The Pleasure Garden', *The Throne and Country*, (29-8-1908), p.393.
(19) 'Refinement versus Coarseness', *The Sweet Pea Annual*, (1909), p.17.
(20) 'Recent developments among sweet peas', *The Sweet Pea Annual*, (1909), p.31.
(21) DA14/100/3, Minute Book, Wem Urban District Council, (1907-1912), p.147, Shropshire Record Office.
(22) 'Florists' Flowers', *The Gardeners' Chronicle*, (5-2-1910), p.83.
(23) 'Florists' Flowers', *The Gardeners' Chronicle*, (26-3-1910), p.195.
(24) Thomas Stevenson, *The Modern Culture of Sweet Peas*, (1910), p.5.

(25) Ibid, p.64.

(26) Ibid, p.68.

(27) Ibid, p.67.

(28) '£1,000 for a bunch of Sweet Peas', *The Daily Mail,* (20-2-1911), p.7.

(29) Henry Donald, *A Bunch of Sweet Peas*, (1995 edn.), p.14.

(30) 'The Cult of the Sweet Pea', advertisement, *The Daily Mail*, (21-2-1911), p.10.

(31) Henry Donald, op. cit., pp.34-35.

(32) 'The "Daily Mail" Sweet Pea Competition', *The Gardeners' Chronicle*, (5-8-1911), p.90.

(33) Walter P. Wright, *The Perfect Garden*, (1908), p.4.

(34) 'Another £1,000 for Sweet Peas', *The Gardeners' Magazine,* (13-1-1912), p.22.

(35) 'Items of Interest', *Amateur Gardening*, 27-1-1912, p.603.

(36) 'Current topics', *The Journal of Horticulture*, (1-2-1912), p.103.

(37) 'Eckford's £1,000 Sweet Pea Competition', acknowledgment of entry, (1912), author's possession.

(38) 'Sweet Peas', *The Gardener*, (7-9-1912), p.323.

(39) DA14/111/1, Minutes of Committees' Book, (1900-1913), Wem Urban District Council, 1912-1913 passim, Shrops. R.O.

(40) 'National Sweet Pea', *The Gardeners' Chronicle*, (26-7-1913), p.71.

(41) 'National Sweet Pea Society', *The Gardeners' Chronicle*, (20-9-1913), p.205.

(42) 'Perfume in Sweet Peas', *The Gardeners' Chronicle,* (20-9-1913), p.205.

(43) 'Sweet Peas', *The Gardener*, (9-11-1912), p.423.

(44) Marion Cran, *The Garden of Ignorance*, 1913, (1924), p.251.

(45) 'What Scotsmen have done for the Sweet Pea', *The Sweet Pea Annual,* (1917), p.52.

(46) 'First award of the Henry Eckford Memorial Medal', *The Gardeners' Chronicle*, (22-10-1921), p.204.

(47) 'A Memorial to Henry Eckford', *The Sweet Pea Annual*, (1921), p.30.

(48) 'Wedding', *The Shrewsbury Chronicle*, (22-10-1922), p.8.

(49) 'Wedding', *The Shrewsbury Chronicle*, (26-6-1925), p.10.

(50) J.C. Loudon, *Encyclopaedia of Gardening*, (1822), p.941.

(51) Charles W. J. Unwin, *Sweet Peas: Their History, Development, Culture*, (1926), p.9.

(52) Marion Cran, *I Know a Garden*, (1933), pp.35, 163.

(53) 'Salop Bankruptcy Court', *The Shrewsbury Chronicle,* (8-6-1934), p.9.

(54) 'Bankruptcy Discharge', *The Shrewsbury Chronicle*, (22-2-1935), p.14.

(55) *Kelley's Directory of Shropshire*, (1937), p.293.

(56) Letter from Dorothy Fairweather, née Eckford, to Bernard Jones, (3-11-1979).

(57) Eckford's seed catalogue, (1943), p.10, author's possession.

(58) Marion Cran, *Hagar's Garden*, (1941), pp.211-212.

(59) M.Higgs, ed., *Popeye: The 60th Anniversary Collection,* (1989), passim.

(60) 'Wem', *The Shrewsbury Chronicle,* (27-3-1942), p.5.

(61) 'Wem', *The Shrewsbury Chronicle,* (3-4-1942), p.5.

(62) 'Wem', *The Shrewsbury Chronicle,* (23-10-1942), p.5.

(63) 'Wem', *The Shrewsbury Chronicle,* (28-4-1944), p.5.

(64) 'Death of Mrs. A. Eckford at Tilley', *The Shrewsbury Chronicle*, (12-5-1961), p.13.

(65) Gilbert Summers, *Historic Edinburgh,* (1986), no paging.

(66) 'The duties and wages of a gardener', *The Gardeners' Chronicle*, (26-11-1898), p.390.

(67) 'The Education of Gardeners', *The Gardeners' Chronicle*, (27-7-1912), p.74.

(68) From 'Inversnaid' by Gerald Manley Hopkins, in W. Scammel, ed., *This Green Earth*, (1992), p.47.

(69) R.W. Emerson, *English Traits*, 1856, (1902 edn.), p.168.

(70) 'Fragrant Memories of Old Wem', *The Shropshire Journal*, (5-1-1968), p.6.

(71) G. Summers, op.cit., quote.

(72) 'Sweet Memories', *Traditional Homes*, (11-1988), p.95.

APPENDICES

Were I to detail the books which I have consulted, and the inquiries which I have found it necessary to make by various channels, I should probably be thought ridiculously ostentatious.

James Boswell,
Advertisement, The Life of Samuel Johnson, 1791, London.

APPENDIX A.
Plants originated by Henry.
Lathyrus odoratus, Sweet Peas.

The dates after the name refers, as far as can be ascertained, to their first mention in the horticultural press. Although Henry died in 1905, Eckford varieties up to 1909 have been included, they taking a while to 'fix'.

G.C.	The Gardeners' Chronicle.
G.L.	Garden Life.
G.M.	The Gardeners' Magazine.
H.E.S.C.	Henry Eckford Seed Catalogue.
J. of H.	Journal of Horticulture.
M.G.G.	Market Growers' Gazette.
J.R.H.S.	Journal of the Royal Horticultural Society.
S.P.B.C.R.	Sweet Pea Bicentenary Celebration Report.
S.P.A.	Sweet Pea Annual.
T.G.	The Gardener.
T.Gdn.	The Garden.
T.G.H.	The Garden Home.

A

Apple Blossom	6-8-1887	G.C.
Alice Eckford	4-10-1894	J. of H.
Agnes Johnston	3-1-1903	G.L.
Achievement	27-10-1904	J. of H.
Agnes Eckford	26-7-190	J. of H.
Annie B .Gilroy	1909	H.E.S.C.

B

Bronze Prince	10-8-1882	J. of H.
Blue King	2-9-1882	G.C.
Blue Beauty	1-9-1883	G.C.
Boreatton	8-8-1889	J.of H.
Blushing Beauty	1-9-1892	J. of H.
Blanche Burpee	12-8-1893	G.C.
Black Knight	10-12-1898	G.M.
Black Michael	27-10-1904	J. of H.

C

Cardinal	26-7-1884	G.C.
Charmer	1-8-1885	G.C.
Charming	1-8-1885	G.C.
Captain of the Blues	8-8-1889	J. of H.
Countess of Radnor	1-9-1892	J. of H.
Countess of Aberdeen	12-8-1893	G.C.
Captivation	4-10-1894	J. of H.
Coquette	29-2-1896	G.M.
Countess of Shrewsbury	29-2-1896	G.M.
Crown Jewel	29-2-1896	G.M.
Coccinea	1898	J.R.H.S.
Colonist	10-12-1898	G.M.
Calypso	1900	S.P.B.C.R.
Chancellor	1900	S.P.B.C.R.
Countess Cadogan	1900	S.P.B.C.R.
Countess of Powis	2-6-1900	G.C.
Countess of Lathom	21-7-1900	G.C.

D

Duchess of Albany	2-9-1882	G.C.
Duchess of Edinburgh	1-8-1885	G.C.
Delight	6-8-1887	G.C.
Dorothy Tennant	13-2-1892	G.C.
Duke of Clarence	12-8-1893	G.C.
Duchess of York	4-11-1893	G.C.
Duke of York	4-11-1893	G.C.
Duchess of Sutherland	29-2-1896	G.M.
Duke of Sutherland	29-2-1986	G.M.
Duchess of Westminster	1898	J.R.H.S.
Duke of Westminster	10-12-1898	G.M.
Dorothy Eckford	2-8-1902	G.C.
David R. Williamson	20-1-1906	G.L.
Dodwell F. Browne	1909	H.E.S.C.

E

Emperor	30-8-1883	J. of H.
Empress of India	1-9-1883	G.C.
Emily Eckford	1-9-1892	J. of H.
Eliza Eckford	12-8-1893	G.C.
Excelsior	12-8-1893	G.C.
Earl Cromer	11-7-1906	M.G.G.

F

Fascination	1-9-1883	G.C.
Firefly	12-8-1893	G.C.

G

Grandeur	2-9-1882	G.C.
Gaiety	1-9-1892	J. of H.
George Gordon	28-7-1900	G.C. & G.M.
Gracie Greenwood	4-1-1902	G.C.

H

Hon. F. Bouverie	10-12-1889	G.M.
Her Majesty	5-9-1891	G.C.
Hon. Mrs. E. Kenyon	28-7-1900	G.M.
Henry Eckford	1906	J.R.H.S.
Horace Wright	14-7-1906	T.G.
H.J.R. Digges	18-8-1906	T.G.H.

I

Indigo King	1-9-1883	G.C.
Imperial Blue	26-7-1884	G.C.
Isa Eckford (Two varieties)	26-7-1884	G.C.
Ignea	5-9-1891	G.C.
Isa Eckford	1908	S.P.A.

J

Jeannie Gordon	3-8-1901	T.G.
James Grieve	1908	S.P.A.

K

King Edward VII	3-1-1903	G.L.

L

Lavender Gem	1-9-1883	G.C.
Leviathan	1-9-1883	G.C.
Lottie Eckford (Two varieties)	1-9-1883	G.C.
Lady Penzance	1-9-1892	J. of H.
Lemon Queen	12-8-1893	G.C.
Lottie Eckford	12-8-1893	G.C.
Lady Beaconsfield	4-10-1894	J. of H.
Lady Harlech	4-10-1894	J. of H.
Little Dorrit	4-10-1894	J. of H.
Lady Grisel Hamilton	29-2-1896	G.M.
Lovely	29-2-1896	G.M.

Lady Mary Currie	10-12-1898	G.M.
Lady Nina Balfour	10-12-1898	G.M.
Lady Skelmersdale	10-12-1898	G.M.
Lord Kenyon	28-7-1900	G.M.
Lady Margaret Ormsby Gore	4-8-1900	G.M.
Lord Rosebery	4-1-1902	G.C.

M

Mrs. Eckford (Two varieties)	1-8-1885	G.C.
Maggie Ewing	6-8-1887	G.C.
Mauve Queen	6-8-1887	G.C.
Miss Hunt	6-8-1887	G.C.
Mrs. Gladstone	8-8-1889	J. of H.
Mrs. Sankey	5-9-1891	G.C.
Monarch	1-9-1892	J. of H.
Mrs. Eckford	1-9-1892	J. of H.
Meteor	12-8-1893	G.C.
Mrs. Joseph Chamberlain	12-8-1893	G.C.
Mrs. Dugdale	4-10-1894	J. of H.
Maroon	1895	J.R.H.S.
Mikado	30-3-1895	G.C.
Mars	29-2-1896	G.M.
Mrs. Fitzgerald	1898	J.R.H.S.
Miss. Willmott	28-7-1900	G.M.
Mrs. Walter Wright	3-1-1903	G.L.
Mrs. Knights-Smith	26-9-1903	G.L.
Maud Guest	26-1-1906	J. of H.
May Perrett	11-7-1906	M.G.G.
Marchioness of Cholmondeley	1908	S.P.A.
Mima Johnston	1908	S.P.A.
Mrs. Charles Masters	1909	H.E.S.C.

N

Novelty	12-8-1893	G.C.

O

Orange Prince	30-8-1883	J. of H.
Othello	10-12-1889	G.M.

P

Princess	2-9-1882	G.C.
Princess of Wales	26-7-1884	G.C.
Purple King	26-7-1884	G.C.
Primrose	6-8-1887	G.C.
Purple Prince	5-9-1891	G.C.
Peach Blossom	1-9-1892	J. of H.
Princess May	12-8-1893	G.C.
Princess Victoria	12-8-1893	G.C.
Perfection	4-10-1894	J. of H.
Prince of Orange	1895	J.R.H.S.
Prima Donna	29-2-1896	G.M.
Prince Edward of York	29-2-1896	G.M.
Prince of Wales	21-7-1900	G.C.
Princess Maud of Wales	14-7-1906	G.C.
Purple King	18-8-1906	T.G.H.
Primrose Waved	1908	S.P.A.

Q

Queen of Roses	1-9-1883	G.C.
Queen of the Isles	26-7-1884	G.C.
Queen of England	8-8-1889	J. of H.
Queen Victoria	29-2-1896	G.M.
Queen Alexandra	1906	S.P.A.
Queen of Spain	14-7-1906	G.C.

R

Rosalind	1-8-1885	G.C.
Royal Robin	1-9-1892	J. of H.
Romolo Piazzani	27-10-1894	G.C.
Royal Robe	30-3-1895	G.C.
Royal Rose	1896	H.E.S.C.

S

Salmon Queen	1-9-1883	G.C.
Splendour	6-8-1887	G.C.
Stanley	1-9-1892	J. of H.
Senator	12-8-1893	G.C.
Salopian	30-3-1895	G.C.
Shahzada	29-2-1896	G.M.
Sadie Burpee	10-12-1898	G.C.
Scarlet Gem	26-9-1903	G.L.
Sybil Eckford	26-7-1906	J. of H.

T

The Belle	12-8-1893	G.C.
The Doctor	4-10-1894	J. of H.
The Bride	1895	J.R.H.S.
The Queen	1895	J. R.H.S.
Triumph	29-2-1896	G.M.

V

Victoria	1-9-1883	G.C.
Venus	12-3-1892	T. Gdn.

W

Waverley	6-10-1894	J. of H.
White Waved	1908	S.P.A.

Verbenas.

Ace of Trumps, Anna Keynes, Bravo, Christine, Conspicua, Countess of Radnor, Dictator, Earl of Radnor, Earl Russell, Eclipse, Fanny Martin, Fanny Purchase, Firebrand, George Brunning, George Peabody, Goliath, Grand Duke, Grand Monarch, Hugh Low, Isa Brunton, Isa Eckford, King Charming, Kingcraft, Lady Ann Spiers, Lady Braybrooke, Lady Edith, Lady Folkestone, Lady Gertrude, Lavender Queen, Lord Derby, Lord Folkestone, Lotty Eckford, Magdala, Master Jacob, Mauve Queen, Memorial, Miss

Charlotte Mildmay, Mr. Brunton, Mr. Ellice, Mr. Laing, Mrs. Bouverie, Mrs. Dodds, Mrs. Eckford, Mrs. Knight, Mrs. Levington, Pearl, Peter William, Pluto, Polly Perkins, Rose Imperial, Royalty, Sensation, Star, Stuart Low, The King.

Pisum sativum. Garden Peas.

Ambassador, Amorial, Aston Gem, Censor, Centenary, Chieftan, Colossus, Consummate, Copious, Critic, Dawn, Diamond, Duke of Connaught, Dwarf Monarch, Eckford's No. 1, Empress, Epicure, Essential, Fame, Gem, Heroine, Ideal, Invincible, Jubliee, Juno, Magi, Magnificent, Memorial, New Wrinkled, Perpetual, Pioneer, Potentate, Prior, Progress, Record, Renown, Rex, Royalty, Shropshire Hero, Superabundant, Superiority, The Bruce, The Clive, The Don, The Echo, The Sirdar, Victor, Wem, Wem Wonder.

Dahlias.

Ace of Trumps, Achievement, Christine, Conspicua, Cremorne, Earl Russell, Eliza, Firebrand, General Gordon, Isa Key, Julius, Lady Folkestone, Lord Derby, Maggie Gerring, Master Jacob, Memorial, Mr. Brunton, Mr. Laing, Mrs. Eckford, Novelty, Royal Queen, Sensation, The Hon. Mr. Bouverie, The Hon. Mrs. Bouverie, Willy Eckford.

Others.

Aster novi-belgii. Michaelmas Daisy.
Dwarf Triumph.
Cucumis sativus. Cucumber.
Coleshill.
Pelargonium x hortorum. The Zonal or Bedding Geranium.
Coleshill.
Primula sinensis. Chinese Primula.
Effulgent, Elegance, Salopia.
P.s. Fimbriata.
Eckford's Crimson, Eckford's Lilac, Snow Flake.
Viola x wittrockiana. Pansy.
Bronze Prince.

This is likely to be an incomplete list. I have left out any very doubtful or short - lived names. There were references in the horticultural press to unnamed Pelargonium hybrids and to 'strains' and 'races' of Henry's pansies, Primulas and Cinerarias. He must have worked on other genera too, of which there appears to be no record.

APPENDIX B
Bibliography
Periodicals

The dates refer to the periods covered in the text.

Amateur Gardening, 1912.
American Journal of Botany, 1938.
Botanical Journal of the Linnean Society, 1996.
Bromley and District Times, 1906.
Curtis's Botanical Magazine, 1793-1855.
Der Naturforschende Verein in Brünn,(Moravia), 1865.
Garden Life, 1903.
Gardening Illustrated, 1889-1900.
Gloucestershire Life, 1973.
Horticulture, (U.S.A.), 1908.
Hutchins' Sweet Pea Annual for 1897.
International Journal of Plant Sciences, 1994.
Journal of the Royal Horticultural Society, 1872-1906 & 1965.
Journal of the Society for the Bibliography of Natural History, 1968.
L'llustration Horticole, (Belgium), 1892.
Swansea Herald and Neath Gazette, 1886.
The Agricultural Economist, 1904-1905.
The Cottage Gardener, 1854. (Becomes The Journal of Horticulture in 1861).
The Daily Mail, 1900, 1911 & 2008.
The Dalkeith Advertiser, 1894.
The Dundee Advertiser, 1864.
The Florists' Exchange, (U.S.A.), 1905.
The Garden, 1892-1905.
The Garden Home, 1906.
The Gardener, (Scottish), 1875-1879.

The Gardener, (English), 1904-1912. (Becomes Popular Gardening in 1920).
The Gardeners' Chronicle, 1841-1935.
The Gardener's Magazine, 1832-1842.
The Gardeners' Magazine, 1893-1912.
The Gardening World, 1889-1905.
The Horticultural Advertiser, 1907.
The Horticultural Trade Journal, 1906.
The Journal of Horticulture, 1889-1912.
The Mendel Journal, 1909-1912.
The New York Times, 1880.
The Salisbury Journal, 1862.
The Scottish Gardener, 1853-1863. (Becomes The Gardener in 1866).
The Shrewsbury Chronicle, 1888-1961.
The Shropshire Journal, 1968.
The Shropshire Magazine, 1956-1958.
The Sweet Pea Annual, 1906 – 1922.
The Throne and Country, 1908.
The Times, 1900.
The Wem Herald, 1912.
Traditional Homes, 1988.

Books, including inventories, registers, catalogues and reports.

Abercrombie, J., (Thomas Mawe's) *The Universal Gardener and Botanist...* 1778, 2nd edn. 1798, G. G. & J. Robinson, London.

Accession Register, Vols. 1865, 1866, 1867, The Linnean Society, London.

Anderson, R. D., *Education and the Scottish People 1750-1918,* 1995, Clarendon Press, Oxford.

Ash, R., Higton, B., *Gardens in Art,* 1994, Aurum Press, London.

Austin, A., *The Garden That I Love,* 1906, Adam and Charles Black, London.

Balfour, M., *Britain and Joseph Chamberlain,* 1985, Allen & Unwin, London.

Bamford, M., *A Short History of Sandywell Park,* c.1966, M. Bamford, Gloucester.

Barker, C. M., *Flower Fairies of the Garden,* 1944, 2002 edn., Frederick Warne, London.

Bartholomew, J., *Yew and Non-Yew,* 1998, Arrow Books, London.

Barton, D.A. *Discovering Chapels and Meeting Houses,* 1975, 1990 edn., Shire,

Princes Risborough.

Bateman, J., *The Great Landowners of Great Britain and Ireland,* 1878, 4th edn. 1883, reprint 1971, Leicester University Press, Leicester.

Battiscombe, G., *Mrs. Gladstone: The Portrait of a Marriage,* 1956, Constable, London.

Bauhin, J., *Historiae Plantarum,* 1651, Vol. 2, (no publisher given), Yverdon (Switzerland).

Beckinsale, R. P. *Companion into Berkshire,* 1952, 1972 edn., Spurbooks, Buckinghamshire.

Beeton's All About Gardening, c .1890, Ward, Lock, London.

Best, G., *Mid-Victorian Britain, 1851-75,* 1971, Weidenfeld and Nicholson, London.

Billing, M., *Billing's Directory and Gazetteer of Berkshire and Oxfordshire,* 1854, Martin Billing, Birmingham.

Bly, J., *Discovering English Furniture,* 1976, reprint 1993, Shire, Princes Risborough.

Boom, B.K., Kleijn, H., *The Glory of the Tree,* 1966, George G. Harrap, London.

Boswell, J., *The Life of Samuel Johnson,* 1791, 1906 edn., Vol.2, J.M. Dent & Co., London.

Boyd, E., *British Cookery,* 1976, 2nd.edn. 1988, Helm, Bromley (Kent).

Brimble, L. J. F., *Intermediate Botany,* 1936, 4th edn. 1962, Macmillan, London.

Bronowski, J., *The Ascent of Man,* 1973, reprint 1974, B.B.C., London.

Brown, D., *Four Gardens in One,* 1992, H.M.S.O., Edinburgh.

Brown, J., *The Pursuit of Paradise,* 1999, HarperCollins, London.

Brown, Y., *Boreatton Park,* 1989, Y.Brown, Ruyton XI Towns.

Bryant, J., Colson, C., *The Landscape of Kenwood,* 1990, English Heritage, London.

Bryant, K., Giles R., eds., *Queen Camel, Now and Then,* 1990, Queen Camel Playing Field Committee, Queen Camel.

Bryson, W., *A Short History of Nearly Everything,* 2003, Doubleday, London.

Burke, B., *A genealogical and heraldic dictionary of the peerage and baronetage of the British Empire,* 59th edition, 1897, H.Colburn, London.

Burnby, J.G.L., Robinson, A.E., *And They Blew Exceeding Fine,* 1976, Edmonton Hundred Historical Society, Enfield.

Burnby, J.G.L., Robinson, A.E., *Now Turned into Fair Garden Plots,* 1983, Edmonton Hundred Historical Society, Enfield.

Campbell-Culver, M., *The Origin of Plants,* 2001, Headline, London.

Catalogue of Scientific Papers, Vol.8, 1864-1873, (1879), The Royal Society, London.

Cervantes Saavedra, M.de, *Don Quixote,* 1604-5 and 1615. Translated by John Rutherford. 2002 edn., reprint 2003, Penguin, London.

Chadwick, O., *The Victorian Church, Part 2, 1860-1901,* 1970, A.&C. Black, London.

Chandler, J., *Salisbury : History and Guide,* 1992, Allan Sutton, Stroud.

Chekhov, A., *Stories,* Vol.1. Translated by Constance Garnett. 1968, Heron Books, London.

Christiansen, R., *The Visitors: Culture Shock in Nineteenth-Century Britain,* 2000, Chatto & Windus, London.

Clark, G.K., *The Making of Victorian England,* 1962, 1966 edn., Methuen & Co., London.

Clarke, A., *Finding Out About Victorian Schools,* 1983, Batsford, London.

Clayton-Payne, A., Elliott, B., *Victorian Flower Gardens,* 1988, Weidenfeld & Nicholson, London.

Coats, A. M., *Flowers And Their Histories,* 1956, Hulton, London.

Cobbett, W., *Rural Rides,* 1830, 1967 edn., reprint 1985, Penguin, London.

Cobbett, W., *The English Gardener,* 1829, 1996 edn., Bloomsbury, London.

Countess of Ashburnham, *Lady Grisell Baillie: A Sketch Of Her Life,* 1892, R.&R. Clark, Edinburgh.

Commelin, C., *Horti Medici Amstelodamensis,* Vol.2, 1701, Lugdani Batavorum, Amsterdam.

Cox, E.H.M., *A History of Gardening in Scotland,* 1935, Chatto & Windus, London.

Cox, K.N.E., Curtis-Machin, R., *Garden Plants for Scotland,* 2008, Frances Lincoln, London.

Cran, M., *Hagar's Garden,* 1941, Herbert Jenkins, London.

Cran, M., *I Know A Garden,* 1933, Herbert Jenkins, London.

Cran, M., *The Garden of Ignorance*, 1913, 1924 edn., Herbert Jenkins, London.

Cupani, F., *Hortus Catholicus*,1696, Franciscus Benzi, Neapoli.

Curtis, S. J., Boultwood, M.E.A., *An Introductory History of English Education Since 1800*, 1960, 4th edn. 1966, University Tutorial Press, Foxton (Cambridgeshire).

Darwin, C., *The Variation of Animals and Plants Under Domestication*, Vol.2, 1868, John Murray, London.

David, V., ed., *Brno*, c.1972, Rapid, Brno.

Davies, J., *Saying It With Flowers*, 2000, Headline, London.

Davies, J., *The Victorian Flower Garden*, 1991, BBC Books, London.

Davies, J., *The Victorian Kitchen Garden*, 1987, BBC Books, London.

Davies, S., *A Century of Troubles*, 2001, Pan Macmillan, London.

Dean, R., ed., *The Sweet Pea Bicentenary Celebration*, 1900, R. Dean, Ealing.

Degler, C.N., *Out of Our Past: The Forces That Shaped Modern America*, 1959, 3rd edn. 1984, Harper & Row, New York.

Desmond, R., *Dictionary of British and Irish Botanists and Horticulturists*, 1977 & 2nd. 1994 edns., Taylor & Francis, London.

Diary of a Victorian Gardener: William Cresswell and Audley End, 2006, English Heritage, Swindon.

Dickens, C., *Reprinted Pieces*, 1898 edn., Chapman and Hall, London.

Dickson-Scott, J., *Scottish Counties*, 1937, reprint 1950, Thomas Nelson, Edinburgh.

Donald, H., *A Bunch of Sweet Peas*, 1988, reprint 1995, Canongate, Edinburgh.

Duthie, R., *Florists' Flowers and Societies*, 1988, Shire, Princes Risborough.

Edwards, P., *Rural Life*, 1993, B.T. Batsford, London.

Elliott, B., *The Royal Horticultural Society; A History: 1804-2004*, 2004, Phillimore & Co., Chichester.

Elliott, B., *Victorian Gardens*, 1986, Batsford, London.

Emerson, R. W., *English Traits*, 1856, reprint 1902, Unit Library, London.

Ensor, R.C.K., *England, 1870-1914*, 1936, reprint 1949, O.U.P., London.

Evans, H., Evans, M., *The Victorians At Home And At Work: As Illustrated By Themselves*, 1973, David & Charles, Newton Abbot.

Famous German Novellas of the 19th Century, 2005, Mondial, New York.

Ferenbach, C., *Annals of Liberton,* 1975, Howie & Seath, Edinburgh.

Fitter, A., *An Atlas of the Wild Flowers of Britain and Northern Europe,* 1978, Collins, London.

Forging the Modern Age:1900-14, 1999, The Reader's Digest, London.

Forman, H.B., ed., *Keats's Poetical Works,* 1924, O.U.P., London.

Fox, G., ed., *Coleshill 2000,* 2000, A. Buratta, Coleshill (Oxfordshire).

Fraser, D.D., *Sweet Peas: How To Grow The Perfect Flower,* 1913, The Daily Mail, London.

Galthorne-Hardy, R., *The Berkshire Book,* 2nd.edn. 1951, Berkshire Federation of Women's Institutes, Reading.

Genders, R., *Scented Flora of The World,* 1978, Granada, Frogmore (Hertfordshire).

George, F.P., *The London Conductor,* 1851, reprint 1984, Boethus Press, Kilkenny.

Gillespie, C.C., ed., *Dictionary of Scientific Biography,* 1974, Scribner, New York.

Good, G., *Liberton in Ancient and Modern Times,* 1893, Andrew Elliott, Edinburgh.

Goode, C.T., *Trentham – The Hall, Gardens and Branch Railway,* 1985, (no publication details).

Gorer, R., *The Growth of Gardens,* 1978, Faber & Faber, London.

Grant, J., *Old and New Edinburgh,* Vol.3, 1882, Cassell, Petter & Co., London.

Greenslade, M. W., Stuart, D.G., *A History of Staffordshire,* 1965, Darwen Finlayson, Beaconsfield.

Groome, F.H., *Ordnance Gazetteer of Scotland,* Vol.1, 1882, Vol.3, 1883, Vol.5, 1884, W. Mackenzie, London.

Grossmith, G., Grossmith, W., *The Diary of a Nobody,* 1892, 1945 edn., Penguin, Harmondsworth.

Hadfield, M., *A History of British Gardening,* 1960, 1969 edn., Spring Books, London.

Harris, J., ed., *The Garden,* Exhibition held at the V.&A., London, 1979, (exhibition catalogue), Mitchell Beazley, London.

Hart-Davis, A., *What the Victorians did for us*, 2001, Headline, London.

Harvey, J., *Early Nurserymen,* 1974, Phillimore & Co., Stroud (Gloucestershire).

Hellyer, A.G.L., ed., *The Gardeners' Golden Treasury,* 1966, W.H.& L. Collingridge, London.

Henig, R.M., *A Monk and Two Peas,* 2000, Weidenfeld & Nicholson, London.

Heslop, J.A.B., Unpublished mss. on Coleshill House and Gardens, c.1990, Berkshire National Trust.

Hibberd, S., *The Amateur's Flower Garden,* 1871, Groombridge & Sons, London.

Hibbert, C., *The English,* 1987, reprint, 1988, Guild Publishing, London.

Higgs, M., ed., *Popeye: The 60th Anniversary Collection,* 1989, Hawk Books, London.

Hill, G., *Canaan,* 1996, Penguin, London.

Hilliers' Manual of Trees and Shrubs, 1975 edn. David & Charles, Newton Abbot.

Holden, E., *The Nature Notes of an Edwardian Lady,* 1989, Webb & Bower, Exeter.

Horn, P., *Labouring Life in the Victorian Countryside,* 1976, Gill & Macmillan, Dublin.

Howitt, W., *The Rural Life of England,* 1838, 1845 edn., Longman, Brown, Green & Longman, London.

Huggett, F. E., *Victorian England as seen by Punch,* 1978, Book Club Associates, London.

Hughes, A.M.D., *Cobbett,* 1923, reprint 1946, O.U.P., London.

Hughes, M. V., *A London Family, 1870-1900,* 1946, O.U.P., London.

Hunter, T., *Woods, Forests and Estates in Perthshire,* 1883, Henderson, Robertson & Hunter, Perth.

Inventory of Gardens and Designed Landscapes in Scotland, Vol.4 Tayside, Central and Fyfe; Vol. 5, Lothians and Borders, Countryside Commission for Scotland, 1987, Battleby.

Jalland, P., *Death in the Victorian Family,* 1996, 1999 edn., O.U.P., Oxford.

Janes, E.R., *Sweet Peas,* 1953, reprint 1961, Ward Lock, London.

Johnson, G.W., ed., *The Gardeners' Dictionary,* 1856, reprint 1877, George Bell, London.

Jones, R.N., Karp A., *Introducing Genetics,* 1986, John Murray, London.

Kauvar, G.B., Sorensen, G.C., *The Victorian Mind,* 1969, Cassell, London.

Keane, W., *The Beauties of Middlesex...* 1850, Chelsea.

Keay, J., Keay, J., eds., *Encyclopaedia of Scotland,* 1994, 2nd. edn. 2000, Harper Collins, London.

Kelley, E.R., *The London Post Office Directory,* 1857, Kelley's Directories, London.

Kelley, E.R., *The Post Office Directory of... Berkshire,* 1869, Kelley's Directories, London.

Kelley, E.R., *The Post Office Directory of Shropshire,* 1885, 1891, 1895, 1905 & 1937 edns., Kelley's Directories, London.

Kenicer, G., *Unpublished Doctoral Thesis on the Genus Lathyrus,* 2007, The Royal Botanic Garden, Edinburgh.

Kennedy, J., *Page's Prodromus,* 1818, John Murray, London.

Keynes, G., *Hazlitt's Selected Essays,* 1930, 1970 edn., The Nonesuch Press, London.

Knight, C., *Pictorial Half Hours of London Topography,* 1851, 1984 edn., Boethus Press, Clifden (Co. Kilkenny).

Lambert, D., Goodchild, P. & Roberts, J., *Researching a Garden's History,* 1995, Landscape Design Trust, Reigate.

Lamborn, E.A.G., *History of Berkshire,* 1909, Clarendon Press, Oxford.

Lang, J., *An Assemblage of 19th Century Horses & Carriages,* 1971, Perpetua Press, Farnham (Surrey).

Leapman, M., *The Ingenious Mr Fairchild,* 2000, Headline, London.

Levitt, I, Smout, C., *The State of the Scottish Working Class in 1843,* 1979, Scottish Academic Press, Edinburgh.

Life in the Victorian Age, 1993, reprint 1994, The Reader's Digest, London.

Loudon, J.C., *An Encyclopaedia of Gardening,* 1822, Longman, Hurst, London.

Loudon, J.C., *An Encyclopaedia of Plants,* 1829, Longman, Rees, Orme, Brown & Green, London.

Loudon, J.C., *The Suburban Gardener and Villa Companion,* 1838, reprint 1982, Garland, New York.

Mabey, R., *Flora Britannica,* 1996, Sinclair-Stevenson, London.

Mak, G., *In Europe : Travels through the Twentieth Century,* 2007, Vintage, London.

Martin, A.M., *Let The Trees Be Your Guide,* 2002, A.M. Martin, Haddington.

Matalová, A., *Mendelianum,* 1990, Moravian Museum, Brno.

Matthew, C., ed., *The Nineteenth Century,* 2005, O.U.P., Oxford.

Maxwell, H., *Flowers: A Garden Note Book,* 1923, Maclehose, Jackson, Glasgow.

McWilliam, C., *The Buildings of Scotland: Lothian Except Edinburgh,* 1978, Penguin, London.

Mee, A., *The King's England: Gloucestershire,* 1966, Hodder & Stoughton, London.

Midwinter, E.C., *Victorian Social Reform,* 1968, reprint 1971, Longman, London.

Miller, P., *The Gardener's Dictionary,* 1st edn. 1731 and 8th edn. 1768, P. Miller, Chelsea.

Mingay, G.E., *Rural Life in Victorian England,* 1976, Book Club Associates, London.

Mitchison, R., *Life in Scotland,* 1978, Batsford, London.

Mitford, M., *Our Village,* 1824-1832, 1997 edn., Folio Society, London.

Moore, G., ed., *The Penguin Book of American Verse,* 1977, reprint 1979, Penguin, Harmondsworth.

Morgan, J., Richards, A., *A Paradise Out of a Common Field,* 1990, Harper & Row, New York.

Morris & Co's *Commercial Directory & Gazetteer of Gloucestershire...* 1876, Morris & Co., Nottingham.

Neale, J.P., *Views of the Seats of Noblemen and Gentlemen in England, Wales, Scotland and Ireland,* 2nd series, Vol.2, 1825, Vol.4, 1828, Sherwood, Gilbert & Piper, London.

Olby, R.C., *Origins of Mendelism,* 1966, 2nd edn. 1985, University of Chicago Press, Chicago.

Osborne, B.D., Armstrong, R., *Scotch Obsessions, 1996,* Birlinn, Edinburgh.

Page, W., ed., *A History of Berkshire,* (Vol.4, The Victoria History of the Counties of England), 1924, St. Catherine Press, London.

Pardon, G.F., *The London Conductor,* 1851, 1984 edn., Boethus Press, Clifden, Co. Kilkenny.

Parker, R., *The Common Stream,* 1975, Collins, London.

Paterson, M., *Life in Victorian Britain,* 2008, Constable & Robinson Ltd., London.

Pignatti, S., *Flora D'italia,* Vol. 1., 1982, Edagricole, Bologna.

Pigot & Co's *New Commercial Directory of Scotland... 1825-1826,* 1825, Pigot & Co., London.

Plukenet, L., *Almagesti Botanici Mantissa,* 1700, Sumptibus Autoris, London.

Pols, R., *Looking at Old Photographs,* 1998, Federation of Family History Societies, Bury.

Priestley, J.B., *Victoria's Heyday,* 1972, 1974 edn., Penguin, London.

Quest-Ritson, C., *The English Garden,* 2001, reprint 2003, Penguin, London.

Raven, P., Evert, R. & Curtis, H., *Biology of Plants,* 2nd edn. 1976, Worth, New York.

Reader, W. J., *Victorian England,* 1974, Batsford, London.

Rice, G., *The Sweet Pea Book,* 2002, B.T. Batsford, London.

Robinson, F., *The Country Flowers of a Victorian Lady,* 1999, Apollo, London.

Robinson, W., *The English Flower Garden,* 1883, 9th edn. 1905, John Murray, London.

Royal Society of Edinburgh, *Transactions of the Royal Society of Edinburgh,* Vol.39, 1900, J. Dickson, Edinburgh.

Scammell, W., ed., *This Green Earth,* 1992, Ellenbank Press, Maryport (Cumbria).

Scott-James, A., *The Cottage Garden,* 1981, Allen Lane, London.

Shakespeare, W., *Complete Works,* 1980 edn., O.U.P., Oxford.

Sitwell, S., Blunt, W., *Great Flower Books 1700-1900,* 1956, 1990 edn., H.F. & G. Witherby, London.

Smout, T.C., *A Century of the Scottish People, 1830-1950,* 1986, reprint 1997, Fontana Press, London.

St. John Mildmay, H.A., *A Brief Memoir of the Mildmay Family,* 1913, John Lane, London.

Stace, C.A., *Hybridisation and the Flora of the British Isles,* 1975, Academic Press, London.

Staley, A., *The Pre-Raphaelite Landscape,* 1973, 2001 edn., Yale University Press, New Haven.

Stevenson, T., *The Modern Culture of Sweet Peas,* 1910, The Cable Printing and Publishing Company, London.

Stuart, D., *The Garden Triumphant,* 1988, Viking, London.

Sudell, R., ed., *The New Illustrated Gardening Encyclopaedia,* 1932, Odhams Press, London.

Summers, G., *Historic Edinburgh,* 1986, Jarrold, Norwich.

Sweet. M., *Inventing the Victorians,* 2001, Faber & Faber, London.

Tait, A.A., *The Landscape Garden in Scotland 1735-1835*, 1980, Edinburgh University Press, Edinburgh.

Taylor, G., *The Victorian Flower Garden,* 1952, Skeffington, London.

Tennyson, A., *In Memoriam,* 1850, 2004 edn., W.W. Norton, New York.

Thacker, C., *The Genius of Gardening,* 1994, Weidenfeld and Nicholson, London.

The Book of Revelation, Authorised version of the Bible.

The Church of Scotland, *Parochial Schools (Scotland) Bill,* Education Committee Report, 1871, Edinburgh.

The New Statistical Account of Scotland, Vol.I Edinburgh Shire, Vol. X Perth, Vol. XIV Inverness & Cromarty, 1845, William Blackwood & Sons, Edinburgh & London.

The Post Office Annual Directory and Calendar for Edinburgh and Leith, 1838-39, 1838, Edinburgh.

Thomas, D.M., *The Collected Stories of Dylan Thomas*, 1983, J.M. Dent, London.

Thompson, F., *Lark Rise to Candleford*, 1945, reprint 1948, The Reprint Society, London.

Thomson, A. B., Mean Winter Temperatures in Edinburgh, 1764/65 – 1962/63, 1971, The Meterological Office, London.

Thomson, D., *England in the Nineteenth Century*, 1950, reprint, 1969, Penguin, Harmondsworth.

Thomson, D., *Handy Book of the Flower Garden,* 5th edn., 1887, reprint 1893, William Blackwood, Edinburgh.

Trevelyan, G.M. *English Social History*, 1944, reprint 1948, The Reprint Society, London.

Trinder, B., *A History of Shropshire,* 1998 edn., Phillimore, Stroud (Gloucestershire).

Turner, W.J., ed., *The Englishman's Country,* 1945, Collins, London.

Unwin, C.W. J., *Sweet Peas: Their History, Development, Culture,* 1926, W. Heffer & Sons, Cambridge.

von Haller, A., (H.B.Rupp's) *Flora Jenensis*, 1718, (1745 edn.), C.H.Cunonis, Jena.

Waterson, M., *The Servants' Hall: A Domestic History of Erddig,* 1980, Routledge & Kegan Paul, London.

White, W., *History, Gazetteer and Directory of Staffordshire,* 1851, W. White, Sheffield.

White, W., *History, Gazetteer and Directory of the County of Essex,* 1848, W.White, Sheffield.

Wilkinson, A., *The Victorian Gardener,* 2006, Sutton, Stroud (Gloucestershire).

Willats, E.A., *Streets with a Story: The Book of Islington,* 1988 edn., Islington Local History Education Trust, London.

Williams, A., *Round About the Upper Thames,* 1922, Duckworth & Co., London.

Williams, C., ed., *A Companion to Nineteenth-Century Britain,* 2004, Blackwell, Oxford.

Wilson, A.N. *The Victorians*, 2002, Hutchinson, London.

Wood, A., *Nineteenth Century Britain, 1815-1914*, 1960, 2nd edn. 1982, reprint 1984, Longman, Harlow.

Woodward, I., *The Story of Wem and its Neighbourhood,* 1952, 1976 edn., Wildings, Shrewsbury.

Wright, H.J., Wright, W.P., *Beautiful Flowers and How to Grow Them,* 1922, T.C. & E. Jack, London.

Wright, W.P., *The Perfect Garden,* 1908, Grant Richards, London.

Yesterday's Britain, 1998, The Reader's Digest, London.

Yurdan, M., *Oxfordshire and Oxford,* 1988, Shire, Princes Risborough.

Acknowledgements

This work, begun in Aberystwyth in mid-Wales in 1994 and finished in Selkirk in the Scottish borders, has hugely indebted me to numerous people and organisations. Firstly, it would never have begun if it hadn't been for the financial help of my late mother, Joyce Edith Martin and of the Stanley Smith (U.K.) Horticultural Trust at Cambridge. Thanks also to the Eckford Sweet Pea Society of Wem. I am deeply grateful to my guiding mentors, in the early years especially: Dr Martin Fitzpatrick and Caroline Kerkham, both formerly of the Department of History at the University of Wales, Aberystwyth. In those early years, too, my research was greatly helped by the kindness of the family of the late sweet pea expert Bernard Jones, who sent me his collection of Eckford memorabilia.

Thanks too, to Dr Greg Kenicer of the Royal Botanic Gardens, Edinburgh, for the use of his unpublished doctoral thesis on Lathyrus, in chapter ten, especially. Thanks go to Philip Lusby OBE, also of 'the Botanics' for his detailed critique of the draft manuscript and discussions of garden history and all things horticultural on our long bus journeys to and from Edinburgh and the Borders. I am grateful to the professionals involved, like those who typed up the manuscript. I particularly wish to acknowledge the help of the many friends and relatives who helped in so many ways, great and small, some of them no longer with us.The great and small who donate their family papers to record offices and libraries provide an indispensable service to those like myself, and I am grateful to those such as Lord Clerk of Penicuik, Lord Stair at Castle Kennedy and to the Countess of Mansfield for their permission to publish from their estate papers.

The expertise of the staff of the following principal organisations was invaluable: The Central Library, Edinburgh City Archives; The Library, Royal Botanic Garden, Edinburgh; Library Headquarters, Mid-Lothian

Community Services, Loanhead; The National Library of Scotland, including the Map Library, Edinburgh; The A.K. Bell Library, Perth and Kinross Council Archive; Dundee Central Library; The National Records of Scotland in H.M. Register House, East Register House and West Register House, Edinburgh; The British Library in Central London and at Colindale, Middlesex; Hackney Archives Department; Wiltshire and Swindon Archives, Trowbridge; The William Salt Library, Staffordshire Records Office, Stafford; English Heritage; Camden Local Studies & Archives; Berkshire Record Office, Reading; The Lindley Library, The Royal Horticultural Society, London; The Library, The Linnean Society, London; The Central Library, Bromley, Kent; Islington Local History Centre; The Somerset Record Office, Taunton; Banbury Museum; The National Trust, High Wycombe; Cambridge University Library; The Mendelianum, The Moravian Museum, Brno, Czech Republic; Scottish Borders Library and Information Service, Selkirk; Gloucestershire Archives, Gloucester; Shropshire Archives, Shrewsbury; The National Library of Wales, Aberystwyth; The London Metropolitan Archives, City of London; The Lompoc Valley Botanical and Horticultural Society, California, U.S.A.; The Lompoc Museum, California, U.S.A.

Picture Credits

The portrait of Grisell Baillie, Lady Hamilton, by James Rannie Swinton in 1844 (ref. B/633) is from the collection of the Earl of Haddington. The painting of the 1847 curling match at Penicuik by Jemima Wedderburn is courtesy of Lord Clerk of Penicuik. The 1845 Fingask time sheet is by kind permission of Mr M. Murray Thriepland (Perth & Kinross Archive, ref. MS169). The 1860 Belcher & Co. bill heading and the 1855 list of men filling the Coleshill ice house are reproduced by permission of the Berkshire Record Office (refs. D/EPb A28 Bd.4 & D/EPb A29). The view of Sandywell Park and the head gardener's house there are from Gloucestershire Archives (RF109/2GS). The Samuelson's advertisement of 1861 is from the Banbury museum. The portraits of Ernst and Fritz Benary and their workplace, c.1900 are courtesy of Ernst Benary Samenzucht GmbH, Germany. Colonel Baker's house is with permission of Salisbury Museum, St Ann Street Salisbury is courtesy of Wiltshire and Swindon archives. The photo of Wem High Street is from *Wem Now & Then* by J. Whitehead, The Adams Press, 1992, p.30. The three-quarter length portrait photo of W. Atlee Burpee and Henry Eckford is from The Burpees' *Thirtieth Anniversary Supplement,* 1906, p.64.

The following images are all either courtesy of the Lindley Library of the Royal Horticultural Society, or are the author's own, including images from the Bernard Jones' collection: the painting of Eckford's sweet peas, *L'illustration Horticole,* Belgium, 15-1-1892 (no paging); photo of William Cuthberston and Richard Dean, *Journal of Horticulture,* 26-7-1900, p.75; Eckford advertisement, *Gardeners' Magazine,* 23-12-1899 (no paging); coloured drawing of nine sweet peas, *Gardeners' Magazine,* 9-3-1901, between pp. 146-147; sweet pea 'Henry Eckford', *The Garden*, 3-2-1906, between pp.72-73; the wild sweet pea, *Curtis's Botanical Magazine,* 1796, (no paging); portrait of George Fleming, *Cottage Gardener,* 1854; glasshouses at Trentham gardens, *Gardeners' Chronicle,* 13-4-1872, p.507; George Eckford's grave, St. Michael's church, Lower Dowdeswell (author's photo); Charlotte Eliza Eckford's grave, All Saints church, Baschurch (author's photo); Henry Eckford's memorial challenge cup, *National Sweet Pea Annual,* 1906, p. viii; covers of Eckford seed catalogues for 1896, 1898 and 1899; Henry, John and Isabella Eckford and W.A. Burpee in the Wem sweet pea grounds, *National Sweet Pea Annual,* 1906, p.60; photo of Charlotte Eckford, née Stainer, 1850's; photo of Henry Eckford as a young man, 1840's; the Eckford's house at Wem (author's photo, 2012); the despatch room, Market Street, Wem, Henry Eckford centre, c.1900; Henry Eckford's grave (author's photo, 2012); photos of Henry Eckford, 1873 and Rev. W.T. Hutchins, c.1896, *Hutchins' Sweet Pea Annual for 1897,* pp.1 & 8; Henry Eckford Testimonial Address, 1905; Henry Eckford's *Curriculum Vitae,* undated; certificate for sweet pea 'Prince Edward of York', c. 1896; certificate 'for new sweet peas', 5-6-1899; R.H.S. 1st class certificate for garden pea 'Empress', 20-7-1886; R.H.S.1st class certificate for Verbena 'Fanny Purchase', 4-9-1872; R.H.S. 1st class certificate for Verbena 'Mrs Eckford', 1-9-1868; R.H.S. 1st class certificate for Verbena 'Anna Keynes', 15-9-1868; painting of Eckford's sweet peas, *The Garden,* 12-3-1892, pp.232-233; portrait of Henry Eckford from an original glass photo, c.1896; photo of Henry Eckford and bicycle, c.1896; the Crystal Palace interior, *The Journal of Horticulture,* 26-7-1900, p.73; the Clive seed grounds, *The Gardeners' Magazine,* 10-12-1898, p.799; hybridisation conference portraits, *The Gardeners' Magazine,* 15-7-1899, between pp.428-429; 'sweet peas of the century', *Gardeners' Chronicle,* 3-8-1901, p. 87; painting of sweet peas, *The Garden,* 13-2-1909, between pp. 78 – 79; sweet pea celebration portraits, Crystal Palace, *The Gardeners' Magazine,* 28-7-1900, p. 475; caption to go with portraits, above: top row left to right, J. Fraser, F.L.S., R.W. Ker,

E. Beckett, S. B. Dicks, W. P. Wright; second row, Richard Dean, V. M. H., George Gordon, H. J. Jones; third row, R. Sydenham, Rev.W. T. Hutchins, J. F. McLeod; bottom row, John S. Eckford, Horace Wright, H.A. Needs, C. H. Curtis, W. Cuthbertson.

Quotes

I am grateful to the publishers of the following books, detailed in the bibliography, for allowing me to quote from them: Graham Rice's *The Sweet Pea Book,* John Harvey's *Early Nurserymen,* Gilbert Summers' *Historic Edinburgh,* Anne Wilkinson's *The Victorain Gardener, The Country Flowers of an Edwardian Lady, Diary of a Victorian Gardener,* Bill Bryson's *A Short History of Nearly Everything,* Ruth Duthie's *Florists', Flowers and Societies,* Colin Matthew (ed.) *The British Isles 1815 – 1901*. Every effort has been made to contact all copyright holders. Those unmentioned are requested to contact the author for acknowledgment in subsequent editions.